Design Methodology for Future Products

Dieter Krause • Emil Heyden
Editors

Design Methodology for Future Products

Data Driven, Agile and Flexible

Springer

Editors
Dieter Krause
Board Member of the Scientific Society for Product Development (WiGeP)
Hamburg University of Technology (TUHH), Institute of Product Development and Mechanical Engineering Design (PKT)
Hamburg, Germany

Emil Heyden
Assistant to the Board of the Scientific Society for Product Development (WiGeP)
Hamburg University of Technology (TUHH), Institute of Product Development and Mechanical Engineering Design (PKT)
Hamburg, Germany

ISBN 978-3-030-78367-9 ISBN 978-3-030-78368-6 (eBook)
https://doi.org/10.1007/978-3-030-78368-6

This Springer imprint is published by the registered company Springer Nature Switzerland AG.
The registered company address is: Gewerbestrasse 11, 6330 Cham, Switzerland

Foreword

It is a great pleasure for me to write the foreword for this book. First, this publication comes out almost exactly a decade after the publication of the book I edited, *The Future of Design Methodology*. This is a consistent continuation of the intention to show promising perspectives for research and application in the field of design methodology. On the other hand, my much-valued colleague Prof. Dieter Krause publishes this book as a publication of the *Scientific Society for Product Development—WiGeP*. This society represents, besides an influential industrial circle, the leading representatives of professors in the German-speaking area who have taken up the subject of *development and design* as an object of research. I feel particularly connected to both goals.

In recent decades, research in the field of product development has grown worldwide in width and detail, which is certainly due to the diversity and heterogeneity of the associated topics and issues. For me and many of my colleagues, it has always been a special challenge as well as a satisfaction to make a contribution to this still young and rapidly growing scientific discipline. The fact that Prof. Krause has taken upon himself to present a summary of promising work with different perspectives in a book deserves great recognition. There is a great risk of getting lost in ramified specializations of the almost endless topic of product development. It is therefore worthwhile to focus on what is common and overarching.

I am also very pleased that Prof. Krause is publishing this summary as a *WiGeP* publication. As an honorary member of this society, formerly active and now as a retired member, I have made every effort to support this society. It is with great pleasure that I see how an equally successful, respected, and collegial community has developed here, which has a national and international impact and plays a highly active role in research and industry. The fact that leading representatives of *WiGeP* have conversely agreed to publish in this book their latest research results and perspectives for future work in the field of development and design underlines the quality of this book as well as the impression of being linked together within *WiGeP*.

I am convinced that this book will attract special attention in the research community, both nationally and internationally. It is a great pleasure for me to see how committed and top-class colleagues are working here in pioneering directions of product development research. After all, in addition to all the scientific findings, they also make a significant contribution to the success of our industrial location and the prosperity of our society.

Technical University of Darmstadt
Institute for Product Development and Machine Elements
Darmstadt, Germany

Herbert Birkhofer

Preface and Acknowledgments

The *Scientific Society for Product Development* (*Wissenschaftliche Gesellschaft für Produktentwicklun—WiGeP*) is an association of leading professors and representatives of the industry in the German-speaking area in the field of product development. The *Scientific Society for Product Development* understands itself as a link between universities and industry with a competence network for the promotion of product innovations in mechanical engineering and related industries. The group of experts *Methods and Processes of Product Development* of the *Scientific Society for Product Development* is particularly committed to the future of design methodology in universities and industries and provides most authors of this book. Dieter Krause is the spokesman of this expert group and member of the board of *WiGeP*. He acts as editor together with Emil Heyden, who is assistant of this group and the board.

This book aims to contribute in the field of design methods and their implementation for innovative future products. A data-driven, agile, and flexible way of working is becoming increasingly essential in the development process of future products. The contributions of this book aim to reveal the strengths and weaknesses in product development. Strengths have to be maintained while overcoming the weaknesses.

Four sections are presented, *Methods for Product Development and Management*, *Methods for Specific Products and Systems*, *Facing the Challenges in Product Development*, and *Model-Based Engineering in Product Development*.

The book starts with the agile strategic foresight of sustainable mechatronic and cyber-physical systems and moves on to the topics of system generation engineering in development processes, followed by the technical inheritance in data-driven product development. Product improvements are shown via agile experiential learning based on reverse engineering and via combination of usability and emotions. Furthermore, the development of future-oriented products in the field of biomechatronic systems, sustainable mobility systems, and in situ sensor integration is shown. The overcoming of challenges in product development is demonstrated through context-adapted methods by focusing on efficiency and effectiveness, as well as designer-centered methods to tackle cognitive bias. Flow design for target-oriented availability of data and information in product development is addressed. Topics of model-based systems engineering are applied to the function-driven product development by linking model elements at all stages and phases of the product. The potential of model-based systems engineering for modular product families and engineering of multidisciplinary complex systems is shown.

The innovative contribution of all the authors in the field of design methods in product development is deeply acknowledged. Needless to say, a big thank you goes to their book contributions. Special gratitude goes to Herbert Birkhofer for his collective work *The Future of Design Methodology*. He provided the inspiration for this work and wrote the foreword for this book.

Board Member of the Scientific Society for Product Development (WiGeP)
Hamburg University of Technology (TUHH),
Institute of Product Development and
Mechanical Engineering Design (PKT)
Hamburg, Germany
April 2021

Dieter Krause

Assistant to the Board of the Scientific Society for Product Development (WiGeP)
Hamburg University of Technology (TUHH),
Institute of Product Development and
Mechanical Engineering Design (PKT)
Hamburg, Germany
April 2021

Emil Heyden

Contents

Part I

Methods for Product Development and Management

From Agile Strategic Foresight to Sustainable Mechatronic and Cyber-Physical Systems in Circular Economies

Iris Gräßler and Jens Pottebaum

Abstract

Today's entrepreneurial decisions have to be taken under the conditions of volatility, uncertainty, complexity and ambiguity. Future holistic product creation therefore requires suitable methods, tools and models from initial strategic steps to product's end of life. In order to improve competitiveness and to realize sustainability, common guidelines for management and operations have to be set in company-specific Product Creation Systems (PCS). Entrepreneurial activities at all management and operational levels must be aligned towards short, middle and long-term horizons. In order to provide a holistic framework for such joint orientation, a generic Product Creation System (gPCS) corresponding to the approach of Lean Production Systems is proposed in this chapter. Mechatronic and Cyber-Physical Systems are key elements in large-scale connected systems. They are complemented by smart functions and concepts like digital twins to enable digital business models and to facilitate intelligent use. Innovative digital business models treat sustainability as beneficial objective and thus lay the foundation for Circular Economies. Product engineers are supported in their tasks by means of Digital and Virtual Product Creation. Typically, they are confronted with uncertainty in Product Creation. For instance, engineers need to gather and interpret uncertain information in Strategic Planning about circularity needs and business potentials. Furthermore, potential effects of alternative design concepts on circularity have to be simulated and weighed up in advance during engineering. Therefore, Agile Strategic Planning, Resilient Requirements Engineering as well as Digital Worker and Learning Assistance are identified as key techniques to be merged with Model-

I. Gräßler (✉) · J. Pottebaum
Paderborn University, Heinz Nixdorf Institute - Chair for Product Creation, Paderborn, Germany
e-mail: iris.graessler@hni.upb.de

D. Krause, E. Heyden (eds.), *Design Methodology for Future Products*,
https://doi.org/10.1007/978-3-030-78368-6_1

Based Systems Engineering (MBSE). Traceable integration into system lifecycle management is key to utilize sustainability potentials along the entire lifecycle.

1.1 Introduction

Mechatronic and Cyber-Physical Systems are key elements in large-scale connected systems. Mechatronic systems provide functionality by synergetic integration of mechanical, electric, electronic and software elements (Isermann 2005). The behavior of a basic mechanic system is controlled by acquiring information through sensors, information processing by software algorithms and application of forces and movements with the help of actuators. While mechatronic systems are defined by their functional and/or spatial integration, Cyber-Physical Systems (CPS) are further characterized by the fact that they are networked with the cyber world, i.e., the Internet of Things and Services (Gill 2006). Based on such kind of systems, socio-technical systems can be built scaling up towards companies of Things and Services (see Fig. 1.1). The functionality of CPS comprises communication capacities and the integration of services like high-performance computations, Artificial Intelligence or image processing. Thus, they include smart functions and concepts like digital twins to enable digital business models and to enable intelligent use.

CPS are enablers of large systems-of-systems, like smart cities or smart transportation hubs. Consequently, business models are changed significantly. Enterprises recognize the potentials of combining tangible commodities with services into hybrid performance

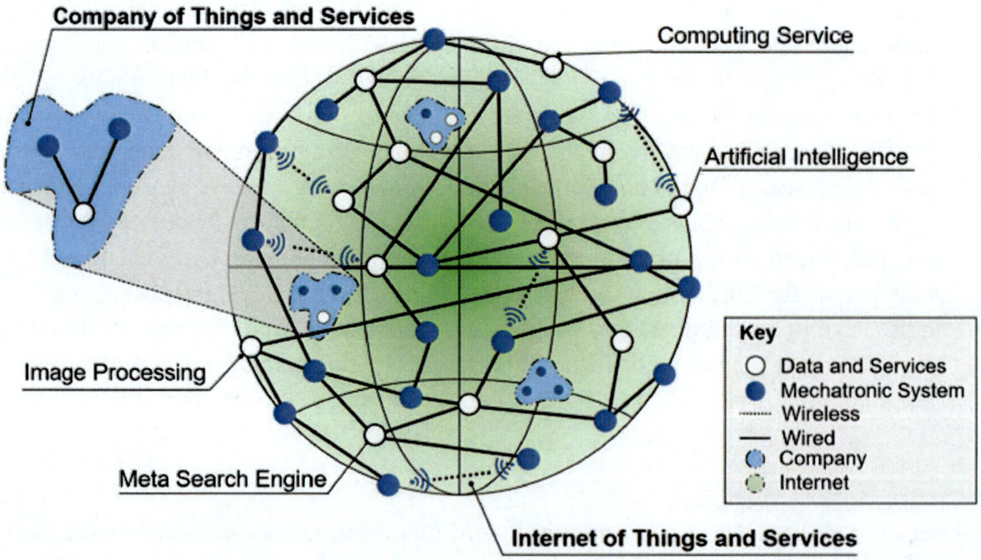

Fig. 1.1 CPS in a connected world (based on (Graessler and Hentze 2020)

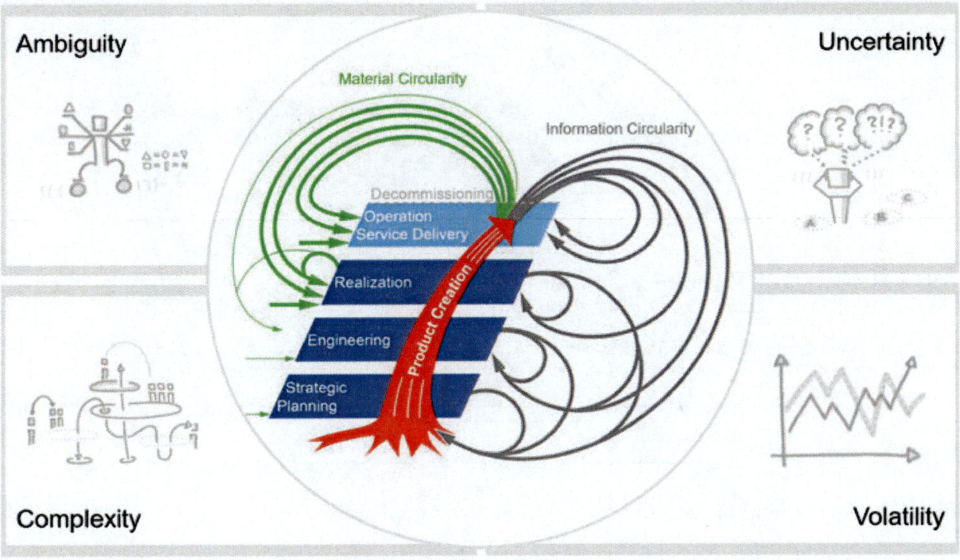

Fig. 1.2 Generic Product Lifecycle (gPLC) in a VUCA world (based on Gräßler and Pottebaum 2021)

bundles as a special type of product (Meier and Uhlmann 2012). Especially in English literature, they are often entitled product-service systems, where products are tangible by definition (Baines et al. 2007; Mont 2002). Like other types of products, they follow the generic Product Lifecycle (gPLC, see Fig. 1.2): Their business model is a result of strategic planning and gives orientation for the entire lifecycle. In product development, the product is specified in a way that reproducible realization is feasible. Properties are specified and validated regarding stakeholder interests. Realization incorporates production and is highly dependent on disciplines. Product use is supported by services like Maintenance, Repair and Overhaul (MRO). In holistic product creation, the End-of-Life phase means an interface towards next product generations and resource re-use. Closing loops in terms of knowledge management, competency building and resource circularity are key challenges of future product creation.

The Action Field of Product Creation is defined and determined in Fig. 1.3. Accordingly, Product Creation comprises the generation of promising product ideas as well as their functional and manufacturing-related realization. Based on the strategic orientation of a company, the most promising ideas for product innovations are identified in Strategic Planning and Innovation Management. As market-driven inputs, future customer requirements of existing markets (market pull) as well as potential new markets (blue ocean strategy) are anticipated. New basic developments in product and production technologies are a further source of new product ideas (technology push). Systems Engineering and Engineering Management form the core of the Action Field of Product

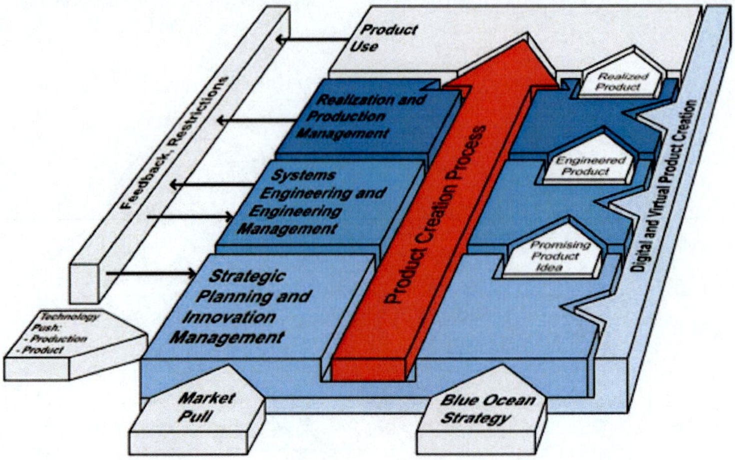

Fig. 1.3 Action field of product creation (based on Gräßler 2015b)

Creation. In case of Mechatronic and Cyber-Physical systems, a product idea is engineered in a multidisciplinary way searching for an overall optimum across all disciplines. Its maturity is continuously increased into an engineered product. This product—as a class—is rolled out into envisioned markets. Individual products—as instances—are operated by or for customers. In parallel to Systems Engineering, suitable realization and production concepts including preparation of intangible product components as well as logistics concepts, production line design and shop-floor layout are planned within Realization and Production Management. Resulting restrictions are fed back to Systems Engineering. The element "Realization and Production Management" in the Action Field of Product Creation also incorporates the interface to current production systems. Digital and Virtual Product Creation serves as enabler of all other elements in the action field. While Product Creation explicitly subsumes the first three phases of the gPLC, later phases are included implicitly, as constraints and opportunities of use cases, MRO procedures and End-of-Life scenarios are considered as restrictions.

Product engineers have to understand the actual and future context of product use. Engineering in general has to consider possible futures to ensure that a business model is resilient with regard to various influence factors. The world evolves into increasingly volatile, uncertain, complex and ambiguous states (VUCA-world, cf. Fig. 1.2). In volatile environments, everything is subject of continuous change. Global supply chains result in risks for small suppliers competing against companies all over the world. Even though access to communication networks is globally managed today, there might be technical or political limitations in the future. Complexity describes a high number of networked elements with partially unknown cause-effect relationships. As an example, autonomous driving can be engineered in a way that cars cause less damages than today—but there will be situations which cannot be anticipated. Ambiguity adds the notation of different meanings. The amount of information increases daily, but the challenge is to extract meaning from information resources. A holistic framework of Product Creation therefore needs to be designed in the view of VUCA challenges. VUCA thus stands for the changed conditions under which entrepreneurial decisions have to be taken today.

Besides technological advances, sustainability plays a major role in engineering future products. Sustainability demands a balanced approach to create value without restricting people in fulfilling their needs in the future. The Brundtland commission postulated the need for a balanced approach towards economic, ecological and social values (Brundtland 1987). This was adopted in 2015 by an international consensus regarding 17 sustainable development goals (United Nations 2015). Concepts like Design for Environment (DfE) and Eco-Design were consequently evolved to change product engineering from a passive to a proactive and integrative approach (McAloone and Pigosso 2021). Studies have shown that sustainability and reliability criteria must be taken into account as part of product design in order to develop more environmentally friendly products without sacrificing performance. (Ostertag et al. 2020). While early concepts focused on single benefits like saving waste in production, concepts like Cradle-to-Cradle (McDonough and Braungart 2002) and Circular Economy (CE) (Ellen MacArthur Foundation 2015) are based on the assumption that sustainability implies its own business value. The vision of Circular Economy is that resources are fully re-used at the product's End-of-Life by product life extension, redistribution, remanufacturing or recycling (Raabe et al. 2019). The pattern of product-service systems means a change of business models in a way that service and performance are provided instead of a material product to be consumed. Therefore, they carry significant potential for implementation of CE (Baines et al. 2007; Kjaer et al. 2019).

1.2 Generic Product Creation System

In the course of introducing lean production in the 1990s, production systems based on the Toyota Production System (Ohno 1988) were established in German companies. This was due to the observation that not the use of individual methods and tools lead to success, but rather the selection and synchronization of guiding principles. Further, if interaction of principles with company goals and with the methods and tools is understood, accepted and implemented by employees at all levels of the company, this leads to sustainable success (Dombrowski and Mielke 2015; VDI 2012). With the guideline VDI 2870, a generic production system was made available for the first time. It can be tailored to suit a specific company's needs. Production systems, however, typically do not contain the enablers incorporated in the Action Field of Product Creation, although the foundations for competitiveness and profitability are laid here. Therefore, the basic idea of a production system was first transferred to Product Creation in (Graessler 2004) with the aim of mapping cause-effect relationships in the early phases and aligning all business activities accordingly. In this first version of the Generic Product Creation System (gPCS), specific conditions of Product Creation Processes were considered. As an example, order fulfilment within Product Creation is mainly characterized by information flows. Material flows only play a subordinate role, for instance, representing result status using physical prototypes. Further, multi-disciplinarity was covered by the first gPCS paying tribute to continuous evolution from original mechanical engineering towards mechatronic systems engineering (Gräßler 2015a). Today, multi-disciplinarity is even more important due to the increasing spread of Cyber-Physical Systems.

Therefore, an enhanced gPCS is proposed in this chapter encompassing CPS in a VUCA world as modern conditions for entrepreneurial activity (Fig. 1.4 and Table 1.1). Furthermore, the social mandate for a Circular Economy (cf. Kohl et al. 2020) is treated now equally with the competitiveness and profitability of a company. Accordingly, the consideration is extended to all phases of the gPLC. The cascade of goals, principles and building blocks of the enhanced gPCS is presented in Fig. 1.4. Using guiding principles (classification system), employees and executives are able to measure their actions and decisions against their contribution to the corporate goals such as Circular Economy or Excellence in Product Creation (value system). The guiding principles are operationalized using standardized building blocks in terms of models, methods and tools. These building blocks enable the implementation of the principles and form the working system. In the proposed gPCS, the working system embraces new models, methods and tools provided by research. In a company-specifically customized PCS, the working system includes best practices of models, methods and tools. Standardization of the included models, methods and tools is a prerequisite for mastering and continuously improving product creation (Gräßler 2015b).

Table 1.1 shows a consolidated list of generally recommendable guiding principles. In an individual company's situation, appropriate guiding principles have to be selected and prioritized out of this enumeration based on company-specific goals (Graessler 2004). For this, the concerned product division and the whole company must be analysed and taken into account. By selecting and defining the guiding principles, appropriate methods and tools are selected and it is ensured that a coherent overall framework with similar or linked methods and tools is created.

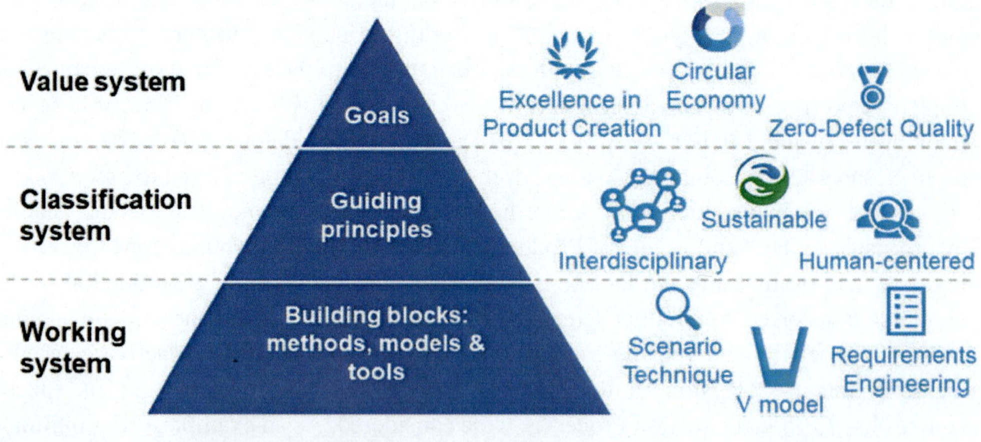

Fig. 1.4 Structure of gPCS (based on Graessler 2004) inspired by Ankele et al. (2008)

Table 1.1 Guiding principles of gPCS

Guiding principles	Definition
Result orientated	Focus on economic performance variables such as rate of return, time and cost compliance
Transparent	Performance, quality, milestone status, implementation difficulties are always known
Reproducible and resilient	Repeatability and insensitivity to disturbances
Open to change	Managing uncertainty and flexibility towards environment or customer induced change requests
Fast	Short lead times from strategic business area planning to market launch
Test driven	Model based verification and validation of the respective result levels based on simulations and experiments
Interdisciplinary	Cooperation and information driven cross linking of involved disciplines with the aim to generate a higher-level optimum
Promoting innovations	Systematic generation and realization of promising product ideas
Application orientated	Consequent orientation of product creation on use cases and experience worlds
Standardized	Using scale and learn effects by reuse of products and processes
Human-centered	Considering employee perception and needs with the goal of high identification as well as the customer's view during the whole product creation
Continuously improving	Continuous improvement of employee participation, technology, IT-tools, organization, processes and methods
Sustainable	Continuous and meaningful use of resources, generated knowledge and thereof acquired competences
Holistic	Overall approach which especially considers the interaction of processes, resources, information and stakeholders

To achieve sustainability, circularity must find its way into requirements, for instance, by Design-for-X approaches or by including circularity as an integral part of a business model. Traceable integration into System Lifecycle Management is key to utilize sustainability potentials in all lifecycle phases. For example, the goal of Circular Economy is coupled with the guiding principles "holistic" and "sustainable". Holistic demands an overall approach comprising all phases of gPLC. Sustainable describes continuous and meaningful use of resources, generated knowledge and thereof acquired competences. Pursuing the goal of circularity, sustainability evolves from a simple trade label towards an integral part of sustainability-oriented business models (Lüdeke-Freund et al. 2018). For instance, in mobility many approaches are focused on saving energy consumption (cf. Chap. 7), different types of simulations are integrated to assess solution alternatives (Göhlich et al. 2021). Regarding End-of-Life approaches, one could ask whether recycling measures are understood as value preservation or even value creation (Leder et al. 2020):

The vision of sustainable development is that products are designed, produced and used without restricting people in fulfilling their needs in the future.

Following the idea of Circular Economies, strategic product planning needs to focus on sustainability when defining the business model of a product (Nußholz 2017). The business model is the entry point of product engineering according to VDI 2206 and other methodologies. When creating product ideas, this should cover a value creation (or at least preservation) at the End-of-Life. Product life extension, redistribution, remanufacturing and recycling are possible measures that can be applied on each system level, from complete product to single components or even material. Changing business models from providing a product as a resource to be consumed towards providing the functionality as a service is one possible trigger to enhance the economic value of sufficiency.

1.3 Building Blocks of Holistic Product Creation

In the following sections, models, methods and tools recommended by the gPCS are outlined. The conditions of a VUCA world induce the need for new basic scientific methods. Ambiguity of predecessor-successor relationships between activities are structured by the V-Model. The inherent concern logic of the V-Model gives orientation within uncertainty and provides clear alignment to entrepreneurial actions (Sect. 1.3.1). Complexity is handled and changes are made traceable by Model-Based Systems Engineering (MBSE, Sect. 1.3.2). Innovative digital business models must answer under volatile conditions, how and with which value proposition to earn money. Agile Strategic Planning, Resilient Requirements Engineering as well as Digital Worker and Learning Assistance are identified as key techniques to be merged with Model Based Systems Engineering. Traceable integration into system lifecycle management is key to utilize sustainability potentials along the entire lifecycle. Agile Strategic Planning treats sustainability as beneficial objective and thus lays the foundation for Circular Economies (Sect. 1.3.3). Rapidly and unpredictably changing specifications are addressed by Resilient Requirements Engineering (RRE, Sect. 1.3.4). Product engineers are supported in their tasks by means of Digital Worker and Learning Assistance (Sect. 1.3.5).

1.3.1 New V-Model for Mechatronic and Cyber-Physical Systems

Necessary pre-condition for the success of entrepreneurial ambitions is an appropriate engineering reference model. Because of its multi-disciplinarity, the V-Model is particularly suitable for Mechatronic and Cyber-Physical Systems. Due to empirical experience and increased requirements, many derivatives of the V-Model appeared in industrial practise and current research. The New V-Model for Mechatronic and Cyber Physical Systems takes up these derivatives and harmonizes them into one coherent approach

(Graessler and Hentze 2020; VDI 2020). It represents the idea of interlinking all disciplines involved in engineering tasks. To prevent the potential misunderstanding as a time sequence, the graphical layout of the V-Model represents the logical sequence of tasks. The key advantage of this content-related logical networking of tasks lies in independence from the chosen form of project organization. Thus, the V-Model can be applied in classical project management as well as in engineering projects run by agile principles (Graessler and Hentze 2020).

The V-Model consists of three strands (Fig. 1.5). The central strand in orange describes the core tasks of engineering from requirements elicitation up to system transition. The inner, yellow strand describes handling and work with requirements, in order to emphasize the importance of changing requirements in engineering practice. The outer, blue strand represents modelling and analysis activities and thus claims a model-based engineering approach. Each of the three strands graphically contains strongly interwoven disciplines, such as mechanics, electrics, electronics, software as well as others, e.g. pneumatics, hydraulics and optics (Graessler and Hentze 2020). The checkpoints one to six shown left and right of the V-Model in Fig. 1.5 support the user in tracking the progress of his development project (Gräßler 2018). Checkpoints provide exemplary incentives to check which results and work content the user may still be missing, which he should have processed, and serve to substantiate the content-related logical networking of tasks in system development. In contrast to “gates” or “milestones” (Cooper 1990), established in

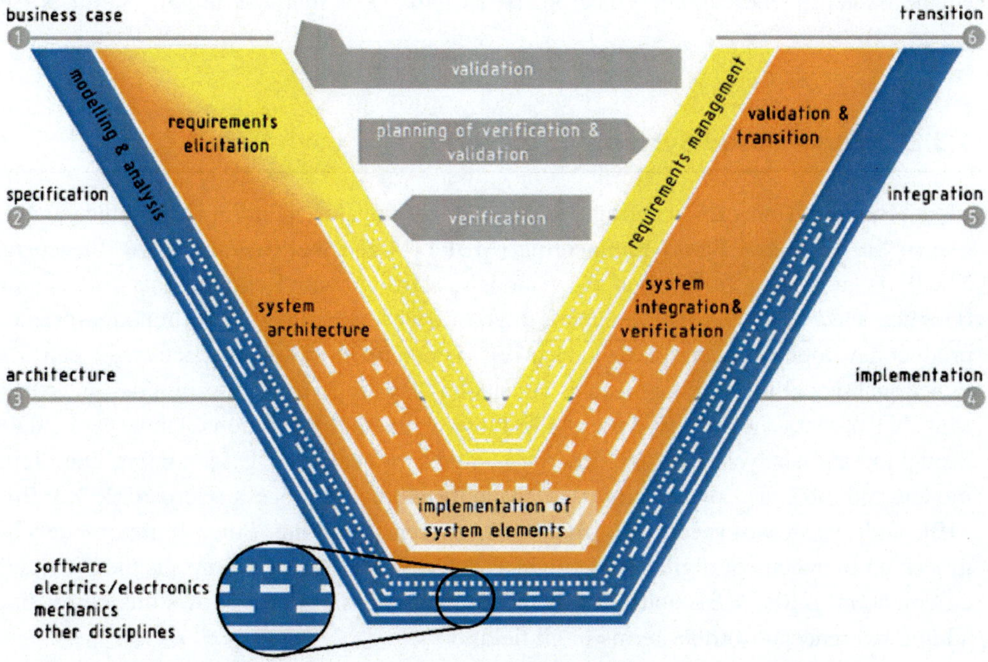

Fig. 1.5 New V-model for Mechatronic and Cyber-Physical Systems (VDI 2020)

stage-gate models, checkpoints neither correlate with a point of time within the development project, nor do they fulfil a release function.

Nestings of V-Models appear on different system hierarchy levels, parallel system elements and sub-systems as well as on different maturity levels (Graessler and Hentze 2020). According to several Systems Engineering approaches, a set of three arrows is exemplarily representing continuous planning, verification and validation to ensure that the requirements for the fulfilment attributes are achieved (Dazer et al. 2020). Although the system is verified and validated several times along the "V", the arrows are included only once and exemplarily, in order to simplify the illustration and achieve good clarity (Graessler and Hentze 2020). Information and communication technology, organization and human beings with their skills, competencies, convictions and emotions are represented by coupling the V-Model with the gPLC (see Fig. 1.2) (Gräßler 2018; Gräßler et al. 2018a).

Systems Engineering (SE) is a structured, multi-disciplinary engineering approach for technical systems, targeting at a cross-disciplinary optimum within a given time frame and budget. For this purpose, the disciplines are structured and networked with each other using models (Gräßler 2015b). The New V-Model explicitly supports Systems Engineering to achieve a cross-disciplinary optimum with regard to safety and reliability, complexity management, user experience, lead-time reduction and cost savings. Systems Engineering balances technical, organizational and managerial activities along the entire system lifecycle (INCOSE 2014). It is adopted, for instance, to utilize system knowledge and enable System Generation Engineering (see Chap. 2) or modular product families (see Chap. 14).

1.3.2 Model Based Systems Engineering

Model Based Systems Engineering (MBSE) extends the aforementioned systemic perspective of SE by Model-Based Engineering (MBE) (Madni and Sievers 2018). In general, MBSE is motivated by the value of models, starting with Requirements Engineering (Göhlich and Fay 2021). Along the left thigh of the V-Model (Fig. 1.5), function-oriented product development is emphasized as a key to open up the solution space (see Chap. 13). Functional modelling is recognized as an enabler, e. g., in agile lightweight design (Albers et al. 2019). Meta-models serve for applicability in diverse applications (Drave et al. 2020). Modelling and analysis are supported by Digital Engineering (DE), assuming that digital models and tools are consistently used in the development process (Gerhard 2020). Both MBE and DE are not specifically focused on systems thinking. Since formal models are processed by means of digital technologies (INCOSE 2018), MBSE means the synergetic combination of SE, MBE and DE. The Venn diagram in Fig. 1.6 shows this correlation, adding references to further terms in all fields.

Problem-solving activities of engineers can always be treated as a model-based activity, starting with mental models (Meboldt 2008). In MBE, specification, design, integration and

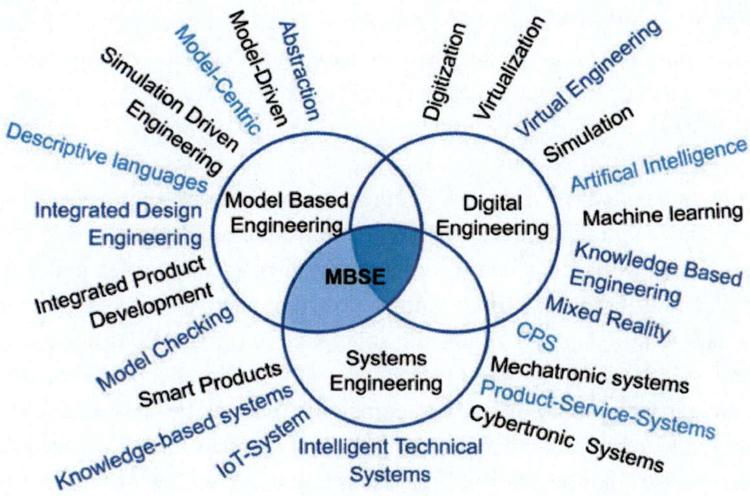

Fig. 1.6 MBSE: exploiting synergies of model-based, digital and systems engineering

validation are based on semi-formal and formal models (Eigner et al. 2017; Estefan 2007). The formalization is a fundamental requirement to apply concepts like data-driven product development (see Chap. 3). If models are determined as the goal of each activity, this is referred to as "model-centric" development differentiating clearly from document-centric development (Bayer et al. 2010; Elaasar et al. 2019). Further, the term "model-driven" development captures algorithms being applied for the automated transformation of (partial) models into products (France and Rumpe 2007; Schmidt 2006).

MBSE is specifically connected with the digital representation of models, for example, to ensure the traceability of dependencies. The system model is an integrated, coherent, and consistent model subsuming the totality of partial models of the system under consideration (Hick et al. 2019). For Mechatronic and Cyber-Physical Systems, system modelling is strongly driven by embodiment design supported by a variety of product modelling options (Matthiesen et al. 2019). For formal and consistent modelling, a modeling tool, a modeling method and a modeling language are combined (Delligatti 2014). In addition to the content of the models, views and viewpoints are defined for the stakeholders who can simultaneously access the system model. By these means, system models and underlying data can be integrated into data and information flow in product development (see Chap. 11).

MBSE carries the potential that Systems Engineers become capable of recognizing and understanding these relationships and to calculate effects of design decisions based on formalized partial models, including structural and behavioral aspects (Hick et al. 2020; see Sect. 1.3.5). In its deep interpretation, MBSE covers the concepts of Simultaneous Engineering, i.e., integrating development of products and Cyber-Physical Production Systems (Gerhard 2017; Vogel-Heuser et al. 2019). While recognizing this potential, the introduction of MBSE means a significant challenge regarding methods, tools and data. From a methodological point of view, the application is linked into business processes like Engineering Change Management through System Lifecycle Management approaches (Konrad et al. 2019). From an organizational point of view, the application of MBSE is scalable to different industries, disciplinary specifics and company sizes (cf. Wilking et al. 2020).

In its core, it means holistic measures in terms of Systems Engineering, Model-Based Engineering and Digital Engineering. When introducing MBSE, three dimensions of internal roles need to be considered: organizational roles, user roles and implementation roles (Gräßler et al. 2021). Broadening this perspective by external relationships, companies act within supply chains in roles like Original Equipment Manufacturers (OEMs) and suppliers on different tier levels. Due to Intellectual Property Protection, access to all partial models at partner companies cannot be assumed on model layer. Usually only document exchange is implemented. Offering a parameter space for joint model management (middle layer in Fig. 1.7) is one approach to overcome this barrier (see Fig. 1.7).

Adopting MBSE concepts, complexity of the system and its environment is not reduced, but handled. Therefore, management platforms like Product and Application Lifecycle Management (PLM/ALM) are utilized for model management. While M-CAD, E-CAD and CASE are available as highly integrated authoring systems interconnected with PLM/ALM, modelling tools for MBSE are still stand-alone products. Advancements are ongoing to enable sophisticated model management on system level, including formalized and traceable links into partial models.

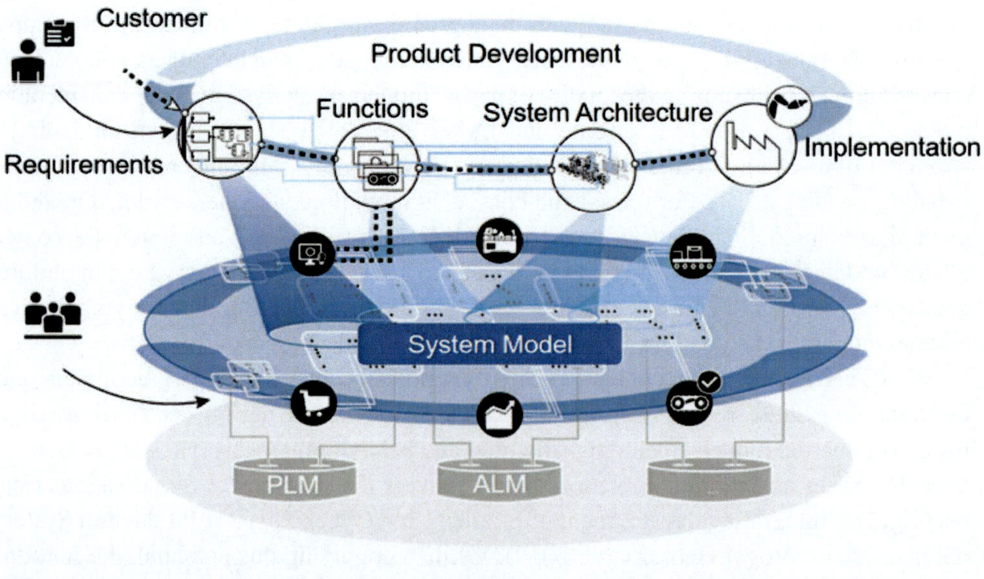

Fig. 1.7 Integrative model-based systems engineering approach of the German research project ImPaKT—ICT-Enabled Model-Based Impact Analysis in Product Engineering

1.3.3 Agile Strategic Planning

For long-term economic success, it is indispensable to always position oneself favourably in comparison to one's competitors. It is necessary to plan new products for the long term and to identify future customer needs at an early stage. Thus, the product development process as a core element of creating new value propositions must be flanked by an innovation and strategic planning process. New product ideas must be continuously examined with regard to feasibility and fit on the basis of internal and external inputs. Promising ideas are investigated further and processed for product development by means of market studies, for example, while other ideas are put on hold to be re-evaluated at a later date. Likewise, planning of release strategies requires value-oriented and cross-domain views (Sahin et al. 2020).

Consequently, business models need to be robust regarding future influences. As emphasized in Sect. 1.1, Circular Economies are implemented by innovative business models treating sustainability as a beneficial objective. This is framed by the envisioned value proposition and targeted revenue channels. Assessing future influences on a business model, both internal factors as well as external factors have to be considered. This applies similarly to, for instance, requirements and their likelihood of change. For anticipating future customer needs and market developments, foresight methods such as the scenario technique can be used. The aim is to anticipate alternative pictures of the future that are consistent in themselves and heterogeneous in comparison, without considering probability of occurrence. Benefits of agility should be made available. Following the values and principles stated in the Agile Manifesto (Beck et al. 2001), several approaches were established in software engineering. Besides clear indications of positive effects, the extension of agility to products which are based on physical core products implies significant challenges regarding cultural and organizational challenges (Atzberger et al. 2020).

In comparison to sequential and workshop-based foresight procedures, the Scenario-Technique is advanced from an instrument for consultants in strategic planning towards an agile tool to be used by the companies themselves for a variety of use cases in early phase of product creation and later in product management (Gräßler et al. 2020b). Agile Scenario-Technique enables iterations in early phases as well as controlling premises of decisions in later phases of the lifecycle. Agility is enabled through (a) the agile nature of the proceeding, (b) the integration of the Integrated Scenario-Data Model (ISDM), (c) the targeted iteration of individual steps and (d) the partial adjustment of the premises (Fig. 1.8). As Agile Scenario Technique is based on the consistency-based school of Scenario-Technique (Reibnitz 1992), it results in consistent scenarios without probability. Clustering enables quick but sound scenarios based on consistency analysis. Traceability and partial automation is implemented by maintaining data across projects using the ISDM (Gräßler et al. 2017a). Therefore, scenarios which were used as a basis for (strategic) decisions can be reflected in time intervals to enable adaptation to new environments. By this measure, disruptive events can be taken into account immediately from a methodological perspective

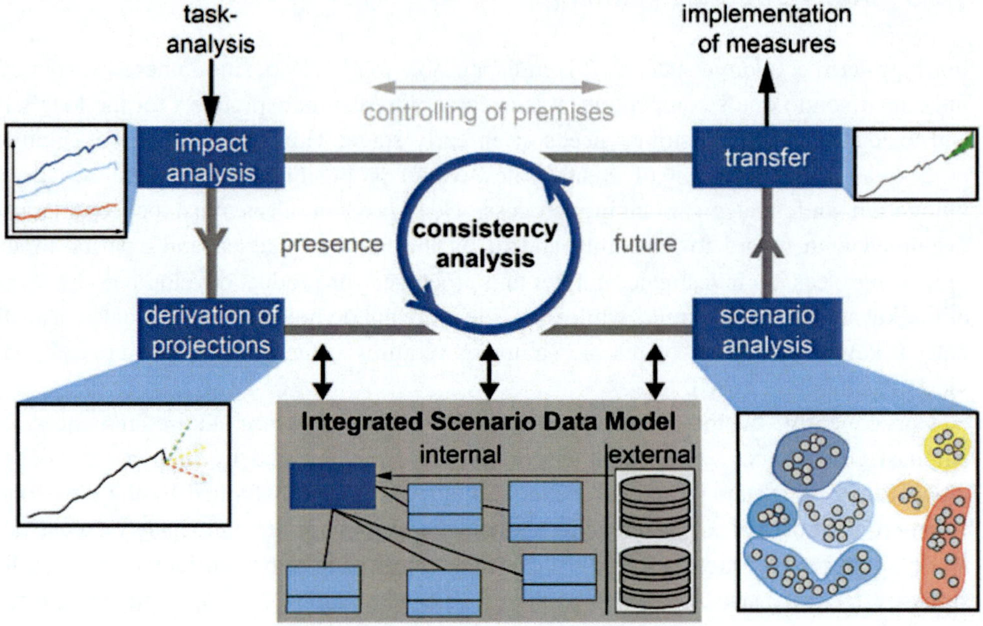

Fig. 1.8 Agile Scenario-Technique based on automated data analysis

and used as chances. Additionally, the impact of possible disruptive events on scenarios can be assessed mathematically (Gräßler et al. 2019b).

Based on a task analysis, the presence is analyzed to identify the key factors of the object under consideration (influence analysis), and the consistency of possible projections into future is checked (consistency analysis). Thus, generated scenarios contain a precise projection of each key factor free from contradictions. By generating two to three scenarios which are as different as possible (scenario analysis), a wide range of strategy and technology options are considered. Thus, alternative strategies can be developed and implemented in addition to the primarily applied strategy in order to react to changing business conditions (transfer). Besides scenario data management, information about generic influence factors can be acquired through the ISDM. It is designed as a knowledge base allowing semantic search. By connectors to IT systems like statistic databases (for instance Statista), Enterprise Resource Planning (ERP) and Product Lifecycle Management (PLM), data can be integrated to define influence factors and to derive reasonable projections.

1.3.4 Resilient Requirements Engineering (RRE)

Strategic Planning paves the way into product development including continuous Requirements Engineering (RE). RE is one of the key activities in Systems Engineering, requirements are treated as core artefacts in MBSE. Deriving necessary functions from requirements and defining logical system elements for these functions builds up technical baselines. Integrating scenarios from strategic planning into the system model extends typical MBSE approaches. While agility can also be supported by human models (see Chap. 6), the integration of scenario-based approaches means an explicit step towards integration of stakeholder needs being the fundamental background of the system model.

By stakeholder requirements, engineers capture the needs of future users and transfer them into system requirements. By nature, requirements are strongly dependent on volatility, uncertainty, complexity and ambiguity. Therefore, resilience of requirements definition has to be ensured in holistic Product Creation. According to Pariès et al., resilience is “the intrinsic ability of a system to adjust its functioning prior to, during, or following changes and disturbances, so that it can sustain required operations under both expected and unexpected conditions.” (Pariès et al. 2017: xxxvi). Building up on strategic input, Resilient Requirements Engineering (RRE, Fig. 1.9) describes a building block which is characterized by intense interactions between stakeholders. In RRE, decisions should be traceable to avoid ineffective iterations. Solving that challenge means to use potentials along the entire product lifecycle (Königs et al. 2012). Cause-effect chains can only be analyzed when relationships are traceable from physical system structure back to functions and even requirements (Eigner et al. 2019; Eigner 2021). Evaluation criteria can be derived in later engineering phases (Horber et al. 2020). In terms of changing requirements, RRE concepts need to be operationalized in anticipation of requirements modification, extension or deletion in every lifecycle phase. Stability of requirements simplifies economic considerations. At the same time, it is essential to be able to advance requirements as a key for customer satisfaction respectively user experience. While formal requirements

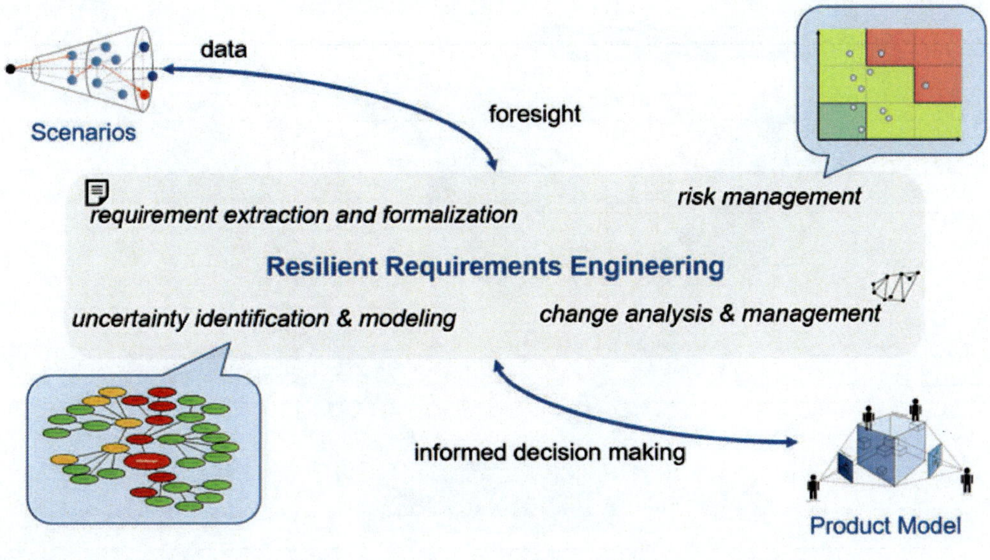

Fig. 1.9 Resilient Requirements Engineering: managing VUCA in early gPLC phase

specification is an enabler for traceability analysis and model-driven generation of artefacts, for instance code fragments, the effort of formalization often is a barrier in practical application of RE. Therefore, resilience needs to be tackled by both methods and supportive IT tools (Graessler et al. 2020; Gräßler et al. 2018b, 2019a).

Besides its primary use case in Strategic Planning, the Agile Scenario-Technique (see Sect. 1.3.4) can be used to anticipate future application scenarios as a background for use case analysis (Gräßler et al. 2017b). Requirements extraction and formalization should be combined with concepts of risk management, assessing the likelihood of requirement changes as well as their propagation within entire requirement sets (Graessler et al. 2020). In RRE, the semantics of risk and the prioritization of requirements are represented by explicit, formal annotations regarding uncertainty (Pottebaum and Gräßler 2020). Uncertainty metadata allows informed decision making, including systematic re-assessment along the product creation process.

1.3.5 Digital Worker and Learning Assistance

The term Digital and Virtual Product Creation is introduced to emphasize (a) the combination of Digital Engineering (DE) and Virtual Engineering (VE) and (b) the inclusion of the entire product creation process (Fig. 1.10). It involves the early, continuous application of digital tools and design-oriented models to support Product Creation for a continuously networked and integrated digital process chain.

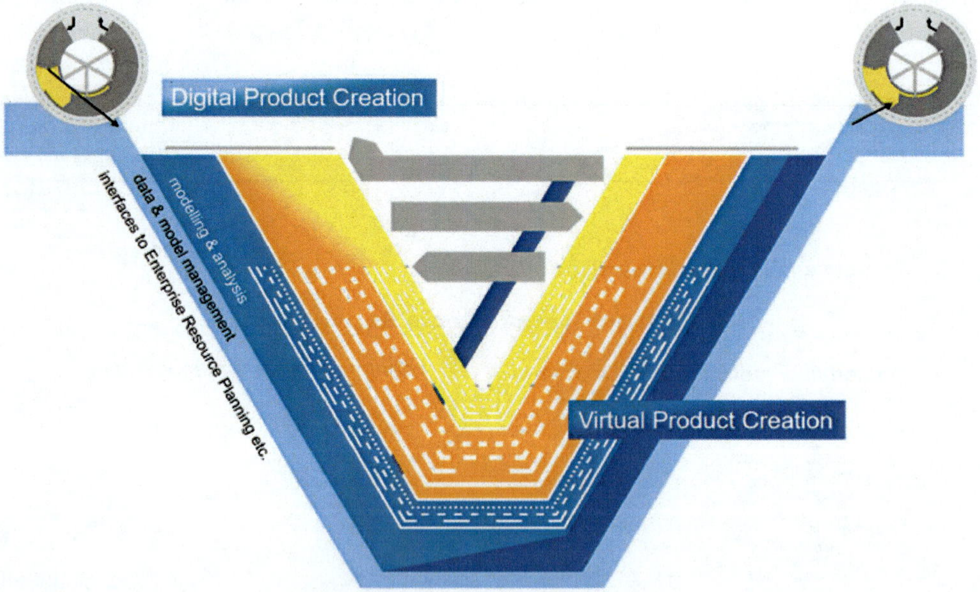

Fig. 1.10 Digital and virtual product creation supporting data and model management

Products under development are represented by digital representations (Anderl 2018). Emphasizing the shift towards model-based engineering, an evolution took place from integrated product models (Grabowski et al. 1993) to system models (Hick et al. 2019). Digital Engineering (DE) describes the early, continuous application of digital tools and design-oriented models for a continuously networked and integrated digital process chain (Gerhard 2020). Virtual Engineering (VE) is a section of Digital Product Creation. According to Ovtcharova, it is characterized by mapping the behavior of technical systems and interaction possibilities in virtual space (Ovtcharova 2020). On a generic layer, tasks are differentiated with regard to data generation and data management (Ovtcharova 2010). Lashin and Stark emphasize the importance of activities that are supported by digitalization and virtualization (Lashin and Stark 2021). Eigner proposes an engineering framework which aligns Systems Engineering with lifecycle management processes and tools (Eigner 2021). While DE is typically dedicated to product characteristics specified by engineers (cf. Rieg et al. 2019), VE enables the analysis of both directly defined as well as resulting product behaviors.

In terms of Digital and Virtual Product Creation, traceability needs to be implemented in the entire Digital Twin. The Digital Twin supports correlated service provision and subsequently extended business models. Digital Twins are, by nature, implementing the concept of traceability into product instances by combining digital master, digital shadow and relationships among these two building blocks (Stark et al. 2020; Stark and Damerau 2019; Schleich et al. 2017) including uncertainty considerations (Hausmann et al. 2021). While emphasizing obvious benefits, the development of a Digital Twin needs to be considered as a software engineering activity requiring intense competency with regard to the product's structure and behavior. Figure 1.11 visualizes the traceability with regard to the example of a turbine. This product is often cited for the evolution from consumed products to new, sustainability enabling hybrid performance bundles (cf. Baines et al. 2007).

Besides implementation to achieve circularity in material flow, MBSE can be utilized as a backbone of human-centered workplaces in production. While methods and tools for Agile Strategic Planning and RRE are designed to tackle challenges in early phases of future product creation, sustainability with its social dimension needs to incorporate humans along the entire process. An approach of experiential learning to support informed product development is presented in Chap. 4. In production, future work places are dedicated to those tasks which cannot be substituted by robotics. Human competency is required in these tasks. Future production must be flexible and changeable. One way to realize this is the human-centric production through context-sensitive digital assistance systems (Gogineni et al. 2019). Through a learning process, workers adapt to new conditions and tasks (see Fig. 1.12).

To improve experiential learning, adaptive assistance systems can be used that customize the support to the produced product and the individual worker (Gräßler et al. 2020c). Considering legal and ethical restrictions, Human Factors can be used to determine an individual worker profile. Human Factors elements can be understood as the interaction in

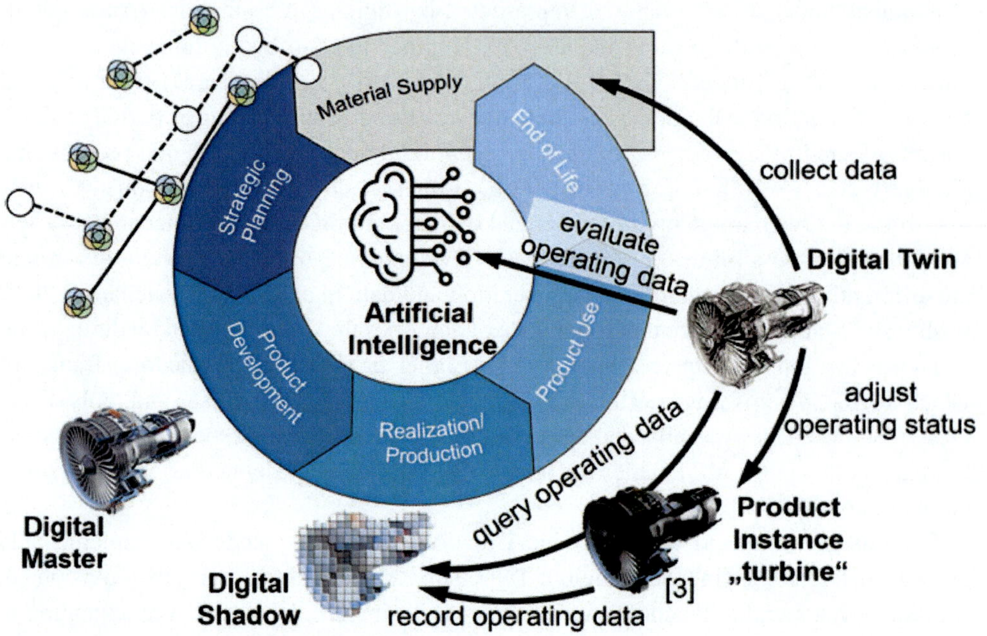

Fig. 1.11 Traceability based on MBSE enabling Digital Twins

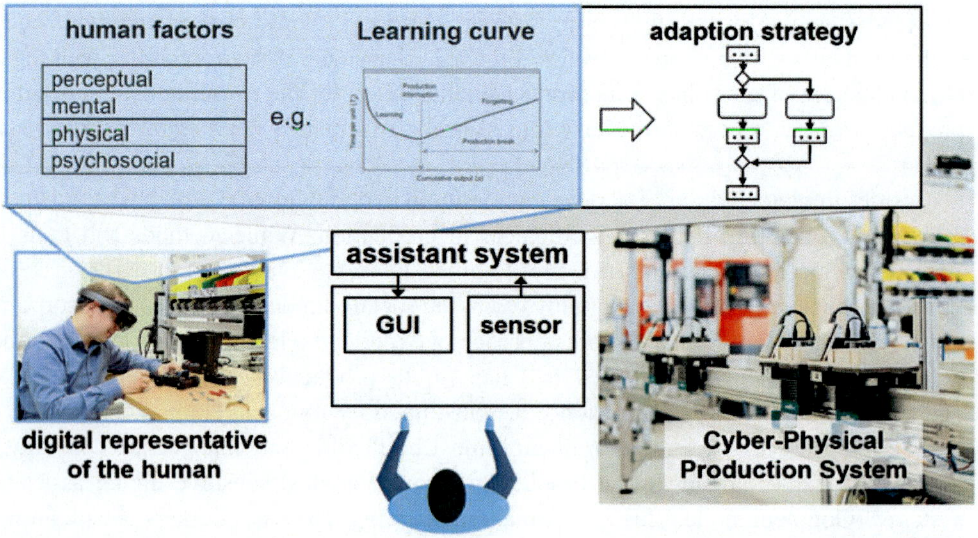

Fig. 1.12 Digital worker and learning assistance (Smart Automation Laboratory at Heinz Nixdorf Institute, Paderborn)

a human-system relationship. Roughly, these can be divided into perceptual, physical (for instance, posture), mental (like skills) and psychosocial aspects (including motivation). All aspects directly influence system outcomes. Specific Human Factors can be measured by sensors of the assistance system. These can be inputs from the worker or recorded process data. By using historical data, it is possible to develop a suitable digital representation of the worker (Gräßler and Pöhler 2017). The data obtained can then be used to adapt the output of the assistance system depending on the situation for a specific worker by using data mining methods and prognosis models such as learning curve models. This can be used to improve the learning experience (Gräßler et al. 2020a).

1.4 Summary and Outlook

The term "holistic" addresses various dimensions in Product Creation. Firstly, sustainability covers the whole range of economic, ecological and social impacts. Sustainable development needs to consider efficiency, consistency and sufficiency as primary values, while at the same time enabling new and innovative business models. Holistic Product Creation avoids defensive entrepreneurial positions, but targets sustainability in terms of value proposition. Secondly, within that frame, enterprises need to find solutions for tasks and problems in a volatile, uncertain, complex and ambiguous environment. Engineers have to be capable of understanding this environment, pinpointing actual needs to be addressed and opening up solution spaces based on an overwhelming breadth of technologies. Thirdly, enterprises need to adopt methodologies that enable multidisciplinary, collaborative work to handle the complexity of Cyber-Physical Systems. Instead of decomposition into disciplinary system elements, solution principles in all fields of Mechatronic and Cyber-Physical Systems need to be considered based on requirements and functional modelling. By means of training and working culture, approaches like Model-Based Systems Engineering have to be rolled out across disciplines and across stakeholders along the entire product lifecycle.

Agile Strategic Planning, Resilient Requirements Engineering as well as Digital Worker and Learning Assistance are specific directions of future scientific developments. Agile Scenario-Technique and RRE represent approaches of intensified front-loading, which is strongly emphasized in Systems Engineering in general and in the new V-Model for Mechatronic and Cyber-Physical Systems. Worker assistance is enabled by information derived from system models and secondary artefacts like digital twins. Utilizing the concepts of MBSE, the intention is to actively manage complexity instead of reducing it in these early phases. The aim is to enable guiding principles in an agile and traceable way. System models are enlarged by artefacts with inherent uncertainty, like projections in Scenario-Technique and application scenarios in RE.

Transferred to multidisciplinary product creation, such a set of building blocks enables a significant increase in competitiveness and sustainability. As a result, Product Engineering Systems have gradually gained acceptance in industrial practice. In order to implement the proposed generic Product Creation System (gPCS) in a company, the generic structure and elements must be tailored and prioritized according to specific restrictions of the company, the regarded branch of industry, product segment and application, where necessary.

References

Albers A, Matthiesen S, Revfi S, Schönhoff C, Grauberger P, Heimicke J (2019) Agile lightweight design—the extended target weighing approach in ASD—agile systems design using functional modelling with the C&C 2 -approach. In: Wartzack S, Schleich B, Guerreiro Gonçalves M, Eisenbart B, Völkl H (eds) Responsible design for our future: Proceedings of the 22nd international conference on engineering design (ICED19), vol 1. Cambridge University Press, Cambridge, pp 2667–2676

Anderl R (2018) Virtuelle Produktentstehung. In: Grote K-H, Bender B, Göhlich D (eds) Dubbel—Taschenbuch für den Maschinenbau. Springer, Berlin, pp 2012–2028

Ankele A, Staiger TJ, Koch T (2008) Chefsache Produktionssystem: Pfade zum Erfolg. LOG_X Verl., Stuttgart

Atzberger A, Nicklas SJ, Schrof J, Weiss S, Paetzold K (2020) Agile development of physical products: a study on the current state of industrial practice. Universität der Bundeswehr München, Neubiberg

Baines TS, Lightfoot HW, Evans S, Neely A, Greenough R, Peppard J, Roy R, Shehab E, Braganza A, Tiwari A, Alcock JR, Angus JP, Bastl M, Cousens A, Irving P, Johnson M, Kingston J, Lockett H, Martinez V, Michele P, Tranfield D, Walton IM, Wilson H (2007) State-of-the-art in product-service systems. Proc Inst Mech Eng Part B J Eng Manuf 221:1543–1552. https://doi.org/10.1243/09544054JEM858

Bayer TJ, Cooney LA, Delp CL, Dutenhoffer CA, Gostelow RD, Ingham MD, Jenkins JS, Smith BS (2010) An operations concept for integrated model-centric engineering at JPL. In: 2010 IEEE aerospace conference. IEEE, pp 1–14

Beck K, Beedle M, van Bennekum A, Cockburn A, Cunningham W, Fowler M, Grenning J, Highsmith J, Hunt A, Jeffries R et al (2001) Manifesto for agile software development

Brundtland GH (1987) Report of the World Commission on Environment and Development: Our Common Future, Oslo

Cooper RG (1990) Stage-gate systems: A new tool for managing new products. Bus Horiz 33:44–54

Dazer M, Bertsche B, Vajna S (2020) Fulfilment attributes. In: Vajna S (ed) Integrated design engineering. Springer, Cham, pp 383–418

Delligatti L (2014) SysML distilled: A brief guide to the systems modeling language. Addison-Wesley, Upper Saddle River, NJ

Dombrowski U, Mielke T (2015) Ganzheitliche Produktionssysteme. Springer, Berlin

Drave I, Rumpe B, Wortmann A, Berroth J, Hoepfner G, Jacobs G, Spuetz K, Zerwas T, Guist C, Kohl J (2020) Modeling mechanical functional architectures in SysML. In: Syriani E, Sahraoui H (eds) Proceedings of the 23rd ACM/IEEE international conference on model driven engineering languages and systems. ACM, New York, pp 79–89

Eigner M (2021) System Lifecycle Management: Digitalisierung des Engineering. Springer, Berlin

Eigner M, Koch W, Muggeo C (eds) (2017) Modellbasierter Entwicklungsprozess cybertronischer Systeme: Der PLM-unterstützte Referenzentwicklungsprozess für Produkte und Produktionssysteme. Springer, Berlin

Eigner M, Dickopf T, Apostolov H (2019) Interdisziplinäre Konstruktionsmethoden und -prozesse zur Entwicklung cybertronischer Produkte—Teil 2. Konstruktion 2019:69–75

Elaasar M, Rouquette N, Jenkins S, Gerard S (2019) The case for integrated model centric engineering. In: Proceedings of the 10th model-based enterprise summit (MBE 2019). National Institute of Standards and Technology, Gaithersburg, MD, pp 9–16

Ellen MacArthur Foundation (2015) Growth within: a circular economy vision for a competitive Europe. Ellen MacArthur Foundation and McKinsey Center for Business and Environment, Brussels

Estefan J (2007) Survey of model-based systems engineering (MBSE) methodologies. Incose MBSE Focus Group 25(8):1–12

France R, Rumpe B (2007) Model-driven development of complex software: a research roadmap. In: Briand LC, Wolf AL (eds) 29th international conference on software engineering (ICSE 2007): FOSE '07; 23–25 May 2007, Minneapolis, MN, Future of software engineering. IEEE Computer Society, Los Alamitos, CA, pp 37–54

Gerhard D (2017) Product lifecycle management challenges of CPPS. In: Biffl S, Lüder A, Gerhard D (eds) Multi-disciplinary engineering for cyber-physical production systems. Springer, Cham, pp 89–110

Gerhard D (2020) Digital Engineering–Basis für Smarte Produkte und Services. In: Steven M, Dörseln JN (eds) Smart Factory: Einsatzfaktoren—Technologie—Produkte, 1st edn, pp 21–35

Gill H (2006) Cyber-physical systems. NSF

Gogineni S, Exner K, Stark R, Nickel J, Oeler M, Witte H (2019) Semantic assistance system for providing smart services and reasoning in aero-engine manufacturing. In: Garoufallou E, Fallucchi F, William De Luca E (eds) Metadata and semantic research. Springer, Cham, pp 90–102

Göhlich D, Fay T-A (2021) Arbeiten mit Anforderungen: Requirements Management. In: Bender B, Gericke K (eds) Pahl/Beitz Konstruktionslehre. Springer, Berlin, pp 211–229

Göhlich D, Nagel K, Syré AM, Grahle A, Martins-Turner K, Ewert R, Miranda Jahn R, Jefferies D (2021) Integrated approach for the assessment of strategies for the decarbonization of urban traffic. Sustainability 13:839. https://doi.org/10.3390/su13020839

Gräßler I (2015a) Generic product creation system. In: 13th industrial simulation conference (ISC) 2015, pp 173–177

Gräßler I (2015b) Implementation-oriented synthesis of mechatronic reference models. In: Fachtagung Mechatronik 2015, pp 167–172

Gräßler I (2018) Competitive engineering in the age of Industry 4.0 and beyond. In: Proceedings of the 12th international tools and methods of competitive engineering, pp XIX–XXVIII

Gräßler I, Pöhler A (2017) Integration of a digital twin as human representation in a scheduling procedure of a cyber-physical production system. In: Proceedings of 2017 IEEE international

Gräßler I, Pottebaum J (2021) Generic product lifecycle model: a holistic and adaptable approach for multi-disciplinary product-service systems. Appl Sci 11:4516–4540. https://doi.org/10.3390/app11104516

Gräßler I, Pottebaum J, Scholle P (2017a) Integrated process and data model for agile strategic planning. In: Vajna S (ed) 11th international workshop on integrated design engineering

Gräßler I, Scholle P, Pottebaum J (2017b) Integrated process and data model for applying scenario-technique in requirements engineering. In: Design Society (ed) ICED17: 21st international conference on engineering design. Design Society, Vancouver, pp 261–270

Gräßler I, Dattner M, Bothen M (2018a) Main feature list as core success criteria of organizing requirements elicitation. In: R&D management conference, Milan, 2019: R&Designing Innovation: Transformational Challenges for Organizations and Society

Gräßler I, Scholle P, Hentze J, Oleff C (2018b) Semi-automatized assessment of requirement interrelations. In: Marjanović D, Štorga M, Škec S, Bojčetić N, Pavković N (eds) Design 2018: Proceedings of the 15th international design conference. Design Society, pp 325–334

Gräßler I, Thiele H, Oleff C, Scholle P, Schulze V (2019a) Method for analysing requirement change propagation based on a modified pagerank algorithm. In: Wartzack S, Schleich B, Guerreiro Gonçalves M, Eisenbart B, Völkl H (eds) Responsible design for our future: proceedings of the 22nd international conference on engineering design (ICED19), vol 1. Cambridge University Press, Cambridge, pp 3681–3690

Gräßler I, Scholle P, Thiele H (2019b) Strategische Planung in Plattformen und Eco-Systemen mittels Szenario-Technik. Fachtagung Mechatronik 2019. https://doi.org/10.17619/UNIPB/1-794

Gräßler I, Roesmann D, Pottebaum J (2020a) Model based integration of human characteristics in production systems: a literature survey. In: 14th CIRP conference on intelligent computation in manufacturing engineering. Elsevier B.V

Gräßler I, Scholle P, Thiele H (2020b) Scenario technique. In: Vajna S (ed) Integrated design engineering. Springer, Cham, pp 615–645

Gräßler I, Roesmann D, Pottebaum J (2020c) Traceable learning effects by use of digital adaptive assistance in production. Procedia Manuf 45:479–484. https://doi.org/10.1016/j.promfg.2020.04.058

Gräßler I, Wiechel D, Pottebaum J (2021) Role model of model-based systems engineering application. In: 19th drive train technology conference (ATK 2021)

Grabowski H, Anderl R, Polly A (1993) Integriertes Produktmodell, 1st edn. Entwicklungen zur Normung von CIM. Beuth, Berlin

Graessler I (2004) Excellency in industrial product development. A promising way to success. In: Tichkiewitch S, Brissaud D (eds) Methods and tools for co-operative and integrated design. Springer, Dordrecht, pp 233–240. https://doi.org/10.1007/978-94-017-2256-8_20

Graessler I, Hentze J (2020) The new V-Model of VDI 2206 and its validation. at-Automatisierungstechnik 68:312–324. https://doi.org/10.1515/auto-2020-0015

Graessler I, Oleff C, Scholle P (2020) Method for systematic assessment of requirement change risk in industrial practice. Appl Sci 10:8697–8725. https://doi.org/10.3390/app10238697

Hausmann M, Koch Y, Kirchner E (2021) Managing the uncertainty in data-acquisition by in situ measurements: a review and evaluation of sensing machine element-approaches in the context of digital twins. Int J Prod Lifecycle Manag 13(1):48–65. Inderscience Enterprises Limited, ISSN 1743-5110

Hick H, Bajzek M, Faustmann C (2019) Definition of a system model for model-based development. SN Appl Sci 1:15

Hick H, Angel HF, Kranabitl P, Wagner-Skacel J (2020) Decision-making and the influence of the human factor. In: Hick H, Küpper K, Sorger H (eds) Systems engineering for auto-motive powertrain development. Springer, Cham, pp 1–26

Horber D., Schleich B., Wartzack S (2020) Conceptual model for (semi-) automated derivation of evaluation criteria in requirements modelling. In: Proceedings of the Design Society: DESIGN conference, pp 937–946

INCOSE (2014) Systems engineering handbook: a guide for system life cycle processes and activities, San Diego

INCOSE (2018) Integrated data as a foundation of systems engineering: whitepaper by the Requirements Working Group, San Diego

Isermann R (2005) Mechatronic systems: fundamentals. Springer, London

Kjaer LL, Pigosso DCA, Niero M, Bech NM, McAloone TC (2019) Product/service-systems for a circular economy: the route to decoupling economic growth from resource consumption? J Ind Ecol 23:22–35. https://doi.org/10.1111/jiec.12747

Kohl JL, van der Schoor MJ, Syré AM, Göhlich D (2020) Social sustainability in the development of service robots. Proc Des Soc Des Conf 1:1949–1958. https://doi.org/10.1017/dsd.2020.59

Königs SF, Beier G, Figge A, Stark R (2012) Traceability in systems engineering – review of industrial practices, state-of-the-art technologies and new research solutions. Adv Eng Informat 26:924–940. https://doi.org/10.1016/j.aei.2012.08.002

Konrad C, Jacobs G, Rasor R, Riedel R, Katzwinkel T, Siebrecht J (2019) Enabling complexity management through merging business process modeling with MBSE. Procedia CIRP 84:451–456. https://doi.org/10.1016/j.procir.2019.04.267

Lashin G, Stark R (2021) Virtuelle Produktentwicklung. In: Bender B, Gericke K (eds) Pahl/Beitz Konstruktionslehre. Springer, Berlin, pp 1097–1153

Leder N, Kumar M, Rodrigues VS (2020) Influential factors for value creation within the circular economy: framework for waste valorisation. Resour Conservat Recycl 158:104804. https://doi.org/10.1016/j.resconrec.2020.104804

Lüdeke-Freund F, Carroux S, Joyce A, Massa L, Breuer H (2018) The sustainable business model pattern taxonomy—45 patterns to support sustainability-oriented business model innovation. Sustain Prod Consumpt 15:145–162. https://doi.org/10.1016/j.spc.2018.06.004

Madni AM, Sievers M (2018) Model-based systems engineering: Motivation, current status, and research opportunities. Syst Eng 21:172–190. https://doi.org/10.1002/sys.21438

Matthiesen S, Grauberger P, Bremer F, Nowoseltschenko K (2019) Product models in embodiment design: an investigation of challenges and opportunities. SN Appl Sci 1. https://doi.org/10.1007/s42452-019-1115-y

McAloone T, Pigosso D (2021) Ökodesign. Pahl/Beitz Konstruktionslehre. Springer Vieweg, Heidelberg, pp 975–1021

McDonough W, Braungart M (2002) Cradle to cradle: Remaking the way we make things, 1st edn. North Point Press, New York

Meboldt M (2008) Mentale und formale Modellbildung in der Produktentstehung—als Beitrag zum integrierten Produktentstehungs-Modell (iPeM). Dissertation, Karlsruher Institut für Technologie (KIT)

Meier H, Uhlmann E (eds) (2012) Integrierte Industrielle Sach- und Dienstleistungen. Springer, Berlin

Mont O (2002) Clarifying the concept of product–service system. J Cleaner Prod 10:237–245. https://doi.org/10.1016/S0959-6526(01)00039-7

Nußholz J (2017) Circular business models: defining a concept and framing an emerging research field. Sustainability 9:1810. https://doi.org/10.3390/su9101810

Ohno T (1988) Toyota production system: beyond large-scale production. Productivity Press, Cambridge, MA

Ostertag A, Schlegl F, Gienger A, Wagner J, Dazer M, Bertsche B, Albrecht A, Leistner P, Tarín C, Sawodny O (2020) Reliable design of adaptive load-bearing structures with focus on sustainability. In: Proceedings of the European Safety and Reliability Conference and the Probabilistic Safety Assessment and Management Conference, Singapore

Ovtcharova J (2010) Virtual engineering: principles, methods and applications. In: Marjanović D (ed) Design 2010: Proceedings of the 11th international design conference. Design Society, pp 1267–1274

Ovtcharova J (2020) Virtual engineering & remote collaboration. In: Leupold A, Wiebe A, Glossner A (eds) IT-Recht. Recht, Wirtschaft und Technik der digitalen Transformation. C.H. Beck, Munich

Pariès J, Wreathall J, Hollnagel E (2017) Resilience engineering in practice: a guidebook. Ashgate studies in resilience engineering series. Chapman and Hall/CRC, Boca Raton, FL

Pottebaum J, Gräßler I (2020) Informationsqualität in der Produktentwicklung: Modellbasiertes Systems Engineering mit expliziter Berücksichtigung von Unsicherheit. Konstruktion

Raabe D, Tasan CC, Olivetti EA (2019) Strategies for improving the sustainability of structural metals. Nature 575:64–74. https://doi.org/10.1038/s41586-019-1702-5

Reibnitz Uv (1992) Szenario-Technik: Instrumente für die unternehmerische und persönliche Erfolgsplanung, 2nd edn. Gabler, Wiesbaden

Rieg F, Hackenschmidt R, Alber-Laukant B (2019) Finite Elemente—Analyse für Ingenieure. Grundlagen und praktische Anwendungen mit Z88 Aurora. Hanser, München

Sahin T, Huth T, Axmann J, Vietor T (2020) A methodology for value-oriented strategic release planning to provide continuous product upgrading. In: 2020 IEEE international conference on industrial engineering and engineering management (IEEM). IEEE, pp 1032–1036

Schleich B, Anwer N, Mathieu L, Wartzack S (2017) Shaping the digital twin for design and production engineering. CIRP Ann 66(1):141–144

Schmidt DC (2006) Model-driven engineering: guest editor's introduction. Computer 39:25–31. https://doi.org/10.1109/MC.2006.58

Stark R, Damerau T (2019) Digital twin. In: Chatti S, Tolio T (eds) CIRP encyclopedia of production engineering, vol 66. Springer, Berlin, pp 1–8

Stark R, Thoben K-D, Gerhard D, Hick H, Kirchner E, Anderl R, Wartzack S, Krause F-L, Grässler I, Pottebaum J, Schleich B, Stelzer R, Hick H u K, Klein P u K, Saske B u K, Czwick C, Gogineni S, Klimmeck L, Bajzek M, Jacobs G, Berroth J, Zimmermann T, Kranabitl P, Göckel N (2020) WiGeP Positionspapier - Digitaler Zwilling

United Nations (2015) Transforming our world: the 2030 agenda for sustainable development: A/RES/70/1

VDI (2012) VDI 2870 lean production systems: Sheet 1—Basic principles, introduction, and review, Düsseldorf

VDI (2020) VDI 2206 Development of cyber-physical mechatronic systems (CPMS). Verein Deutscher Ingenieure, Düsseldorf

Vogel-Heuser B, Fantuzzi C, Wimmer M, Böhm M, Fay A (2019) Herausforderungen in der interdisziplinären Entwicklung von Cyber-Physischen Produktionssystemen. at-Automatisierungstechnik 67:445–454. https://doi.org/10.1515/auto-2018-0144

Wilking F, Schleich B, Wartzack S (2020) MBSE along the value chain—an approach for the compensation of additional effort. In Proceedings of the IEEE 15th international conference of system of systems engineering (SoSE). IEEE 62020, pp 61–66

Model of SGE: System Generation Engineering as Basis for Structured Planning and Management of Development

Albert Albers and Simon Rapp

Abstract

Providing methods and processes for structured planning and management of the development of new systems requires a description model, which describes fundamental phenomena in the development of new systems. Such a model should be based on the theory of socio-technical systems, it should be applicable to the wide range of different types of development projects which are observable in practice and it should provide formalisms for quantitative empirical studies and computer support. The model of SGE—System Generation Engineering aims at this goal. It describes the development of new systems with two fundamental hypotheses. First, every development of a new system is based on a reference system, consisting of subsystems from already existing systems. Second, based on the reference system, the subsystems of a new system are developed by three types of variation: carryover variation, attribute variation and principle variation. The model finds broad approval in development practice and allows also for the description of development increments as well as for the description of production and validation systems. Variation types and characteristics are key factors for innovation potential and development risk in the development of a new system. They are also important factors for the situation specific methodical support of development activities.

A. Albers (✉) · S. Rapp
Karlsruhe Institute of Technology (KIT), IPEK – Institute of Product Engineering, Karlsruhe, Germany
e-mail: albert.albers@kit.edu

D. Krause, E. Heyden (eds.), *Design Methodology for Future Products*,
https://doi.org/10.1007/978-3-030-78368-6_2

2.1 Motivation and Requirements for a Description Model as Basis for Planning and Management of Development

Formulating and providing methods and processes for a certain purpose requires models to describe the underlying phenomena which are subject to those methods and processes. Hence, methods and processes for planning and management of development require a description model[1] for fundamental phenomena in the development of a new system. To ensure applicability and transferability of development methods and processes along different project stages and across different projects such a description model has to fulfil several requirements:

- In development practice the extent to which a new system is actually newly developed can differ strongly across different development projects, resulting in a wide range of observable shares of new development in different projects. A description model should be applicable across this wide range (Albers et al. 2015).
- For modern development projects, e.g. for mechatronic systems, most often developers with background from different disciplines have to collaborate. A model to describe fundamental phenomena of such projects should thus be applicable across different disciplines. Cross-disciplinary applicability is a main motivation of the technical system theory (Ropohl 1979, 2009) which therefore provides a suitable basis for a description model of system development. At the same time applicability of the description model across several branches is also strengthened by this basis.
- During a development process the system in development is—with iterations—continuously specified and concretized. This results in system models and specifications with different levels of detail at different time points in the development process (see e.g. Ponn and Lindemann 2011). The technical system theory as basis for a description model of this process allows for depicting those different levels of detail within the process (hierarchical concept of the system theory).
- The development of a new system is usually closely linked with other systems and their development, especially a production system and a validation system (Albers 1994; Albers and Meboldt 2007; Albers et al. 2016a). A description model for the development of a new system should cover these interdependencies. Furthermore, methods and processes for planning and management of development can have different purposes, for example assessment of innovation potential and development risk, identification and

[1] We use the term 'description model' to emphasize the purpose of the model. The purpose is the description of fundamental observations in the development of new systems. It is not the purpose of the model as such to explain these observations or to structure engineering processes. The latter can be the purpose of processes and methods developed based on the description model. Hence, the purpose of the description model is to give a concept to describe fundamental observations. However, we intentionally do not use a term such as 'concept' or 'concept model' because 'concept' is usually rather associated with a certain level of maturity in the design of a new system.

design of new product ideas, validation support or knowledge management in general. Common and integrated use of different methods and processes from these different fields can be facilitated, if they are based on a common description model of the fundamental phenomena in the development of a new system.

- Important objectives in the planning and management of system development are innovation potential and development risk. To generally enable the management of system development with respect to these objectives, a description model must depict root causes of innovation potential and development risk with independent model elements.[2]
- The description model should provide sufficient formalisms to measure and track key performance indicators in planning and management of development and to allow for the development of computer-based support.

The model of SGE—System Generation Engineering[3] was developed to address these requirements. The fundamental elements and hypotheses in the model are explained in the next section.

2.2 Fundamental Elements and Hypotheses in the Model of SGE: System Generation Engineering

The first hypothesis in the model of SGE is: **every development of a new system is based on a reference system** (Albers et al. 2015, 2019a). The reference system for a new system is a system whose subsystems originate from already existing or already planned socio-technical systems and the associated documentation and which are the basis and starting point for the development of the new system[4] (Albers et al. 2019a). The reference system can therefore include for example, but neither exclusively or necessarily, preceding

[2]To illustrate this aspect, an example can be given from the field of engineering changes: Langer et al. (2012) define critical engineering change (EC) for a survey as follows: "A critical EC endangers the start of production or the whole project in terms of cost, time, resource involvement, or feasibility (e.g. changing customer requirements, changes for a massive cost reduction)". If the definition is used without the examples at the end and if no further general criteria are given for what makes an EC a critical EC, the only possible way to determine, whether an EC is a critical EC, is an individual case analysis. However, this can be very time consuming and does not provide a process that is transferrable to further cases and projects. The examples mentioned in the brackets might serve as additional criteria for criticality. However, they have a rather exemplary nature and there might for example be cases, where customer requirements change without resulting in a critical EC.

[3]Originating from the field of product development, in research so far, the model is called "model of PGE—Product Generation Engineering". Considering the applicability for different types of systems as well as for example for system-of-systems development it seems more suitable to refer to it as "System Generation Engineering" (see also e.g. Albers et al. 2017a).

[4]Based on the definition in Albers et al. (2019a) the formulation of the definition here is adjusted slightly following the previous remark.

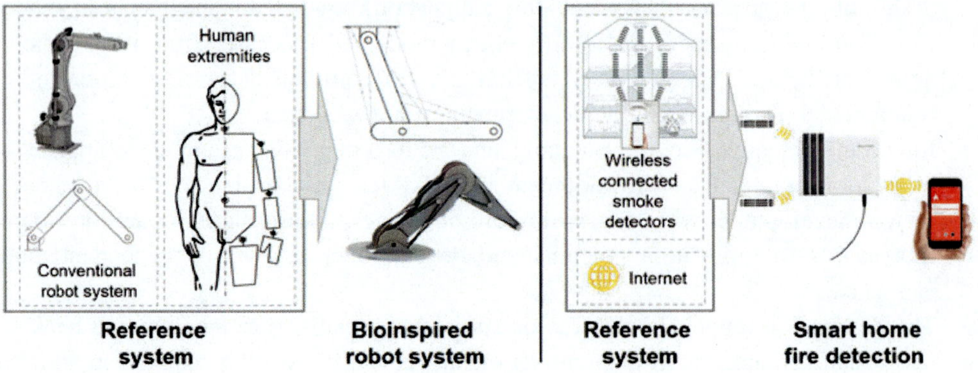

Fig. 2.1 Examples of systems and subsystems of the underlying reference system. Simplified illustration with figures from Bartz et al. (2019), Bartz (2019) and (Hekatron Brandschutz)

products of a company, competitor's products, parts of those products, corresponding partial product models and product documentation. This documentation may comprise for example requirements, models, test reports, technical concepts or descriptions of them, but also corresponding production processes. The reference system is established by developers of a new system by their intention to use selected references as basis and starting point for the development of their own system. Figure 2.1 shows examples of systems and some subsystems of the underlying reference systems.[5]

The shown examples support the hypotheses that every development of a new system is based on a reference system, reaching from cases such as a new generation of an established car product line to cases where there is no specific preceding product generation in the market, in recent years for example the first electric cars in various companies, but also the first smartphones of different brands (for some more examples see also Albers et al. 2020a). In that way every new system can be seen as a new system generation.

Continuously searching for and identifying potential subsystems of the reference system in a development project as well as analysing, evaluating and selecting them is a core task for developers in the project (Albers et al. 2019a).

The second main hypothesis in the model of SGE is: **based on the reference system a new system is developed by a composition of three different types of variation of subsystems: carryover variation, attribute variation and principle variation** (Albers et al. 2015, 2020b). Figure 2.2 illustrates the three different types of variation using the example of the smart home fire detection system.

[5]The term "reference product" is also used in some works based upon the PGE model, especially earlier ones. Reference products are existing products, which are a source for subsystems of the reference system. In general, however, subsystems of the reference system can also come from other sources, not only existing products which are already available in a market.

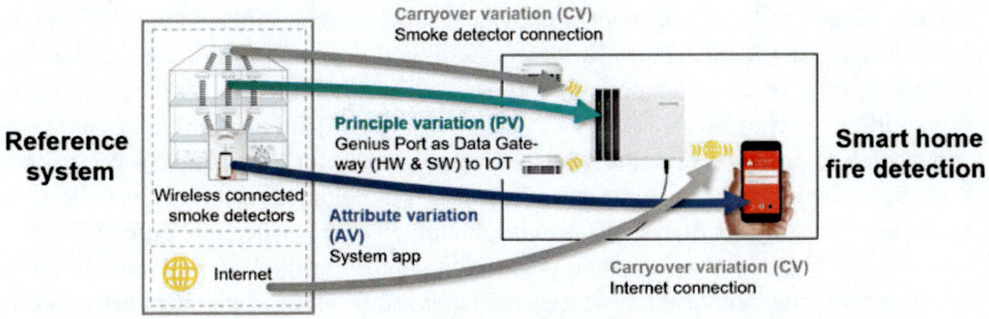

Fig. 2.2 The three different types of variation by the example of selected subsystems of the smart home fire detection system from Fig. 2.1 (simplified illustration)

In general, the three different types of variation are defined as follows (Albers et al. 2015, 2020b)[6]:

- **Carryover variation** (CV) is an activity for the development of a subsystem in a new system generation, where the underlying subsystem from the reference system is carried over and adjustments are made at most at the interfaces of the subsystem due to system integration.
- **Attribute variation** (AV) is an activity for the development of a subsystem in a new system generation, where, starting with a subsystem from the reference system, the elements and links within this subsystem are maintained in principle but their attributes are at least partially altered.
- **Principle variation** (PV) is an activity for the development of a subsystem in a new system generation, where, starting with a subsystem from the reference system, elements and links between them are added or removed.

Principle variation always goes along with attribute variation as the resulting new subsystem structure with its elements and links has to be specified. In the example of the development of zero-emission electric buses by (Göhlich et al. 2018, see also Chap. 7) the electric drive, the battery and the charging system are developed by principle variation, if the underlying subsystems of the reference system are taken from a conventional bus. Other subsystems such as body or interior are developed by attribute variation or carryover variation.

Applying the definitions above to specific domains or views on systems and different types of systems gives criteria for identifying the different types of variation (Albers et al. 2020b). One example is the embodiment-function relation of mechatronic systems, which can for example be modelled using the C&C²-approach (Albers et al. 2002). In case of an attribute variation[7] all structures and contacts (working surface pairs) and their connections

[6]Those definitions follow Albers et al. (2020b) with minor updates.

[7]In this case it can also be referred to as embodiment variation

are maintained but the embodiment of individual structures and surfaces might be changed. In a principle variation there are also structures or working surface pairs added or removed (Albers et al. 2016c). Another possible view on systems to find criteria for the identification of the different types of variation are system properties. Removing or adding properties corresponds to principal variation. Maintaining properties, for example the acceleration of a car, but changing it, for example increasing the acceleration, is an attribute variation.[8] An example for other types of systems than mechatronic systems where the different types of variation are observable is a system of coupled simulation models, for example for the integrated investigation of the production process and resulting component behaviour for fiber reinforced materials. Adding or removing individual models from such a system is then a principle variation (Albers et al. 2020b).

All variations by which a new system is developed together form an operator V that describes the emergence of a new system generation based on the underlying reference system R_i.[9] This can be denoted as follows, using furthermore G_i for a system generation and $i = n$ for the system generation, which is currently in development and nearest to market entry. CS_n are all subsystems within G_n which are developed by carryover variation. AS_n and PS_n are those developed by attribute variation or principle variation, respectively (Albers et al. 2015, 2019a, 2020b):

$$R_n \xrightarrow{V} G_n = CS_n \cup AS_n \cup PS_n$$

Based on this formulation it is also possible to calculate the share of subsystems which are developed by a specific type of variation compared to the overall number of subsystems in the new system generation. The result is a variation share, denoted below for the share of principal variation $\delta_{PV,\ n}$ as an example, working analogously for the share of carryover variation and the share of attribute variation (Albers et al. 2015).

$$\delta_{PV,n} = \frac{|PS_n|}{|G_n|} = \frac{|PS_n|}{|CS_n \cup AS_n \cup PS_n|} [\%]$$

[8]In this case it is an attribute variation of a product property. This is not the same as an attribute variation of a product characteristic. Although product properties and product characteristics are connected by relations in the way that developers define characteristics to realise desired properties (Weber 2005) attribute variations of characteristics and properties can be distinguished. However, as a result of their relation, there is presumably a correlation in the way that attribute variations of properties are often achieved by attribute variations of characteristics. But in general, it is also possible to realise an attribute variation of a property (e.g. acceleration of a car) by a principle variation of a characteristic (switch from combustion engine to electrical drive).

[9]It is important to note, that in general R_i is not the same as G_{i-1}. However, a great amount of elements in R_i might origin from G_{i-1}.

The variation shares are potential key factors for the planning and management of development. This is described in more detail in the next section. The calculation can be refined, for example by using individual weight factors for different subsystems, corresponding to their relative importance for the fulfilment of system functions or also by taking in to account the amount of linkages of a subsystem to other subsystems (Albers et al. 2019b).

The nomenclature can be expanded by further indices to include further information about a system generation, for example about the intended customer or market or whether it is a variant[10] of a product that is already in the market. Some examples are given below based on (Albers et al. 2020c).

$$G_i^{\{product\ line, customer, user, \ldots\}}$$

Beyond modelling the development of a new system and the dependency of that development with other, preceding systems, the model of SGE is also capable of describing the development of an individual system generation more in detail. Different increments, iterations or maturity levels[11] in the development of a new system generation are also the result of a composition of the three different types of variation. From this perspective those increments can be described as generations within the engineering of a system generation and are thus called engineering generations (Albers et al. 2016b). The nomenclature for engineering generations follows the one for system generations, using "E" instead of "G" and an additional index to mark the corresponding system generation. The second engineering generation in the system generation currently in development is therefore denoted as $E_{n,2}^{\{\ldots\}}$, for example. Figure 2.3 shows variation shares along six engineering generations in the development of a new sports car generation.

The model of SGE is also applicable to the description of the development of production and validation systems which are usually closely connected with the development of a new system generation and developed alongside (Albers et al. 2016a).

The next section introduces important aspects for the planning and management of development based on the presented model of SGE.

[10]The development of variants of a product or system in general can also be described by the presented model. Hence, a new system variant can also be understood as a new system generation. Variants are usually characterized by relatively high shares of carryover variation and they are usually at the same time in the market as the system generation from which they are derived (Peglow et al. 2017).

[11]There is a broad variety of approaches to define maturity levels. We are not referring to a particular one here.

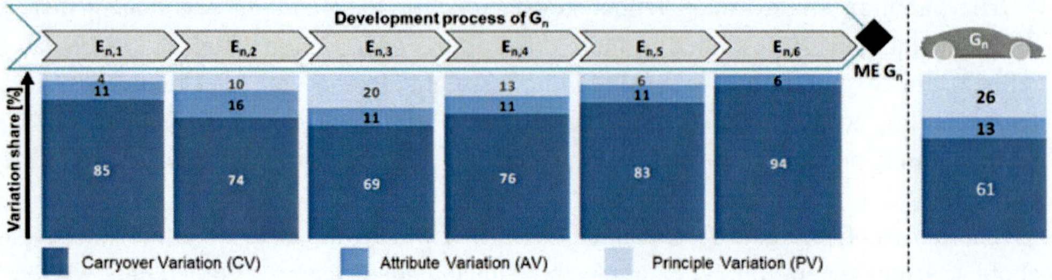

Fig. 2.3 Variation shares along six engineering generations in the development of a new sports car generation (Albers et al. 2019c, translated)

2.3 Variation Types and Reference System Characteristics as Key Factors for Innovation Potential and Development Risks, Planning and Management

Innovation potential and development risk[12] are important factors in the planning and management of development. The phenomena described by the two main hypotheses in the model of SGE are key factors with regard to innovation potential and development risk.

On the one hand, innovation potential usually requires to improve or create new system properties and functions for a set of increased or new customer benefit, user benefit and supplier benefit, compared to existing systems (Albers et al. 2018c). This is, in general, achieved by technical novelty through attribute variation and principle variation but also through the use of subsystems in the reference system that are new to a specific branch. An example for the latter is the successive integration of infotainment technology known from smartphones into cars. On the other hand, variations as well as the selected reference system influence development risks.

An increasing share of attribute variation and principle variation tends to increase development risks due to technical novelty of the new solutions. However, it is important to point out that it is not possible to state in general, that risk increases when switching from carryover variation to attribute variation and then principle variation. Depending on the extent of an attribute variation it is possible to observe attribute variations, which pose more risk than some principle variations. Furthermore, carryover variation can also hide risks, for example if the operating boundary conditions for a subsystem in the new system differ strongly from the original use.

Another source of risk is the reference system with its subsystems. Subsystems of the reference system can be characterized more in detail by different characteristics. One

[12]There is a broad variety of understandings and definitions for both these terms in literature. Regarding innovation we follow the understanding in Albers et al. (2018c) which is based on Schumpeter (1927). The focus for risk is on technical risk and development cost without being strictly limited to those two.

important characteristic in this respect is the organisational origin of subsystems in the reference system (Albers et al. 2016b). The organisational origin usually affects the accessibility of knowledge about a subsystem of the reference system. If a subsystem of the reference systems is from the own company, usually product documentation and even implicit knowledge of the people, who developed it, is accessible. If a subsystem of the reference system is from another company, the access to technical documentation is often limited.[13] If it is from another branch, the developers' understanding of system behaviour and relevant design parameters might be less detailed than it is for systems from their own branch. In addition, required production technology and validation systems might initially not be available. Furthermore, if subsystems of the reference system come from research activities, for example at universities, they might have not yet been proven in the customer market. Other characteristics of the reference system and its subsystems that contribute to development risks are the engineering discipline (for example mechanics vs. electrics) or the age in terms of time that passed since something was engineered as the availability of knowledge might drop over time.[14] The illustrated relations between variations, the reference system and development risks provide a basis for the evaluation of solution concepts.

Because an initial reference system is usually already known in the early stage of a development, such evaluations are possible early. Figure 2.4 shows an approach for such an evaluation using the tendencial influence of variations and the organisational origin of subsystems of the reference system (Albers et al. 2017a). The availability of a reference system from the beginning of a development project is an important aspect for the understanding of early stages in development that results from the model of SGE (Albers et al. 2017a).

Now and then examples can be found in the media where the previously described relations between variations, the reference system and development risks likely contributed to project failure. Selected examples are shown in Fig. 2.5.

The relations between variations, the reference system and development risks also provide a basis for the planning and management of a development project to handle the identified risks. If subsystems in the reference system are from outside the company or another discipline, possible actions are the initialisation of suitable cooperations or building up lacking competencies in the own company, for example by pre-development projects or by hiring corresponding specialists.

For the planning of the development process to implement different variations it is necessary to investigate the relation between variations and specific engineering activities in more detail. Such specific engineering activities are for example (but not only) 'idea

[13]Regarding the knowledge about function-embodiment relations see for example corresponding risk clusters described in Matthiesen et al. (2018).

[14]The identification of further potentially relevant characteristics is subject to current and future research.

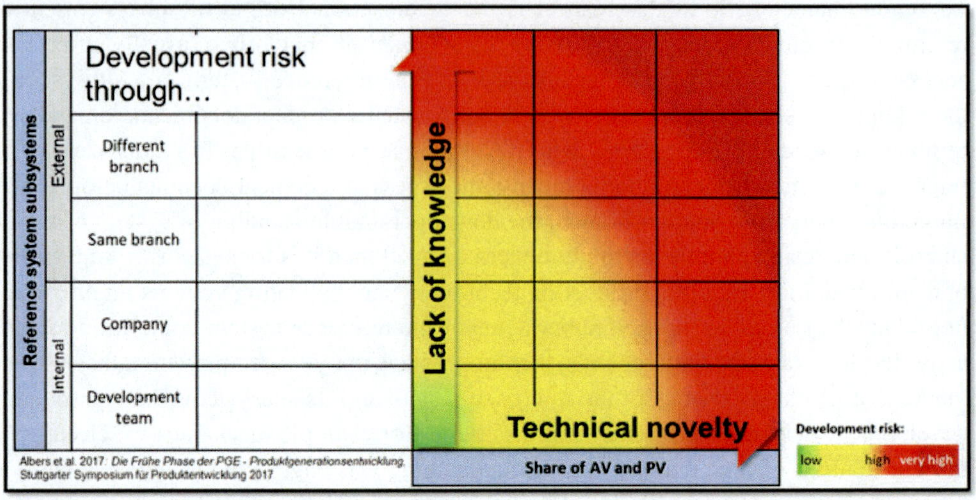

Fig. 2.4 Risk portfolio for the early evaluation of system concepts based on the influence of the reference system and variations on development risk (Albers et al. 2017a)

Fig. 2.5 Examples of projects, where risks and project failure can likely be explained at least partially with variations or the underlying reference system. Left: while today regular wind power plants reach rotor hub heights of more than 100 m, building the GROWIAN project shown in the picture with that height in the mid-80s required a large attribute variation compared to standard power plants in those days. The project is considered to be one of the biggest failures in the history of wind power plants. Middle and right: several companies which were not established car manufacturers tried to develop cars in the field of autonomous driving (Google Car in the middle) or electric vehicles (Dyson, right picture). However, even if the company has expertise in software and algorithms (Google) or battery and electric engine technology (Dyson) building a car requires still a great number of additional subsystems in the reference system, which come from outside the company, if the company is not an established car manufacturer (this explains for example early cooperations of Tesla with Lotus for the first roadster generations). This goes along with a lack of knowledge and engineering competence in the respective areas and thus considerable development risks. The cost of the failed Dyson project is roughly around 500 million pounds (Times 2020). Image source from l. t.r: Weller (1984), N-TV (2016), Dyson (2021)

detection', 'modelling of principle and embodiment', 'build up prototype' or 'validation'. They are used, for example in the Integrated Product Engineering Model (iPeM) (Albers and Meboldt 2007; Albers et al. 2016a) and the VDI 2221 standard based upon that, as generic common building blocks to model development processes, where the uniqueness of each process (Albers 2010) is depicted by the unique combination of these building blocks. The implementation of every variation requires several of these engineering activities. A principle variation might require, for example, idea detection, modelling of principle and embodiment and validation (Albers et al. 2019d). The composition of this "bundle" of activities tends to differ depending on the variation type. Attribute variation and principle variation require for example in tendency more idea detection and modelling of principle and embodiment than a carryover variation (and might therefore be, in tendency, more time consuming and costly).

Based on such relations, challenges in the development of new systems, which are from a company point of view a "first generation" of the respective system type, can presumably be explained by an increased amount of attribute variation and principle variation at late stages in the development process, compared to a development project as illustrated in Fig. 2.3. The required time for those late variations then endangers project schedules (Albers et al. 2020a).

Furthermore, it has to be considered, that variations of individual subsystems can, of course, trigger further variations in a development project. Beyond that variations can also require development activities in the corresponding production and validation system. This can be modelled by variations in the development of a new system triggering variations in the development of the corresponding production or validation system. This relation can be bidirectional. Variations in a production system can also enable new variations in the development of a new system, for example.

Knowledge about the bundles of engineering activities for a certain type of variation, which might even manifest in statistic patterns for individual product types or product ranges,[15] provide a basis for planning the development process.

Considering the perspective on variations in time there are two views to be distinguished overall:

- **A retrospective analysis of variations by comparing a developed system and its subsystems with the underlying reference system and its subsystems.** This approach can be used for empirical investigations of variations, characteristics of the reference system and their impact on the development process. However, within the development process and across several engineering generations the reference system evolves and subsystems might be subject to several variations. This dynamic is linked with different levels of maturity and specification of a new system during its development process and

[15]This has to be subject of future research.

is not always traceable or reconstructable in retrospective analysis. Thus, this analysis can be inaccurate (Rapp et al. 2020).

- **A prescriptive perspective in the development process where, based on subsystems of the current reference system, intended variations for individual subsystems of the new system are planning assumptions and specifications of the design space**[16] (Wessels et al. 2019). For some subsystems carryover variation might be specified as a goal, for others attribute variation is allowed or even principle variation. The basis for these specifications is an analysis on how the current development objective can be achieved, based on the current reference system.[17] If it seems possible to achieve an objective by optimization without a change of the underlying functional structures, attribute variations might be intended. If it seems to be necessary to search for new solutions principle variation[18] might be intended. Those different intended variations then lead to different specific activities, for example optimization simulations for attribute variation or the search for additional potential content of the reference system in case of an intended principle variation. Figure 2.6 illustrates the two perspectives.

An example for such an analysis in which an existing system and its properties are compared to a development objective is found in Schröppel et al. (2019) and Chap. 5. Following an analysis of existing systems and corresponding models, development activities which can be seen as variations, are derived to design user-friendly products for specific target users. This process can be applied spanning across engineering generations as well as system generations.

The time frame for the planning of variations can be defined in terms of engineering generations. A time frame with similar intention in agile development approaches is the Sprint time. Hence, in the agile development of mechatronic systems (Albers et al. 2018a) an engineering generation can correspond to the increment resulting from a sprint.

Beyond planning the development process of a specific system, the model of SGE also provides an ontological basis for the planning of systems across several generations. Examples are modular design for construction kits, for example (Albers et al. 2019e) and

[16]See also Albers et al. (2004) for a generic description of different ways of adapting a system based on the C&C^2 approach.

[17]The comparison between objectives by which the reference system and its subsystems were developed and the objectives for the system in development leads to observations that can also be described as variations. Those are then for example variations of (desired) system properties, such as the acceleration of a car. In this case it is an attribute variation, if a car in development is meant to have a greater acceleration than a car in the reference system. The topic of variations regarding development objectives is for example addressed in the field of requirements managements when investigating changes of requirements (see for example Graessler et al. 2020 and Chap. 1).

[18]A principle variation does often not mean using a new physical principle, but rather a new way of realizing a desired function through designing the embodiment of structures and working surface pairs.

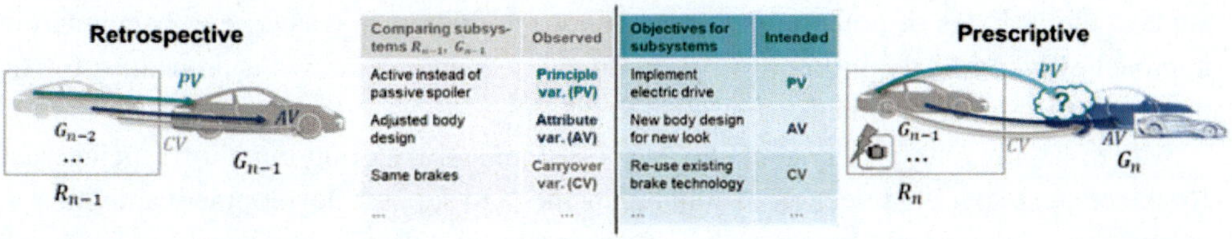

Fig. 2.6 Two perspectives on variation: retrospective analysis and design space specification in process planning (simplified exemplary illustration)

approaches for the planning of product family generations (see for example Küchenhof et al. 2020).

Once the development activities are planned, the different intended variations and the underlying reference system are also important factors regarding the methodical support of development activities. The next section gives a brief introduction in this field.

2.4 Methodical Support of Variations

There are already various methods for the support of product development, which are based on references to some extent. However, they often do not actually refer to them as 'references' as described by the model of SGE.[19] Some refer to other concepts with a similar purpose (but without fulfilling the requirements stated in the introduction of this chapter). Others deal with references without explicitly being built upon a concept that describes them. Probably one of the main challenges in the common and integrated use of those methods, which consider in some way contents of a reference system or variations as thought of here, is the missing common underlying description model. Besides that, it is often possible to link their purpose with the presented model of SGE. A brief overview is given below.

Other methods do neither explicitly nor implicitly consider content of a reference system or variations made on that basis. However, as those phenomena are presumably found in every development project, this might be at least partially the cause, if the use of such methods in practice is limited. In the last part of this section an example is shown, how a method can be adapted by taking the reference system and variations into account.

When looking at existing methods with purposes that can be linked to the model of SGE, a substantial share of approaches in the field of knowledge management can be seen as the management of knowledge about the reference system and (potential) subsystems. Particular examples are various methods of technology scouting, benchmarking and building of analogies. Those methods aim to support the search for potential content for the reference system in the development of a new system generation. They also might

[19]Which is, in 2021, of course due to model being relatively "young".

support the analysis of potential subsystems for the reference system, e.g. benchmarking approaches when analyzing competing products. Chapter 4 shows a procedure for that purpose based on online media, for example.

Moreover, the reference system and its subsystems form an important influencing factor for the context-specific selection of suitable methods to support development activities. An example in this case is the approach by Laukemann et al. (2015), also illustrated in Chap. 9, for method provision based on process similarities to former processes. These similarities are determined, amongst others, by the processed information objects.

There are also approaches from numerous fields that can be understood as support in the idea detection for new technical solutions based on the reference system. One field are works about the role of 'examples' in creativity (see e.g. Herring et al. 2009 for an overview). The work of Weber and Husung (2016) illustrates in a more general manner the importance of solution patterns in the development of new systems. The basis for the description of those pattern is the CPM-approach by Weber (2005, 2014), which describes in a general conceptual way the relation between product characteristics[20] and resulting properties.[21] The role of references to existing systems and solutions in the formulation of e.g. user needs to be addressed by specific new products is also visible in the example of a bicycle cleaning device used by Herrmann et al. (2018) to illustrate the Emoji Method (see also Chap. 9). Another creativity method for the detection of new product ideas, which is explicitly built on the concept of the SGE model is the Innobandit approach by Heimicke et al. (2018). An important basis for variations and thereby new solutions can be data from the lifecycle of subsystems in the reference system. The concept of technical inheritance by Lachmayer et al. (2014)[22] provides a basis for this. A very detailed view on the analysis of a subsystem in an existing system with the aim of improving it is given in Chap. 10, see also Nelius et al. (2020). The work investigates cognitive biases, which can occur in the process and lead to wrong assumptions on function-embodiment relations in the analysed subsystem of a reference system.

Other methods support analysing the potential effects of variations, for example on the structure of a system and its subsystems. Methods for engineering change propagation management serve this purpose, using for example DSMs as basis (Clarkson et al. 2001) or also the CPM approach (Conrad et al. 2007).

Furthermore, a very important activity in the development of every system is validation (Albers 2010). Building early prototypes by using existing cars together with potential engines of new car generations is an established approach of car manufacturers for early engine testing (Albers et al. 2017a). This is an example how the development of validation

[20]The understanding of "characteristic" in that case differs from the understanding of reference system characteristics in the model of SGE.

[21]The CPM/PDD approach can be considered a general description model for a detailed description of the relations and their transformations in the course of variations.

[22]See also Chap. 3.

systems can also be described by the model of SGE. It is also an example for the importance of the systematic use of the reference system for agile development of mechatronic systems (see also Albers et al. 2017b). Other validation approaches building up on the reference system use Augmented Reality, for example Reinemann et al. (2018). Göhlich et al. (2021) addresses the overall system validation in the system generation engineering of the urban mobility system with an integrated assessment approach for strategies for the decarbonization of urban traffic.[23]

To conclude this section, the case of a method adaption is described in the following. This case of Albers et al. (2018b) is taken from a students' development project within a course for mechanical design. The case is about the creation of principle sketches. The aim of principle sketches is to illustrate key aspects of technical solution ideas with little formalization and thus less creation effort than a full specification of the components' embodiment. A principle sketch of a gearbox can e.g. include the arrangement of shafts, bearings and maybe the housing (Kirchner 2007). Figure 2.7 shows two typical examples from students' works with annotations.

The principle sketches enable a first discussion and evaluation of the ideas. The course format was then adjusted by giving the students of a course the development result from the previous course as reference system. To trigger development activities the task for the new course included selected differentiation objectives compared to the previous course, for example making a four-wheel drive powertrain with flexible torque distribution for a vehicle instead of a powertrain with fixed torque distribution in the previous course. As a result of working explicitly based on the reference system and the corresponding documentation, especially drawings, a student team changed the way they created principle sketches as shown in Fig. 2.8. The students used transparent film for their principle sketches. In doing so they could position the sketch over the drawing of the corresponding subsystem from the reference system. This allowed for displaying different system areas with different types of variation, which would be necessary for the realization of a solution idea, and hence illustrated a potential focus of development activities. Furthermore, by moving the film they could switch between different alternative solution concepts. This facilitated a profound discussion of the different solution concepts.

2.5 Conclusion and Outlook

The model of SGE describes fundamental phenomena which are observable in every development of a new system. As it is based on the theory of technical systems it has a broad potential applicability for the description of the integrated development of a system, the corresponding production system and the corresponding validation system. Furthermore, it is applicable across different branches.

[23] See also Chap. 7.

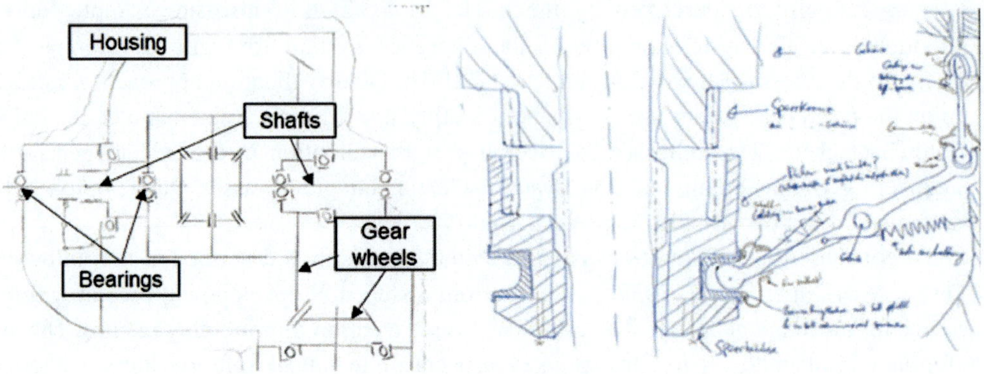

Fig. 2.7 Examples for conventional principle sketches for a differential gear box (left) and an actuation mechanism for a gear wheel coupling (Rapp 2019, presentation slides to Albers et al. 2018b)

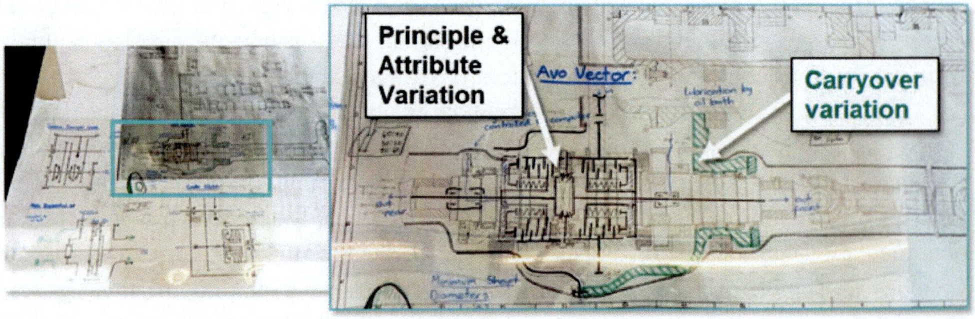

Fig. 2.8 Principle sketches based on the documentation from the reference system. Different system areas with different types of variation are marked individually for a solution concept (Rapp 2019, presentation slides to Albers et al. 2018b)

The fundamental phenomena described by the model of SGE can be linked with innovation potential and development risk in a development project. The approaches to measure and quantify the elements in the model of SGE in terms of variation types, variation shares and characteristics of the reference system therefore provide key factors for the planning and management of development projects with regard to innovation potential and development risk.

Intended variations as well as characteristics of the reference system are furthermore a basis to derive specific necessary development activities as well as measures for handling development risks, e.g. triggering cooperations.

Furthermore, variations as well as characteristics of the reference system or its subsystems are important context factors for the methodical support of development activities and need to be considered in method selection and method adaptions.

The adaption and integration of existing methods based on the model of SGE as a commonly used underlying description model is an objective of further research. The model of SGE provides a basic ontology for a future methodical framework of Advanced Systems Engineering,[24] which will be created by integrating Systems Engineering concepts with further, recent methodological approaches, for example for agile development of mechatronic systems.[25] Future works will also include the further research and evaluation of key performance indicators for planning and management of development based on the model of SGE. An important issue in this field will be the development of computer support and the use of AI algorithms.

References

Albers A (1994) Simultaneous Engineering, Projektmanagement und Konstruktionsmethodik—Werkzeuge zur Effizienzsteigerung. VDI Berichte-Verein Deutscher Ingenieure 1120:73–106

Albers A (2010) Five hypotheses about engineering processes and their consequences. In: Proceedings of the TMCE, April 2010, vol 12, p 16

Albers A, Lohmeyer Q (2012) Advanced systems engineering—towards a model-based and human-centered methodology. In: International symposium series on tools and methods of competitive engineering, TMCE, May 2012, pp 407–416

Albers A, Meboldt M (2007) IPEMM—integrated product development process management model, based on systems engineering and systematic problem solving. In: Bocquet J-C (ed) Proceedings of ICED 2007, the 16th international conference on engineering design, Paris, France, 28–31 July. The Design Society (ICED), Paris, France, pp 611–612 (exec. summ.), full paper no. DS42_P_537

Albers A, Matthiesen S, Lechner G (2002) Konstruktionsmethodisches Grundmodell zum Zusammenhang von Gestalt und Funktion technischer Systeme. Konstruktion (1981) (7–8):55–60

Albers A, Burkardt N, Ohmer M (2004) Principles for design on the abstract level of the contact & channel model. In: Proceedings of the TMCE, pp 87–94

Albers A, Bursac N, Wintergerst E (2015) Product generation development—importance and challenges from a design research perspective. New Dev Mech Mech Eng 16–21

Albers A, Reiß N; Bursac N, Richter T (2016a) iPeM—integrated product engineering model in context of product generation engineering. In: Wang L, Kjellberg T (eds) 26th CIRP Design conference, Stockholm, Sweden (Procedia CIRP, 50), pp 100–105

Albers A, Haug F, Heitger N, Arslan M, Rapp S, Bursac N (2016b) Produktgenerationsentwicklung—Praxisbedarf und Fallbeispiel in der automobilen Produktentwicklung. In: Gausemeier J (ed) Vorausschau und Technologieplanung. 12. Symposium für Vorausschau und Technologieplanung. 8. und 9. Dezember 2016 Berlin. Heinz Nixdorf Institut, Paderborn, pp 227–242

Albers A, Bursac N, Rapp S (2016c) PGE-product generation engineering-case study of the dual mass flywheel. In: DS 84: Proceedings of the DESIGN 2016 14th international design conference, pp 791–800

[24]For a first description of the objective of such a framework see Albers and Lohmeyer (2012).

[25]See for example Albers et al. (2018a).

Albers A, Rapp S, Birk C, Bursac N (2017a) Die Frühe Phase der PGE—Produktgenerationsentwicklung. In: Binz H, Bertsche B, Bauer W, Spath D, Roth D (eds) Stuttgarter Symposium für Produktentwicklung SSP 2017. Stuttgarter Symposium für Produktentwicklung SSP 2017. Fraunhofer-Institut für Arbeitswirtschaft und Organisation IAO, Stuttgart

Albers A, Behrendt M, Klingler S, Reiß N, Bursac N (2017b) Agile product engineering through continuous validation in PGE—product generation engineering. Des Sci 3

Albers A, Heimicke J, Hirschter T, Richter T, Reiß N, Maier A, Bursac N (2018a) Managing systems of objectives in the agile development of mechatronic systems by ASD-agile systems design. In: DS 91: Proceedings of NordDesign 2018, Linköping, Sweden, 14th–17th Aug 2018

Albers A, Matthiesen S, Bursac N, Rapp S, Hölz K, Eisenmann M (2018b) PGE—product generation engineering in engineering education: real-world problems for beginners in student projects. In: DS 93: Proceedings of the 20th international conference on engineering and product design education (E&PDE 2018), London, pp 68–73

Albers A, Heimicke J, Walter B, Basedow GN, Reiß N, Heitger N et al (2018c) Product profiles: modelling customer benefits as a foundation to bring inventions to innovations. Procedia CIRP 70:253–258

Albers A, Rapp S, Spadinger M, Richter T, Birk C, Marthaler F et al (2019a) The reference system in the model of PGE: proposing a generalized description of reference products and their interrelations. In: Wartzack S, Schleich B (eds) Proceedings of the 22nd international conference on engineering design (ICED19), vol 1, Delft, The Netherlands, 5–8 Aug 2019. The Design Society, Delft, The Netherlands , pp 1693–1702

Albers A, Matthiesen S, Revfi S, Schönhoff C, Grauberger P, Heimicke J (2019b) Agile lightweight design-the extended target weighing approach in ASD-agile systems design using functional modelling with the C&C 2-approach. In: Proceedings of the Design Society: international conference on engineering design, vol 1, no 1. Cambridge University Press, pp 2667–2676

Albers A, Haug F, Heitger N, Fahl J, Hirschter T (2019c) Entwicklungsgenerationen zur Steuerung der PGE—Produktgenerationsentwicklung: Von der Bauteil-zur Funktionsorientierung in der Automobilentwicklung. In: Stuttgarter Symp. Für Produktentwicklung (SSP 2019), Stuttgart, Deutschland, pp 253–262

Albers A, Rapp S, Peglow N, Stürmlinger T, Heimicke J, Wattenberg F, Wessels H (2019d) Variations as activity patterns: a basis for project planning in PGE—product generation engineering. Procedia CIRP 84:966–972

Albers A, Bursac N, Scherer H, Birk C, Powelske J, Muschik S (2019e) Model-based systems engineering in modular design. Des Sci 5

Albers A, Ebertz J, Rapp S, Heimicke J, Kürten C, Zimmermann V et al (2020a) Produktgeneration 1 im Modell der PGE—Produktgenerationsentwicklung: Verständnis, Zusammenhänge und Auswirkungen in der Produktentwicklung. KIT Scientific Working Papers, 149. Karlsruher Institut für Technologie (KIT), Karlsruhe

Albers A, Rapp S, Fahl J, Hirschter T, Revfi S, Schulz M et al (2020b) Proposing a generalized description of variations in different types of systems by the model of PGE—product generation engineering. Proc Des Soc DESIGN Conf 1:2235–2244. https://doi.org/10.1017/dsd.2020.315

Albers A, Fahl J, Hirschter T, Endl M, Ewert R, Rapp S (2020c) Model of PGE—product generation engineering by the example of autonomous driving. Procedia CIRP 91:665–677

Bartz M (2019) Transfer des muskuloskelettalen Leichtbaus auf offene kinematische Ketten. Dissertation. Verlag Dr. Hut, München. ISBN: 978-3-8439-3935-5

Bartz M, Uttich E, Bender B (2019) Transfer of lightweight design principles from the musculoskeletal system to an engineering context. Des Sci 5

Clarkson PJ, Simons C, Eckert CM (2001) Predicting change propagation in complex design. In: Proceedings of DETC 01. American Society of Mechanical Engineers (ASME 2001 Design engineering technical conferences and computers and information in engineering conference), Pittsburgh, PA

Conrad J, Deubel T, Köhler C, Wanke S, Weber C (2007) Change Impact and Risk Analysis (CIRA) - Combining the CPM/PDD Theory and FMEA-Methodology for an Improved Engineering Change Management. 16th International Conference on Engineering Design (ICED 07), Paris, France, 28.-31.08.2007. In: Proceedings of ICED 07 (DS 42), pp 9–10 (Executive Summary), Paper no. 549 (Full Paper, CD-ROM). École Centrale Paris. Paper also available online at www.designsociety.org, paper no. DS42_P_549.pdf

Dyson (2021) Das Dyson Elektroauto von James Dyson. https://www.dyson.de/newsroom/overview/features/dyson-battery-electric-vehicle. Accessed 12.04.2021

Göhlich D, Fay TA, Jefferies D, Lauth E, Kunith A, Zhang X (2018) Design of urban electric bus systems. Des Sci 4

Göhlich D, Nagel K, Syré AM, Grahle A, Martins-Turner K, Ewert R, Jahn R, Jefferies D (2021) Integrated approach for the assessment of strategies for the decarbonization of urban traffic. Sustainability 13(2):839

Graessler I, Oleff C, Scholle P (2020) Method for systematic assessment of requirement change risk in industrial practice. Appl Sci 10(23):8697

Heimicke J, Reiß N, Albers A, Walter B, Breitschuh J, Knoche S, Bursac N (2018) Agile innovative impulses in product generation engineering: creativity by intentional forgetting. In: DS 89: Proceedings of the fifth international conference on design creativity (ICDC 2018), University of Bath, Bath, UK, pp 183–190

Herring SR, Chang CC, Krantzler J, Bailey BP (2009) Getting inspired! Understanding how and why examples are used in creative design practice. In: Proceedings of the SIGCHI conference on human factors in computing systems, April 2009, pp 87–96

Herrmann T, Laukemann A, Binz H, Roth D (2018) The emoji method: an approach for identifying and formulating problem ideas. Int J Comput Inform Eng 12(12):1071–1086

Kirchner E (2007) Leistungsübertragung in Fahrzeuggetrieben: Grundlagen der Auslegung, Entwicklung und Validierung von Fahrzeuggetrieben und deren. Springer, Komponenten

Küchenhof J, Schwede LN, Krause D (2020) Planning & tracking the changes-matrix mapping of modular product family generations. In: DS 101: Proceedings of NordDesign 2020, Lyngby, Denmark, 12th–14th Aug 2020, pp 1–11

Lachmayer R, Mozgova I, Reimche W, Colditz F, Mroz G, Gottwald P (2014) Technical inheritance: A concept to adapt the evolution of nature to product engineering. Procedia Technol 15:178–187

Langer S, Maier A, Wilberg J, Münch TJ, Lindemann U (2012) Exploring differences between average and critical engineering changes: survey results from Denmark. In: Marjanovic D, Storga M, Pavkovic N, Bojcetic N (eds) Proceedings of DESIGN 2012, the 12th international design conference. International Design conference; DESIGN 2012, Dubrovnik, Croatia, 21–24 May. The Design Society (DESIGN), Dubrovnik, Croatia, pp 223–232

Laukemann A, Binz H, Roth D (2015) Konzept eines produktentwicklungsspezifischen Wissensmanagementverfahrens für kleine und mittlere Unternehmen. In: Stuttgarter Symposium für Produktentwicklung

Matthiesen S, Grauberger P, Hölz K, Nelius T, Bremer F, Wettstein A et al (2018) Modellbildung mit dem C&C^2-Ansatz in der Gestaltung-Techniken zur Analyse und Synthese. KIT, Karlsruhe

Nelius T, Doellken M, Zimmerer C, Matthiesen S (2020) The impact of confirmation bias on reasoning and visual attention during analysis in engineering design: An eye tracking study. Des Studi 71:100963

N-TV (2016) Kooperation statt Eigensinn. Google nimmt beim Auto-Projekt die Ausfahrt. https://www.n-tv.de/auto/Google-nimmt-beim-Auto-Projekt-die-Ausfahrt-article19312931.html. Accessed 12.04.2021

Peglow N, Powelske J, Birk C, Albers A, Bursac N (2017) Systematik zur Differenzierung von Varianten im Kontext der PGE—Produktgenerationsentwicklung. In: Tagungsband 15. Gemeinsames Kolloquium Konstruktionstechnik, Düsseldorf, Germany

Ponn J, Lindemann U (2011) Konzeptentwicklung und Gestaltung technischer Produkte: systematisch von Anforderungen zu Konzepten und Gestaltlösungen. Springer, Berlin

Rapp S (2019) Reflexion zum Lehrexperiment: PGE—Produktgenerationsent-wicklung in den Workshops zur Maschinenkonstruktionslehre III/IV. Arbeit im Rahmen des HDZ-Zertifikats Baden-Württemberg. IPEK—Institut für Produkt-entwicklung am Karlsruher Institut für Technologie (KIT)

Rapp S, Barg M, Klotz T, Birk C, Albers A (2020) Influencing factors on the retrospective analysis of variation shares with C&C^{2}A-based criteria in Product Generation Engineering. Procedia CIRP 91:781–788

Reinemann J, Hirschter T, Mandel C, Heimicke J, Albers A (2018) Methodische Unterstützung zur Produktvalidierung in AR-Umgebungen in der Frühen Phase der PGE—Produktgenerationsentwicklung. Design for X. Beiträge zum, 28

Ropohl G (1979) Eine Systemtheorie der Technik. Zur Grundlegung der allgemeinen Technologie. Carl Hanser, München

Ropohl G (2009) Allgemeine Technologie: Eine Systemtheorie der Technik. Universitätsverlag Karlsruhe, Karlsruhe

Schröppel T, Miehling J, Wartzack S (2019) Roadmap für die Entwicklung einer Methodik zur dualen Nutzerintegration. In: Stuttgarter Symposium für Produktentwicklung, vol 16

Schumpeter J (1927) The explanation of the business cycle. Economica 21:286–311. https://doi.org/10.2307/2548401

Times (2020) The Sunday Times. James Dyson interview: how I blew £500m on electric car to rival Tesla. The inventor spent a small fortune developing his idea. https://www.thetimes.co.uk/article/james-dyson-interview-electric-car-tesla-tzls09t5m. Accessed 12.04.2021

Weber C (2005) CPM/PDD—an extended theoretical approach to modelling products and product development processes. In: Proceedings of the 2nd German-Israeli symposium on advances in methods and systems for development of products and processes, July 2005, vol 6, no 10. Fraunhofer-IRB, Stuttgart

Weber C (2014) Modelling products and product development based on characteristics and properties. In: An anthology of theories and models of design. Springer, London, pp 327–352

Weber C, Husung S (2016) Solution patterns—their role in innovation, practice and education. In: DS 84: Proceedings of the DESIGN 2016 14th international design conference, pp 99–108

Weller T (1984) Growian in the summer of 1984, at the Kaiser-Wilhelm-Koog near Marne, Schleswig-Holstein, Germany. https://commons.wikimedia.org/wiki/File:Growian001.png. Accessed 12.04.2021

Wessels H, Heimicke J, Rapp S, Grauberger P, Richter T, Matthiesen S, Albers A (2019). Sprintplanung in der Mechatroniksystementwicklung auf Basis von Referenzsystemelementen. In: 17. Gemeinsames Kolloquium Konstruktionstechnik, KT2019, Aachen, Germany, pp 200–211

Technical Inheritance as an Approach to Data-Driven Product Development

Roland Lachmayer and Iryna Mozgova

Abstract

Hallmarks of modern technical products or systems are an accelerated time-to-market by clear modification cycles, the processing of large amounts of data, an increased flexibility as well as a quick reaction to changes in market situations. The monitoring of technical products is state of the art nowadays. Due to new communication possibilities that have emerged, a multitude of data exist that can be transferred into information and knowledge about products through their life cycle. The developing communication possibilities facilitate new innovative approaches for the application of product life cycle data. New methods of data management and data processing are required for cross-generational process analysis as are software and hardware tools. Furthermore, new methodologies for developing technical products are demanded. This chapter describes the Paradigm of Technical Inheritance, which is based on the idea of developing and modifying a new generation of products or services taking into account the information gathered from the life cycles of the previous generations. The basic principles of this approach are outlined, a process model including data collection, monitoring and analysis methods is presented, and application examples for both a generation-oriented development of a single component and for a complex technical system are given.

R. Lachmayer (✉) · I. Mozgova
Leibniz University Hannover, Institute of Product Development, Garbsen, Germany
e-mail: lachmayer@ipeg.uni-hannover.de

D. Krause, E. Heyden (eds.), *Design Methodology for Future Products*,
https://doi.org/10.1007/978-3-030-78368-6_3

3.1 Evolution in Technology and Generation Oriented Product Development

Throughout human history, the progressive development of society is ultimately conditioned by and inextricably linked to the development, improvement and advancement of engineering and technology. There are various concepts of the origin of technology, which see the development of technology from the expedient human activity and the need for rational use of the means of this activity (Kapp 1877). One of the concepts of the emergence of technology was proposed by O. Spengler (Kidd 2012). According to this concept, technology and technical systems are a way of organizing the joint activities of large masses of people. Therefore, a technology should not be seen as a set of tools, but as a way of handling them. A different concept of the origin of technology is offered by L. Mumford (Fortner 2014), who supposes that machine technology is a product of biotechnology, technology arises from the characteristics of human functioning and in its nature is closely related to the nature of peoples.

Analyzing the history of technological development, we can find the following tendency: non-creative aspects of human work functions are gradually transferred to technical devices, while the creative ones remain for men. Since the early 80s of the twentieth century, the theory of industrial society has been replaced by the concept of information society, where a special role is played by information and the ways of its collecting, processing, distribution and usage. Among the famous works of the authors of this concept we would like to mention the works of A. Toffler (1990), who distinguished three waves of the development of society and the works of D. Bell (Waters 2003), in which he describes three technological revolutions. Within the concept of the information society, the main and decisive factor in social development is the production and use of scientific, technical and other information, which contributes to the development of the service economy and the information sector, as well as radically changing production processes, as we see in Industry 4.0 (Anderl 2015).

3.1.1 Evolutionary Processes in Nature and Technology

Nature, objects of the material world, technologies and various fields of knowledge develop according to their specific laws. Technologies develop in close interaction with social developments and nature and are subject to the laws of dialectics. A brief review of the laws of evolution of technical systems is given in (Lachmayer et al. 2014). The laws of technical evolution describe a generalized idealized process of system development, taking into account the static, kinematic, and dynamic aspects of development (Eversheim 2009). The known laws of evolution of technical systems were formulated by G. Altshuller (1984) and his followers. Among the laws should be noted the law of completeness of the parts of the system, which provides the minimal functionality; the law of transition of working parts of a system from macro to micro level; the law of increasing the degree of ideality of the system; the law of transition of quantitative changes into qualitative or the S-curve law, etc. Within studying the development of technical systems, parallels can be drawn between the laws of development of nature and technology, the method of analogy is used. Thus, the S-curve of the development of technical systems or products, often considered, for example, in innovation management, was first studied and substantiated in the study of the evolution of yeast fungus colonies (Kemp et al. 1999; Hughes 1987).

The evolution in nature became a major direction in development of evolutionary theories. Among the most famous are the Lamarck's theory of biological evolution, (Honeywill 2008), Darwin's theory (Storch 2013), modern synthesis evolutionary theory (Fischer 1958) and Neo-Darwinism, created by A. Weismann. Lamarck's theory is based on the inheritance of acquired properties and the inherent commitment of all living creatures to perfection. C. Darvin has put forward the principle of natural selection as a basis of evolution. The modern synthesis evolutionary theory is a doctrine of the evolution of the organic world developed on the basis of modern genetics, ecology and classical Darwinism. R. Fischer was one of the first representatives of the theory. The theoretically described mechanisms of mutation, recombination and selection are the base for, e.g., evolutionary optimization algorithms (Bäck et al. 2000).

3.1.2 The Role of Data in the Development, Monitoring and Analysis of Modern Products

The current state and trends in the evolution of technology and development of technical products imply the rapid and effective creation of products or systems using the experience and accumulated knowledge. Within the framework of the above-mentioned concept of information society, product developers are working in a dynamic and digitized time, where information about technical products and components can be collected and deployed (Kaufmann 2015). In the context of Industry 4.0 technical systems are networked with each other. Due to the modern created communication possibilities, a multitude of data sets exists which, with the right knowledge, methods and tools, can be transferred to information about the products (Abramovici and Lindner 2011; Wuest et al. 2016). Methods for collecting data records during the product life cycle are the so-called monitoring methods and tools for observing specific problems (Haken 2012).

Usually, today's development of technical systems rarely includes completely new developments. According to Albers et al. (2015), the classification according to Pahl and Beitz (Feldhusen and Grote 2013) into new, the adaptation and a variant design is no longer sufficient and refers to the most common type of development projects as product generation development. Nowadays, technical products are mostly developed from known solution principles or their combination and the most common type of development projects can be classified as product generation development or generation oriented product development. In the course of this, it has to be examined which existing data should be collected in the life cycle of a product or technical system so that the information obtained from the

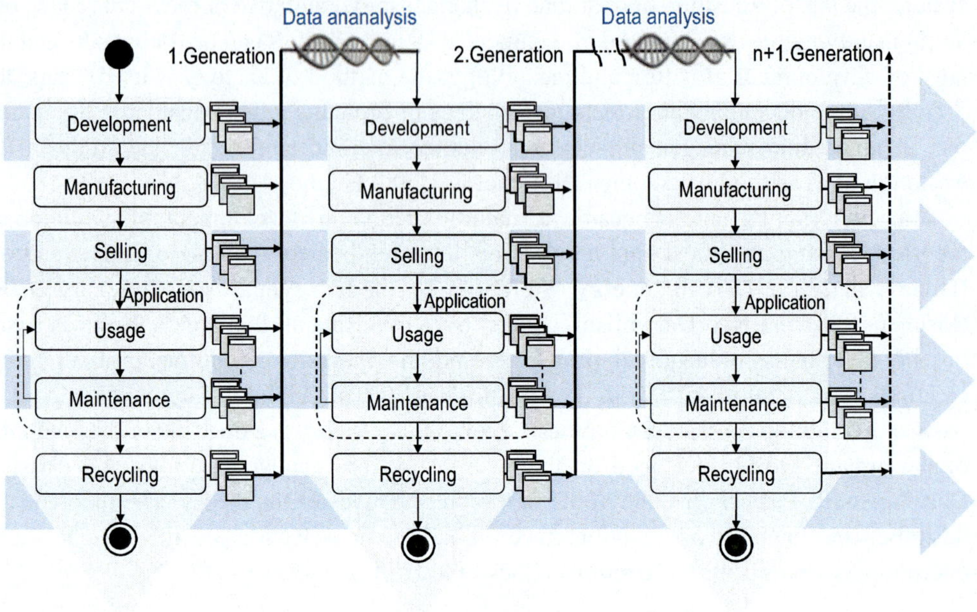

Fig. 3.1 Transfer of life cycle data (Lachmayer and Gottwald 2014)

analysis of this data can be used to effectively develop and adapt a new generation of a product. Therefore, the integration of information as well as the increase of the flexibility of the development processes is necessary. Modern technologies for data collection and data analysis are used to capture relevant information and generate knowledge that can be passed on from one product generation to the next. In this approach, throughout all phases of the product lifecycle, the involved processes must not only be interconnected, but also data acquisition, storage and processing throughout the entire product life cycle should be available. An important aspect is the expansion of the range of services and data standardization. As shown in Fig. 3.1, the process of generation oriented product development is intended to enable the transfer of life cycle data from generation to generation.

Historically, during the life cycle of a product or system, the main sources of data are transactions and operations: order processing, interaction with suppliers and customers, customer service, etc. They can be supplemented with information from surveys and studies. By processing all of this data, manufacturers and suppliers gain mainly insight into consumers, demand and product costs (Ripperda and Krause 2017; Johannknecht et al. 2017), and less insight into how the system or product is used. Nowadays, that smart products and production systems deliver information, unprecedented in volume and variety and in real time, data, along with people (Graessler and Poehler 2019), technology and capital, has become one of the main assets of companies. This new data is valuable on its own, but its value is multiplied when it is combined with other data, such as development, manufacturing, selling, service history and usage patterns. Central importance in the presented process have the large amounts of data collected during the life cycle of the technical product, which form the history of the product, and which are relevant for the development of subsequent generations of this product. For a targeted extraction of information by means of data acquisition in each of the mentioned phases, the use of specially developed or adapted methods or algorithms is necessary (Eigner et al. 2014). In addition to extensive data analysis, the integration of data into the product development process is required.

3.2 Paradigm of Technical Inheritance

A process of technical evolution can be defined as a process that represents the controlled, gradual and continuous change of technical systems, products and processes as well as models with the aim of adapting to environment influences and requirements (Lachmayer et al. 2014). As a rule, a technical system consists of a set of system elements or subsystems and their relationships with each other. The system is delimited or demarcated from the environment by a system boundary. In the process of technical evolution, each system or system element has its own development dynamics (Fig. 3.2).

Each individual subsystem is not exclusively dependent on its parent system; it may be independent of the parent system, or acquired, developed, and operated alone or in conjunction with other systems. System elements or subsystems can be material objects such as components, assemblies, machines, devices, apparatuses, but also immaterial entities such as methods, algorithms, concepts or software.

The Paradigm of Technical Inheritance (TI), which was developed by the project partners in the Collaborative Research Centers 653 "Gentelligent Components in their Lifecycle" (Denkena and Mörke 2017), is based on an algorithmized feedback of information from the life cycle phases of a product into the next product generation. The main idea is the development or modification of a new generation of products or services taking into account the collected information from the life cycles of previous generations of the product. For this, materials, sensors, technologies and methods were developed in order to store knowledge and expand it depending on external loads. An outstanding feature is that the collected data is autonomously captured by intelligent products and stored and processed on them by a genetic code (Demminger et al. 2016; Mozgova et al. 2017).

3.2.1 Evolutionary Mechanisms in Technology

The known evolutionary mechanisms in biology can be tentatively taken as a basis for recognizing and describing the evolutionary processes in technical systems and subsystems. But evolution and hereditary information transfer processes in technology cannot occur exactly as in biological systems. The ideas for integrating evolutionary mechanisms into the product development process can be found in various process models,

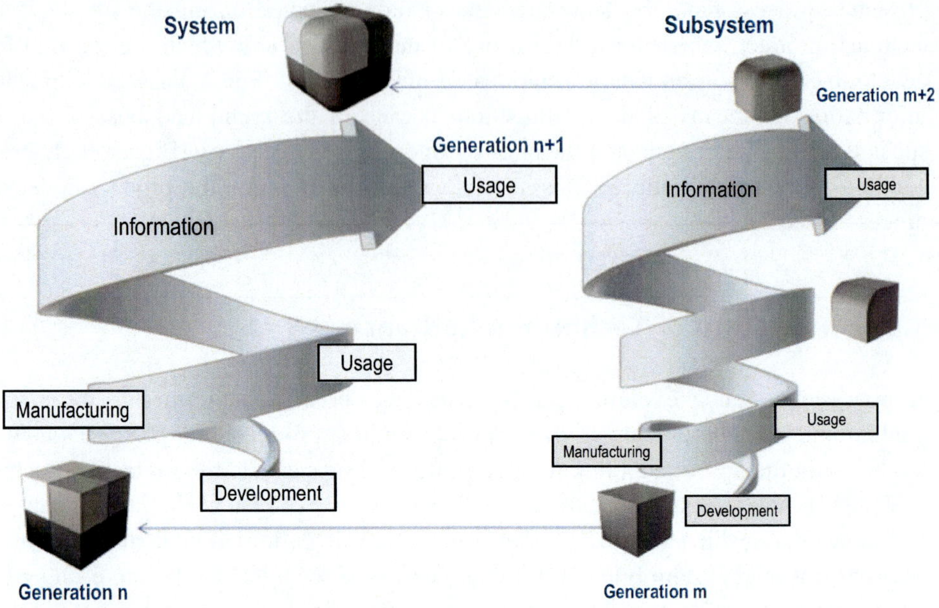

Fig. 3.2 Evolutionary dynamics of systems and subsystems

such as the Munich process model (Lindemann 2009) or the autogenetic design theory (Vajna et al. 2005). The terminology and mechanisms used within the paradigm of TI are given in the Table 3.1.

Some of these terms can be explained at the example of the evolution of a wheel carrier for a race car developed by the Horse Power Team of the Leibniz University Hannover. The development of a first generation of a wheel carrier requires design experience, knowledge of basic design tools, understanding the laws of vehicle dynamics and general information about the expected loads on the component. This includes rough calculations of dimensions for the vehicle as well as multi-body simulations of the racing car to determine loads. As a result, a first generation of the component in the form of a parameterized CAD model is obtained. By testing different scenarios of race car motion, for example, using multi-body simulations, we gain information about the applied loads. By optimizing the resulting model of the component according to the load information computed during simulations of different driving scenarios, we achieve adapted variations of the original parametrized model. In analogy to the biological evolution processes, we name these adapted variations of the component genotypes (Fig. 3.3).

In producing each such variation of the original model, i.e. each genotype, we will obtain real physical components. Each of these produced wheel carriers, despite the common model, will be individual at the physical level. Thus, we are talking about individuals. These physical components must undergo quality control, i.e. selection. Furthermore, in the production process and depending on the parameters of the equipment and

Table 3.1 Terminology of technical evolution

Term	Definition
Technical Evolution	Process of control, stepwise and continuous change of technical systems, products and processes as well as models with the aim to adapt to influences and requirements
Technical Inheritance	Transfer of assembled and verified information from production and application to the next product generation
Individual	An individual is the smallest considered technical system, product, process or model in a population
Generation	A generation is a group of individuals with the same level of development
Population	A population consists of all generations of individuals of a technical system, product and process as well as a model at the current time
Selection	Selection process based on multiple criteria a requirement profile
Mutation	A process with targeted or non-targeted character to create variants with resulting modified properties

Lachmayer et al. (2015)

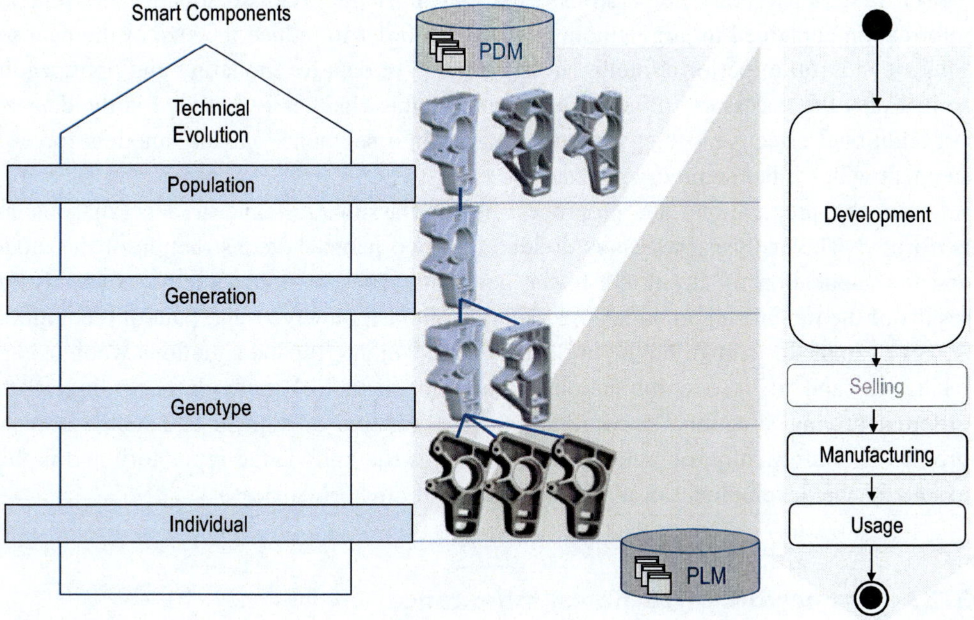

Fig. 3.3 Assignment product life cycle phases and terms in the scope of TI

the production process itself, there can be deviations, for example, from the geometry of the genotype, i.e. there can be some mutations within the genotype. During the operation of the manufactured component, within the framework of the TI, data about the loads on the wheel carrier should be collected. The information obtained from the analysis of this data can be used in the development of the next generations of wheel carrier. All wheel carrier generations represent a population of these components.

Documentation corresponding to the development phase of a component, i.e. corresponding to the population, generation and genotype levels, is supposed to be organized and stored using Product Data Management (PDM) systems. The organization and storage of data and information obtained during the operation of components is achieved using Product Lifecycle Management (PLM) systems (Fig. 3.3).

3.2.2 Process of Information Transfer

For the development of the information transfer process, a goal-oriented algorithmic data feedback is to be realized, which includes statistical methods and operations as well as a design evolution for product adaptations in the development process. Since creating analytical or numerical models for interpreting heterogeneous data is a complicated task, it is useful to apply statistical data analysis that allows structuring and interpretation of data.

The first step is data preparation (Fig. 3.4). It is important to identify the relevant information contained in large amounts of data in order to reduce the size of the data set. Thus, it is useful to perform intelligent aggregation of data for modeling and optimization so that only the necessary information about dynamic changes is included in the data set. An additional effective strategy is to split data into segments and use models for each segment with further summary of results.

After the preparation, the preprocessing, a classification and a data analysis are performed. The analysis process is divided into two parts: the construction of the model and the application of the model to the new data. The developed method of analyzing results of the monitoring includes the methods of cluster analysis and pattern recognition.

For example, as shown below in Sect. 3.3, recognizing typical situations when using a racing car and using a combination of different statistical methods according to the different driving situations show representative patterns of signals and define a driver profile. The information obtained can be stored in the knowledge repository and is thus usable for the development of a new generation of a wheel carrier.

3.2.3 Framework of Technical Inheritance

As indicated in Fig. 3.2, TI takes place in cycles in which information from the life cycle phases of the product under consideration is collected and is available for the development of the next product generation. The central process of information feedback is divided into four phases (Gottwald 2016): identification of the life cycle information; implementation of the monitoring strategy; realization of the data analysis and algorithmized information feedback. The procedure model describes the activities and measures required to set up targeted component monitoring and the feedback of information over the entire life cycle.

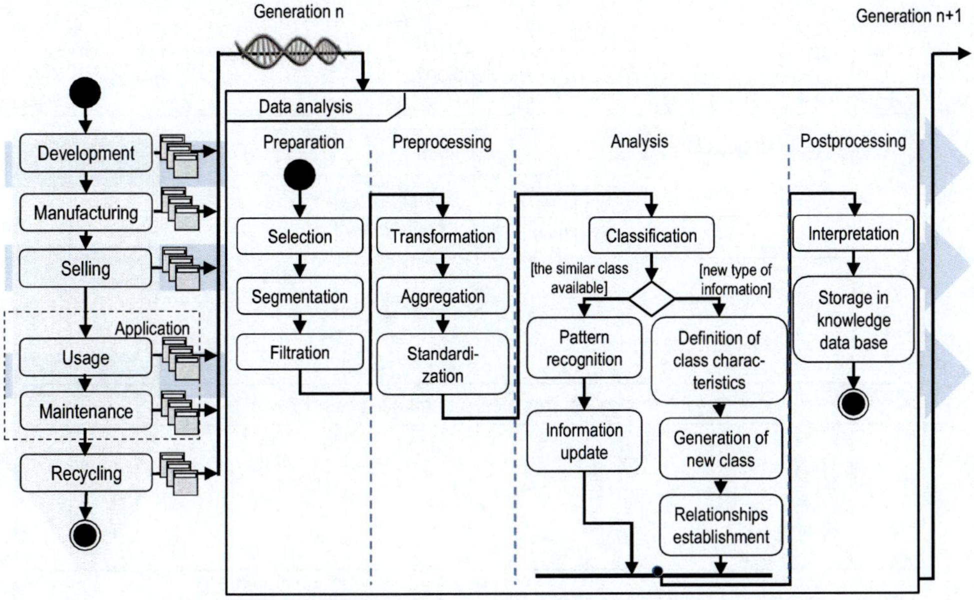

Fig. 3.4 Data analysis (Mozgova 2017)

The approach includes identification of life cycle data and analysis of information flows in the process. For each life cycle stage, four aspects are to be considered: physical principles (stresses), human-related factors (safety), economic aspects and trends. In addition, life cycle information can be important not only for the next generation of a product or system, but can even influence the current development of rework, for example in the case of safety-relevant components or subsystems (Lachmayer et al. 2015).

The four phases of the process of information feedback are part of the methods workflow within the TI paradigm. As shown in Fig. 3.5, the paradigm rests on four pillars: the principle of technical evolution, evolutionary mechanisms, data analysis, and the genetic code of the product. The first three have been described above. A classification of information and Genetic Code (GC) of the component can be described as genetic information of a component and constitutes the basic information which is necessary to identify or reproduce components. This information can be stored as static, unchangeable data in the component and may have been inherited from an older generation of the component. Parts of a GC of a component are described in Mozgova et al. (2017). The implementation of information feedback requires a methodology, a set of methods and tools appropriate to the scope of the paradigm. Methodology in this context refers to a doctrine of scientific methods, totality of all methods applied within a given field. Methodology allows among all methods of data analysis to establish how to implement these through algorithms, how to determine the correct tools for realization of the algorithms and how to describe sequences of methods for an application area, i.e., methods workflow.

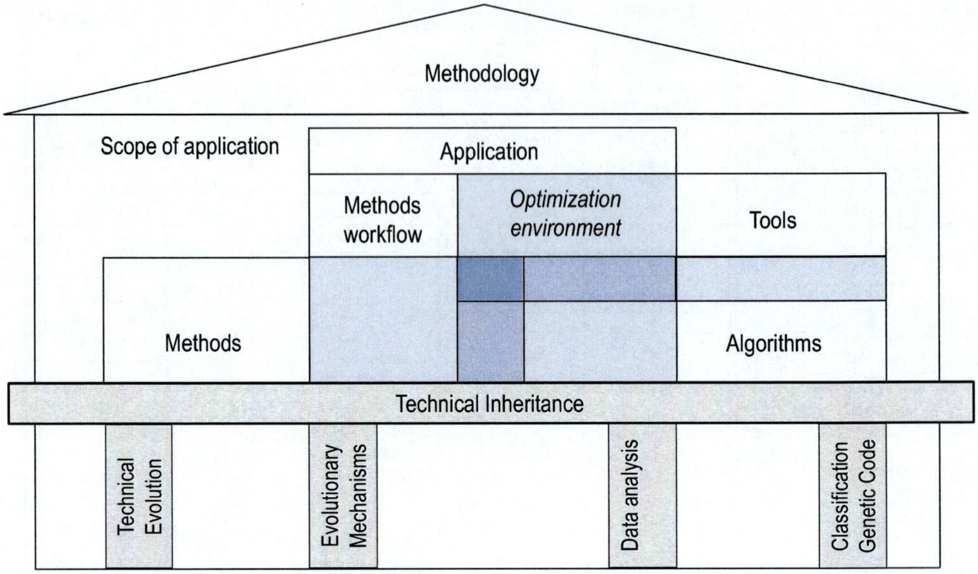

Fig. 3.5 Framework of technical inheritance

The development and adaptation of a new product generation or product genotype is implemented using an automated design, modeling and data processing environment. For the example of a race car wheel carrier, this process is called design evolution (Lachmayer et al. 2013): the adaption of products by analyzing product life cycle while taking evolutionary mechanism into account.

3.3 Application Examples of Algorithmic Data Feedback for Technical Inheritance

For the demonstration of the TI approach, we can address the use of operation data for the development of subsequent generations. The application area includes technical systems with cyber-physical, smart or intelligent products, targeted data feedback and individualization & customization of products. The advantages are linking of life cycle data with the development process and targeted adaptation or configuration of the product based on real operational data. This chapter demonstrates a generation-oriented approach to product development, using the evolution of both the individual component and the system as a whole as an example.

3.3.1 Representation of the Process of Information Feedback for the Development of Structural Mechanical Components Under Dynamic Loading

The effectiveness of the presented approach can be illustrated at the example of an information feedback and the development of a new generation of a wheel carrier. During the development phase different generations of wheel carriers are designed and created, based on the received lifecycle information. The manufacturing uses the information from the development and based on this, individual production and process plans are created, which allows a rerouting of the component and an individual way through the manufacturing. During usage the amount of individual component information increases after the development phase and is saved in a PLM System, meanwhile the amount of generation information is determined after the development phase and stored in a PDM System (Fig. 3.6).

The development process itself is iterative and includes the phases of development and optimization of the component geometry, the selection of a suitable material and the choice of sensors (Mozgova et al. 2017). Each development phase has an iterative character. During developing and optimization of the geometry of the component, properties of the metallic alloy are considered. The selection of the type of sensors depends on the alloy and the geometry of a component. Conditional on the expected loads and simulation results obtained during the geometry development, it might be necessary to change e.g. the properties of the alloy from which the component is made of. That would entail the changes in the selection of sensors. Positions of sensors, for example, on the component surfaces, have to be planned not only taking into account critical or characteristic positions of loading, but also taking into account the availability of locations for reading and writing data and for preventive maintenance, repair or replacement (Mozgova et al. 2018). That can demand a change of the geometry of the component.

The general scheme of the approach to develop a new generation of a wheel carrier based on data collected during the usage of previous generations is depicted in Fig. 3.7. The approach involves manufacturing restrictions in the context of product and component development, analysing data of the component usage (Lachmayer et al. 2013; Lachmayer et al. 2018), monitoring and controlling the current state of the component and analysing the results of usage.

During the development process a selection of suited materials and the type and location of the sensors as well as choosing the data modeling and analysis algorithms is required. Figure 3.8 shows the results of the technical evolution of the wheel carrier.

The criteria of homogeneity of the stress-strain distribution of the component and of the reduction of the component weight were used as the target functions of multicriteria optimization of the component geometry. The simulation of the stress-strain state of the component was performed using the finite element method.

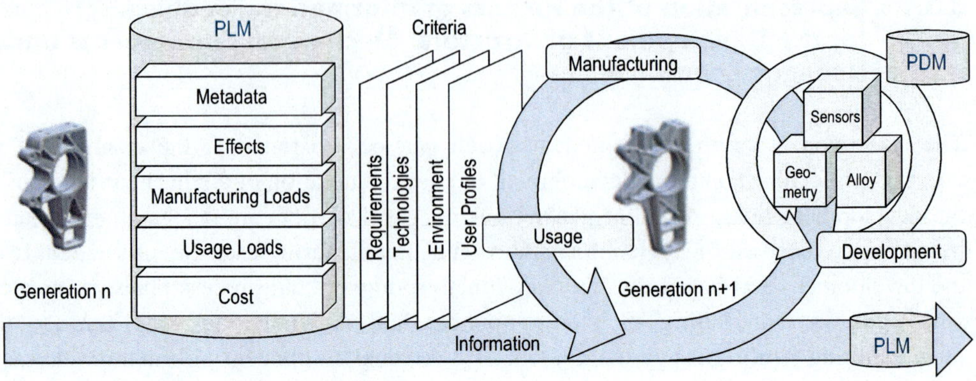

Fig. 3.6 Evolutionary adaptation of the component

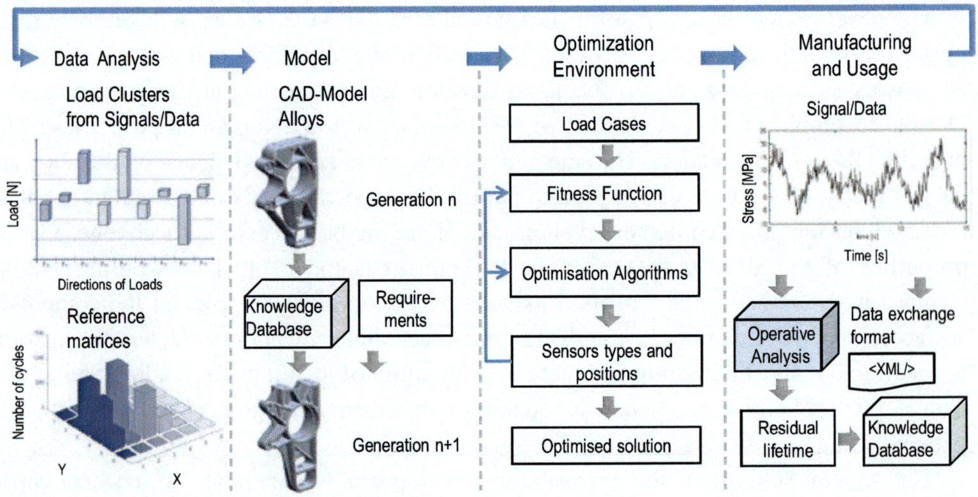

Fig. 3.7 Algorithmized information feedback

3.3.2 Application of Technical Inheritance for the Design, Monitoring and Operation of a Technical System at the Example of an Electronic Sorting Device

The application of the paradigm of technical inheritance (TI) in the system development will be shown at the example of a sorting system. Within the scope of a collaboration with the Nordstadt Clinic Region Hanover, a technical solution for sorting object carriers was to be developed.

The object carriers to be sorted contain tissue sections, which are examined in the histology for diagnosis. After the examination, these are be archived for 20 years so that they can be used for follow-up examinations. In order to be able to guarantee later retrieval

	Development Goal	CAD-Model	Stress-Strain State (FEM)	Development Result
Generation 1	Casting FEM based stress-strain simulation			Weight = 740 g Stress σ_{max} = 132 MPa
Generation 2	Casting FEM optimization Light weight			Weight = 607 g (↓ 18%) Stress σ_{max} = 79 MPa (↓ 40%)
Generation n	3D printing Topology optimization Geometry as an assembly			Weight = 585 g (↓ 21%) Stress σ_{max} = 75 MPa (↓ 43%)

Fig. 3.8 Evolution of a wheel carrier

and error-free assignment of the object carriers, these need to be sorted and filed according to an identification number. The system consists of a table, on which 1000 sorting compartments and 40 compartments for rejects are located, and a gantry robot (Fig. 3.9). The object carriers are to be sorted into the sorting compartments by the gantry robot according to their case number. The object carriers are fed into the system using four magazines. Each of the four magazines has a mechanism for separation, which allows the object carriers to be removed individually from the magazines.

In order to be able to provide all data for a subsequent generation of the system, the CAD models, corresponding technical drawings and the source code of the PLC as well as the requirements list, all data sheets of the purchased parts and the maintenance plan were stored in the PDM system Autodesk Vault (Scheidel et al. 2017). In this way, the PDM system stores the documentation of the first generation of the electronic sorting unit. For all project documents, individual identification number were assigned to the CAD models and drawings according to the developed numbering system. The numbering system corresponds to the concept of the GC.

In developing this technical system within the TI paradigm, an analysis was conducted which data and information was required to monitor the functioning of the system during its operation. During the analysis the following data and information to be monitored was identified: number of slides per day, number of slides per case, waste area capacity, number of used trays, pattern in the sorting, interruptions during operation, system operation times.

It is then determined at which points these data and information are recorded. Preventive periodic maintenance has been defined for the electronic sorting unit, which is carried out at 6-monthly maintenance intervals. This strategy includes a check of all electronic,

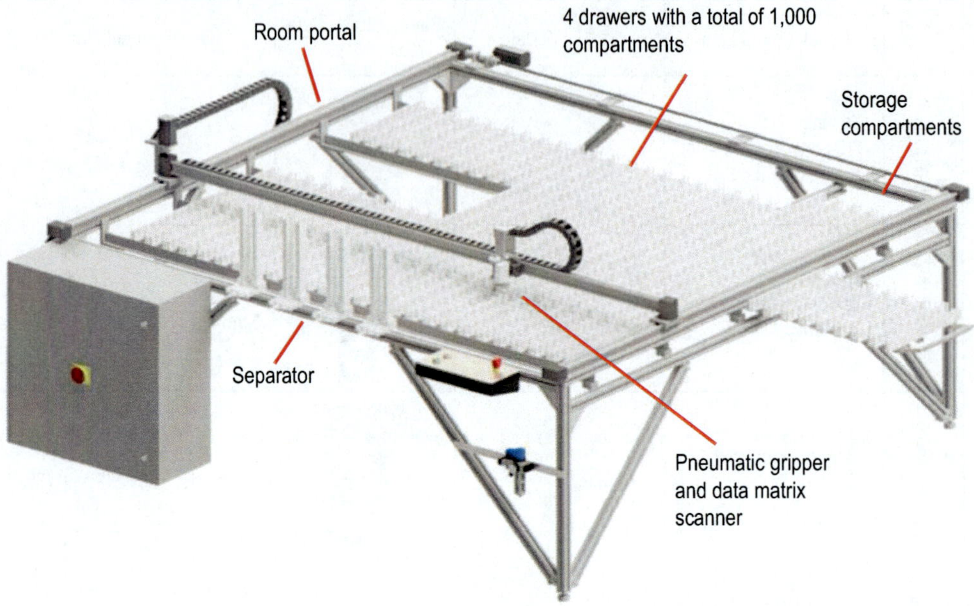

Fig. 3.9 Electronic sorting unit

pneumatic and mechanical components as well as an update of the system code and test runs. All operational data were collected as part of this strategy.

The analysis of collected data and information obtained during the operation of the electronic sorting unit allowed to specify the requirements for the next generation of the system. As a result of monitoring of the system operation the following parameters were analyzed: number of scans, trips of a tray, departures of a tray, timestamp of trips. Based on the obtained information, the following adaptations for the second generation of the system were considered: dimensions and performance of the room portal and adaptation of its capacity, dimensions of the compartment intake, dimensions of the waste area, adjustment the sorting algorithm and adaptation of service activities.

When the waste area is approached, different types of errors can occur and are documented. For example, object carriers with an incorrect location code or an incorrect sorting year can be returned to the sorting process for a later sorting run. During usage was recorded that object carriers were found in the wrong slots and that the design of the carrier holder for the slots was too small. Analysis of the causes of this type of error showed the need to develop an adapted design of carrier holder (Fig. 3.10).

Thus, at the example of this system we can observe different dynamics of development of the whole system and its subsystems, as described in Sect. 3.2: for the first generation of the electronic sorting unit the second generation of the carrier holder is developed. Figure 3.10 also shows the GC obtained according to the numbering created for the electronic sorting unit. Thus, according to the technical inheritance approach, the information obtained during operation was transferred to adapt the second generation of carrier

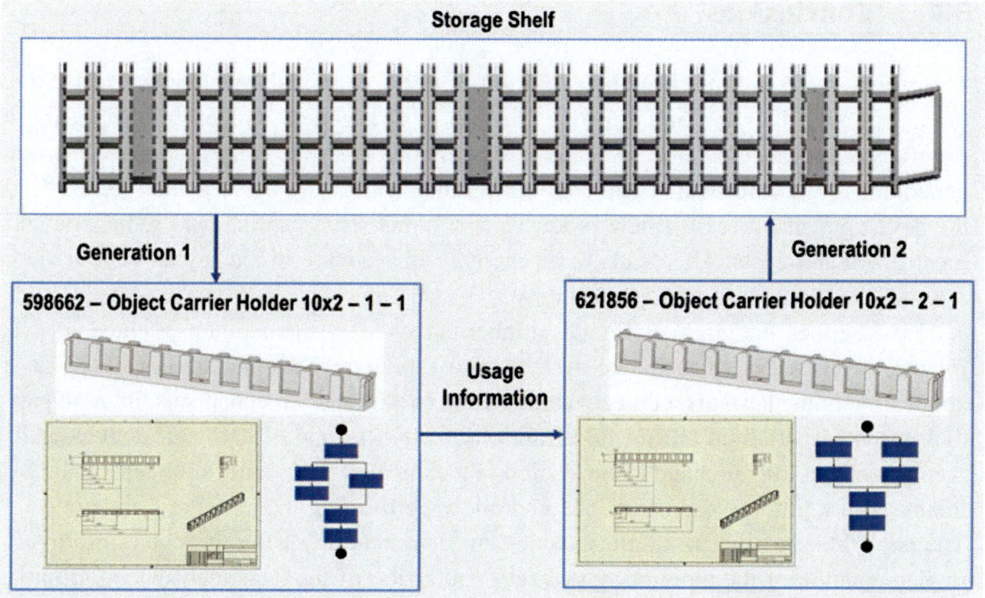

Fig. 3.10 Two generations of the subsystem object carrier holder (Scheidel et al. 2017)

holder. Furthermore, the traversing speeds of the sorting arm determined for the first generation are adjusted based on the number of cycles, which enables the reduction of the size and the use of smaller stepper motors and shorter linear guides.

In general, data processing during device use included summarizing, sorting and formatting process data, filtering errors and duplicates. The data analysis revealed that the size and operating capacity of the second generation of the electronic sorting unit can be optimized. For example, the analysis showed that that only 33.5% object carriers of the original were used is the case and the waste area can therefore be reduced by about 50%. Thus, fewer compartments can be provided for the next generation of the electronic sorting unit, whereby a dynamic assignment of the compartment numbers makes sense here. It must be taken into account that the cases are not distributed randomly over the compartments, but follow a pattern. Thus, for the next generation of the electronic sorting unit, both an adjustment of the number of compartments and a possible adaptation of the sorting algorithm is planned. So, the dimensions of the frame and the height of the compartments are therefore adjusted in this concept. The gantry arm and the associated stepper motors must also be reduced in size. Compared to the gantry arm of the first generation of the electronic sorting unit the reduction in its size results in a saving of 25% of item cost.

3.4 Conclusions

The current state and trends in the evolution of technology and development of technical products imply the rapid and effective creation of new generations of products using the experience and accumulated knowledge throughout the life cycle of previous product generations. The information gathered during the product life cycle is then fed back into the development process of new product generations. This facilitates to reduce development times and costs. This leads to an increase in resource efficiency as well as higher dynamics and thus increased productivity.

The described Paradigm of Technical Inheritance (TI) is a generation-oriented approach for the development of technical systems and products. The paradigm is based on the application of the laws of technical evolution and evolutionary mechanisms. Integral part of the presented paradigm are methods, algorithms and means of data analysis, as well as methods of classification and identification of technical systems and subsystems, which are the basis of a product's genetic code and allow performing unambiguous authentication. The application of the paradigm includes an identification of the life cycle information, implementation of the monitoring strategy, realization of the data analysis, algorithmized information feedback and is carried out within the framework of the TI. Based on a methodology appropriate to the scope of the application, a selection of methods, algorithms and data analysis tools for a technical system or product is performed to adapt a new generation of the system or product.

A methodology for targeted data feedback was presented and it was shown how the operational data can be used for development as well as adaptation of the next generation of a product. At the example of wheel carrier was shown how the paradigm of TI can be used to optimize lightweight constructions with additionally improved stress distribution. The example of the electronic sorting unit demonstrates the application of the TI paradigm for the design of a whole system and individual subsystems.

References

Abramovici M, Lindner A (2011) Providing product use knowledge for the design of improved product generations. CIRP Ann Manuf Technol 60(1):211–214

Albers A, Bursac N, Wintergerst E (2015) Produktgenerationsentwicklung—Bedeutung und Herausforderung aus einer entwicklungsmethodischen Perspektive. In: Binz H, Bertsche B, Bauer W, Roth D (eds) Stuttgarter Symposium für Produktentwicklung (SPP) 2015, pp 1–10

Altshuller G (1984) Creativity as an exact science. CRC, London

Anderl R (2015) Industrie 4.0—technological approaches, use cases, and implementation. At-Automatisierungstechnik 63(10):753–765

Bäck T, Fogel DB, Michalewicz Z (2000) Evolutionary computation 2. IOP, Bristol

Demminger C, Mozgova I, Quirico M, Uhlich F, Denkena B, Lachmayer R, Nyhuis P (2016) The concept of technical inheritance in operation: analysis of the information flow in the life cycle of smart product. Procedia Technol 26:79–88

Denkena B, Mörke T (eds) (2017) Cyber-physical and gentelligent systems in manufacturing and life cycle. genetics and intelligence—keys to Industry 4.0. Academic, London

Eigner M, Dickopf T, Apostolov H, Schaefer P, Faißt KG, Keßler A (2014) System lifecycle management: Initial approach for a sustainable product development process based on methods of model based systems engineering. IFIP Adv Inf Commun Technol 442:287–300. https://doi.org/10.1007/978-3-662-45937-9_29

Eversheim W (2009) Innovation management for technical products. Springer, Berlin

Feldhusen J, Grote KH (2013) Pahl/Beitz Konstruktionslehre: Methoden und Anwendung erfolgreicher Produktentwicklung. 8., Berlin

Fischer R (1958) The genetical theory of natural selection. Dover, New York

Fortner RS (2014) Lewis Mumford. Technics, civilization, and media theory. In: The handbook of media and mass communication theory, pp 210–224

Gottwald PR (2016) Prozess einer generationsübergreifenden Produktentwicklung durch Technische Vererbung. Dissertation, Leibniz University Hannover

Graessler I, Poehler A (2019) Human-centric design of cyber-physical production systems. Procedia CIRP 84:251–256

Haken KL (2012) Grundlagen der Kraftfahrzeugtechnik. 2., aktualisierte und erweiterte Auflage. Carl Hanser, München

Honeywill R (2008) Lamarck's evolution: two centuries of genius and jealousy. Murdoch Books, Sydney

Hughes TP (1987) The evolution of large technological systems. MIT Press, Cambridge, MA, pp 51–82

Johannknecht F, Gatzen M, Lachmayer R (2017) Systematic approach to optimize cost drivers based on life cycle cost modeling. In: Proceedings of the 21st international conference on engineering design (ICED 17), 21–25 August, Vancouver, vol 5, pp 121–130

Kapp E (1877) Grundlinien einer Philosophie der Technik. Braunschweig

Kaufmann T (2015) Geschäftsmodelle in der Industrie 4.0 und dem Internet der Dinge—Der Weg vom Anspruch in die Wirklichkeit. Springer Vieweg, Wiesbaden

Kemp R, Mudler P, Reschke CH (1999) Evolutionary theorising of technological change and sustainable development. Research Centre for Economic Policy, Rotterdam

Kidd IJ (2012) Oswald Spengler, technology and human nature. Eur Legacy 17(1):19–31

Lachmayer R, Gottwald P (2014) Integrated development by the consideration of product experiences. In: Proceedings in 10th international workshop on integrated design engineering, 10–12 Sept 2014, Gommern, pp 73–82

Lachmayer R, Mozgova I, Deiters A, Sauthoff B, Gottwald P (2013) Design evolution by integration of life cycle information in knowledge repositories. In: 5th IASDR 2013 Tokyo, vol 2, pp 309–318

Lachmayer R, Mozgova I, Reimche W, Colditz F, Mroz G, Gottwald P (2014) Technical inheritance: a concept to adapt the evolution of nature to product engineering. Procedia Technol 15:178–187. https://doi.org/10.1016/j.protcy.2014.09.070

Lachmayer R, Mozgova I, Gottwald P (2015) Formulation of paradigm of technical inheritance. In: Proceedings of the 20th international conference on engineering design (ICED15), 27–30 July, Milan, vol 8, pp 271–278

Lachmayer R, Yanchevskyi I, Mozgova M, Gottwald P (2018) Identification of several non-stationary loads applied to an elastically deformed structure. Appl Comput Mech 12(1). https://doi.org/10.24132/acm.2018.365

Lindemann U (2009) Methodische Entwicklung technischer Produkte. Springer, Berlin

Mozgova I (2017) Intelligente Datenanalyse für die Entwicklung neuer Produktgenerationen, Wissenschaftsforum Intelligente Technische Systeme (WInTeSys), 11 und 12.05.2017, Paderborn, vol 369. Heinz Nixdorf Institut, pp 335–346

Mozgova I, Barton S, Demminger C, Miebach, T, Taptimthong P, Lachmayer R, Nyhuis P, Reimche W, Wurz M (2017) Technical inheritance: Information basis for the identification and development of product generations. In: Proceedings of the 21st international conference on engineering design (ICED 17), 21–25 August, Vancouver, vol 6, pp 91–100

Mozgova I, Yanchevskyi I, Gerasymenko M, Lachmayer R (2018) Mobile automated diagnostics of stress state and residual life prediction for a component under intensive random dynamic loads. Procedia Manuf 24:210–215. https://doi.org/10.1016/j.promfg.2018.06.037

Ripperda S, Krause D (2017) Cost effects of modular product family structures: Methods and quantification of impacts to support decision making. J Mech Des Trans ASME 139(2):2593216

Scheidel W, Mozgova I, Lachmayer R (2017) A case study for a unified designation of development information in a PDM system for technical inheritance. Procedia Manuf 24:2–7. https://doi.org/10.1016/j.promfg.2018.06.002

Storch V (2013) Evolutionsbiologie. Springer, Berlin

Toffler A (1990) Power shift: knowledge, wealth, and violence at the edge of the 21st century. Bantam Books, New York

Vajna S, Jordan A, Bercsey T (2005) The autogenetic design theory: an evolutionary view of the design process. J Eng Des 16(4). https://doi.org/10.1080/09544820500267781

Waters M (2003) Daniel Bell. In: Ritzer G (ed) The Blackwell companion to major contemporary social theorists. Blackwell, Malden, MA. https://doi.org/10.1002/9780470999912

Wuest T, Weimer D, Irgens C, Thoben KD (2016) Machine learning in manufacturing: advantages, challenges, and applications. Prod Manuf Res 4(1):23–45

Application of Agile Experiential Learning Based on Reverse Engineering as Support in Product Development

Frank Mantwill and Valentin Multhauf

Abstract

In an information society, there is a trend towards finding solutions by researching online references. This possibility must also be used for the product development process. Consequently, this chapter shows a combination of methods between the product development process and reverse engineering. This approach aims to support product planning, task clarification and conception by use of reference fundamentals. For this purpose, iterative empirical values are generated by the model of experiential learning according to LEWIN. This application has been tested by means of hardware development of a condition monitoring system. The experience gained is documented in terms of benefits and restrictions that arise. A recommended course of action in dealing with online media is shown with an application example as a basis for the reverse engineering process. In conclusion, the statement can be made: Employing reverse engineering, online media can make an effective and early knowledge contribution to the product development process.

4.1 Importance of Product Knowledge in the Early Phase of Product Development

By mere logical thinking we are not able to gain any knowledge about the world of experience; all knowledge about reality starts from experience and flows into it. Purely logically won sentences are completely empty with regard to the real (Albert Einstein).

F. Mantwill (✉) · V. Multhauf
Helmut-Schmidt-Universität, Machine Elements and Computer-Aided Product Development, Hamburg, Germany
e-mail: frank.mantwill@hsu-hh.de

D. Krause, E. Heyden (eds.), *Design Methodology for Future Products*,
https://doi.org/10.1007/978-3-030-78368-6_4

Albert Einstein clarifies with this quotation that every knowledge is traceable to experience. From this it can be deduced that often, especially inexperienced product developers, lack a portfolio of different solution approaches when searching for solution principles (Meboldt et al. 2012). Therefore "the reuse of components, concepts and related knowledge is an important factor in design practice, but also in design education" (Weber and Husung 2016). This is because, in the course of developing a product, people have to generate and process different information through their thoughts and actions so that a functional and manufacturable product can be created from a need or an order (Matthiesen 2021). Even if the information to be created can be as different in detail as the products resulting from a development process, there are at the same time generic building blocks underlying every product development process, which result in particular from the fields of system theory, various model representations as well as from the human abilities of thinking and acting as individuals and in groups (Verein Deutscher Ingenieure 2019). As a result, the range of challenging situations that product developer encounters in the course of his or her professional life is considerable. Even if, due to the existing expertise, these problems are only tasks—and thus not problems—for the experienced development engineer, new elements in the design process occur daily, so that in the current situation there is often not enough knowledge available to find a solution immediately (Badke-Schaub and Frankenberg 2003). The revision of VDI 2221 in November 2019 reinforces the now indispensable use of information and communication technologies to support product development (Verein Deutscher Ingenieure 2019). Accordingly, the topic of this chapter starts here and tries to extend the search process of suitable solution principles using the methodology of reverse engineering based on online media, in order to learn from reference products and consequently to generate product knowledge itself. Section 4.2 therefore proposes a system for a product development process, which takes reference solutions from online portals into account as a central element. The established principles of product development, such as frontloading and the Rule of Ten, will be taken into account. After a preceding analysis (see Sect. 4.3) of online repair portals for the portfolio offered there, quality and usability of the data, recommendations for action are explained in Sect. 4.4 through a use case.

4.2 Integration of Reference Product Knowledge into Product Development

In order to apply the integration of reference product knowledge into the product development process, the reference knowledge must be available, methodically analyzed and systematically incorporated into the product development process. For product knowledge generation, the reverse engineering process is first explained in this subsection. This is followed by a description of the LEWIN experience learning process, which enables the integration of reference product knowledge into the product development process according to VDI 2221 (see Fig. 4.2). Since this combination of methods between reverse engineering and the product development process was tested in the context of the hardware development of the condition monitoring system PANDA I TIMESWIPE, the focus is explicitly on mechanical components.

The process of reverse engineering means, conversely, developing or also reconstructing. Consequently, the process can be interpreted as extracting design elements from an existing system by examining its structures, states, and behaviors. A basic, general definition of reverse engineering is provided by Elliot J. Chickowsky and James H. Cross II in Reverse Engineering and Design Recovery: "Reverse engineering is the process of analyzing a subject system to identify the system's components and their interrelationships and create representations of the system in another form or at a higher level of abstraction" (Chickowsky and Cross II 1990). The procedure of analyzing an existing product (reference products) in order to gain knowledge for its further development originates from mechanical engineering (Chickowsky and Cross II 1990). This means that in the context of this chapter, the definition from the VDI 5620 "Reverse Engineering of Geometry Data" does not apply to the term reverse engineering. For mechanical assemblies, this typically means disassembling and then analyzing, measuring and documenting the parts. The goal of redesign based on existing products is to determine detailed information and specifications of the sub-elements and to further learn about the functionalities and manufacturing processes. Furthermore, interactions of the respective sub-elements are investigated and the corresponding sub-functions are projected into the overall functional structure. In the model of system generation engineering the "reference system for the development of a new product generation is a system whose elements originate from already existing or already planned socio-technical systems and the associated documentation and are the basis and starting point for the development of the new product generation" (Albers et al. 2019). In the case integration of reference product knowledge into product development, a systematic approach has become established for improving one's own product based on the analysis of existing products. Product teardown analysis is used specifically to obtain a starting point for a solution-centric development approach. A basic, general definition of product teardown inclusive main purposes of the methodology supplies Kevin N. Otto and Kristin L. Wood: "Product teardown is the process of taking a product to understand it, and to understand how the company producing the product succeeds. A product teardown serves three primary purposes:

- Dissection and analysis during reverse engineering
- Experience and knowledge for an individual's personal database
- Competitive benchmarking" (Otto and Wood 2001)

Thus, the reference product knowledge is systematically generated by the product teardown analysis as part of the reverse engineering process, which now has to be transformed by analysis into own product experiences. The LEWIN model of experiential learning can be used for this purpose (see Fig. 4.1).

The starting point of experiential learning according to LEWIN is the concrete experience. This is followed by extended observation, data collection and reflection on the

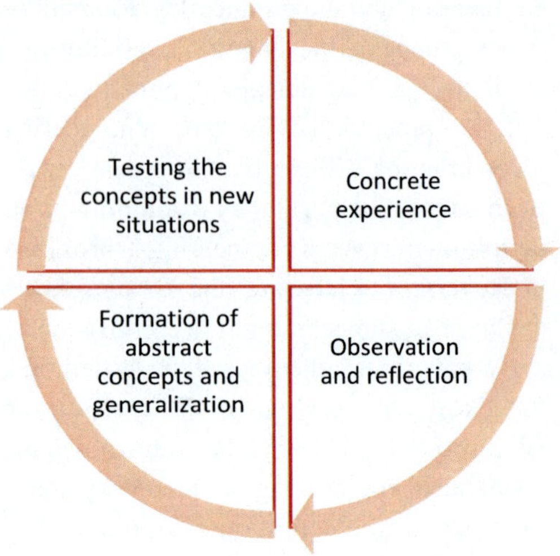

Fig. 4.1 Model of experiential learning according to LEWIN (Gruber 1999)

experience made. Based on this process, new concepts are formed by evaluating the collected data and interpreting the observations made, which in turn lead to new experiences by testing them in new situations. Consequently, the model of experiential learning according to LEWIN can be understood as an iteration loop, which generates incremental experience values from the perspective of a product developer and thus expands the solution principle portfolio. Thus, by testing the concepts in new situations, the external reference product knowledge is transformed into individual's experience values. These new insights must be incorporated into the product development process in a structured manner. To this end, the general procedure for development and design is described below, followed by a demonstration of how the experience values can be incorporated into the development process. The state of the art for the product development process is defined by means of VDI 2221. In 1986, VDI 2221 was the first description of the main phases of development and design derived from systems engineering and is still one of the best-known methodologies for product development. This approach divides the development process into eight activity steps, each of which has a work result. First, according to the guideline, the task is clarified and specified. The result of this first step is a list of requirements, i.e. which customer needs the product must meet and which constraints must be observed. Based on the requirements, the overall functions and the essential subfunctions and their structures are defined in activity step two. The goal of step two is to structure and modularize the problem with the help of a solution-neutral description. In mechanical engineering, this is usually mapped with the help of function structures. In activity step three and four, the principle solutions are worked out, which provide the active structure for fulfilling individual functions in accordance to the requirements. The functional solutions found in each case are broken down into feasible modules in activity step five, which are then specified and implemented in activity step six. The integration of the partial solutions takes place in activity step seven. The documentation of the product and the creation of the production documents are carried out in activity step eight.

Thus, the product development process defined by VDI 2221 can be understood as a "bottom-up principle". This basic idea involves the functional solution of delimited and detailed partial requirements in order to be able to solve hierarchically superimposed functional requirements with their overall function. The individual partial solutions are assembled from "bottom" to "top" until all product requirements are finally met. Inexperienced product developers lack knowledge of the solution-centric development approach. This means that inexperienced product developers do not possess the expertise to accomplish without information bases a strategic product planning, to provide a requirement list, to conceive a function structure and to sketch their construction solutions. The previously described model of experiential learning supports the hypothesis that ultimately all expertise is traceable to experience. Accordingly, LEWIN's experiential learning model is intended to act as an interface between the product development process and the reverse engineering process, and to show how inexperienced product developers can implement a solution-centric development approach based on reference values. In principle, the teardown analysis of competitor products has become established in the business world, but reference product knowledge based on online sources is rarely used systematically as a cross-sectional activity for a company's product development.

As a starting point for a solution-centric development approach, reverse engineering can support the product development process by means of online media (cf. Sect. 4.3). Reverse engineering can be interpreted as a "top-down principle". The process can thus be described as a procedure in which the respective design solutions of the sub-elements, the functional structures and the requirements are extracted from a product that already exists on the market by means of product teardown. As a result, the product developers have an information basis based on existing products, which can make an early and effective contribution concerning the product development process according to VDI 2221. The advantages of the combination of methods can be explained by the opposing vertical principles of "bottom-up" and "top-down" and realized by horizontal information flows (cross-sectional activities).

The requirements of the reference products can be derived based on the product specification description. In commercial online media, a product specification description is already included in the report. For the open content media, a corresponding specification description, exemplified by product data sheets, can be obtained from another online source. The generated information about the requirements of the reference products provides the basis for a horizontal information flow between the reverse engineering process and the product development process (see orange arrow in Fig. 4.2) as the requirements list of existing products plus the systematic decomposition can be associated as the concrete experience according to LEWIN's model of experiential learning. By means of the subsequent observation and reflection of the competitor products, the phase of task clarification according to VDI 2221 can be supported based on the product specification description. In conclusion, the description of competitor products provides a valuable

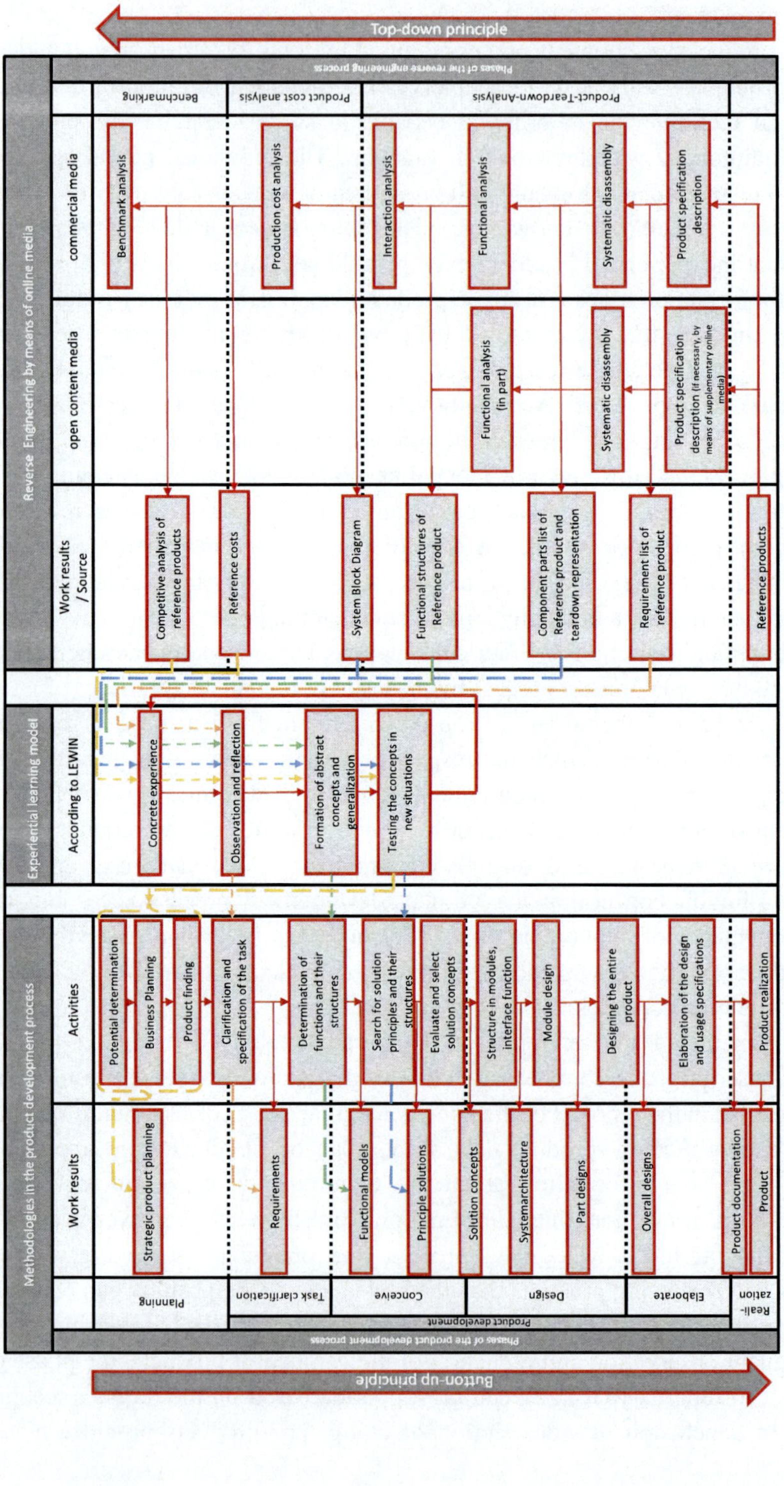

Fig. 4.2 Combination of the product development process and the reverse engineering process through the experiential learning

reference basis for the requirements list of the product to be developed. Through professional observation and reflection, this information basis can make an efficient contribution to the clarification and specification of the task.

The second horizontal flow of information (see green arrow in Fig. 4.2) leads to the determination of functions and their structure based on the function structure of reference products. Using the methodology of the teardown analysis, a function structure of selected reference products can be generated. Again, the report from the commercial online media already includes a detailed functional analysis. As a result, each sub-component can be assigned to its function(s). In the context of the reverse engineering process based on open content media, no direct functional analysis is offered in principle. However, through the presented disassembly and the comments of the "crowd", the function/s of the sub-elements can be determined in many cases. Again, the functional structure of competitor products can provide a positive contribution to the identification of functions and structures for the product development process.

The main contribution of the reverse engineering process to product development is based on the third horizontal information flow (see blue arrow in Fig. 4.2). The work results generated in the phases of product teardown analysis, product cost analysis and benchmark analysis within the reverse engineering process can provide an effective contribution based on reference products for the search for solution principles and their structures. Through the systematic disassembly of already existing products including the functional analysis on module level as well as the further interaction analysis, important insights for requirement-specific design solutions can be gained. This approach is fundamental for an understanding of experience building according to LEWIN and thus has a direct influence on the design approaches for the respective function fulfilment based on reference solutions. These solution approaches of already developed products offer a representation on submodule level, in which all expertise of experienced developers has been incorporated. The active principles of the reference products examined have already established themselves on the market and convinced the respective consumers—otherwise, they would not receive the attention they do in the online media. By means of the teardown representation, functional structures of reference products, as well as the system block diagram, novel concepts and generalizations can be developed through expert reflection. According to LEWIN's model of experiential learning, the fourth phase involves testing the newly developed concepts in a new situation. This means that the requirement-specific design solution approaches, which were generated based on the reference solutions, are implemented in the product environment to be developed. This should be understood as an iterative process since it is rarely possible to develop the optimal solution approaches in the first iteration loop. Therefore, it is important to question the experience of the non-optimal solution approach (observation and reflection), to develop new functional solutions based on the gained expertise and to test them again. The reverse engineering process can provide the first effective basic idea based on reference product knowledge. Concluding, it can be said that reverse engineering and its reflection offers a portfolio of reference design solutions which effectively contribute to the generation of solution principles according to VDI 2221. This purpose is served by both commercial and open content media.

4.3 Online Media as a Source of Product Knowledge

In this section, the online media already mentioned are examined concerning their added value when used for a reverse engineering process. The resulting findings are presented in a clear overview in Table 4.1. In the following section, the open content media are defined and their principles are described. In contrast to the open content media, the commercial media are then described. In the final section, the advantages and restrictions of online media as a basis for reverse engineering are explained.

The term Open Content was coined by the Open Content Initiative in 1998. This resulted in the use of free content in the areas of software (Open Source), technologies (Open Hardware), databases (Open Data), science and education (Open Access) as well as in politics (Open Government). The "Open Source Ecology" initiative is already taking the approach about scaling existing product knowledge. In general, an open content platform can be understood as an information medium by means of which authors pass on their knowledge to third parties based on altruistic motivation. This knowledge can usually be used free of charge. The basis for open content platforms is provided by the phenomenon of crowdsourcing. "Crowdsourcing is an interactive form of service provision that is organized collaboratively or competitively and involves a large number of extrinsically or intrinsically motivated actors of different knowledge levels using modern information and communication systems based on Web 2.0. The object of performance are products or services of different degrees of innovation, which are developed by the network of participants reactively due to external impulses or proactively by self-actively identifying gaps in demand or opportunities" (Martin et al. 2008). In this context, the following Open Content media are based only on collaborative interactions between the intrinsically motivated authors and the platform operators. The open content media iFixit and YouTube were investigated.

iFixit is a wiki-based website with the basic idea: "show people how to fix almost anything". Altruistically motivated authors can create a repair manual for a device on iFixit.com or edit and improve already existing manuals. This website enables interested people to share their technical knowledge with the rest of the world and provides the necessary platform for this. In addition to repair instructions, the medium also provides an ideal source for a reverse engineering process. Analyses of disassembled products can be viewed under the menu item "Teardown" (iFixit 2021).

The video portal "YouTube" enables the uploading of video in order to share one's knowledge with society via contributions. The portal was founded in 2005 and pursued the basic idea of crowdsourcing. That is, intrinsically motivated filmmakers upload their video and thus share their opinion. As a result, users on the portal can watch the video clips for free, rate them, comment on them and in turn upload films themselves. This results in a

Table 4.1 Results of the study of online media

	Open Content media			Commercial media	
	iFixit		YouTube	Tech Insights	IHS Markit
	Teardown Analysis	Repair manual			
Scope of the portfolio	o Smartphone o Smartwatch o Laptop o Smart-speaker o Game Console o Electric bike	o Clothing o Household facilities o Camera o Car parts o Computer-Hardware o Electronic devices o Game consoles o Household appliances o Mac products o Medi Player o PC o Mobile phone o Tablet o Vehicles	o Undefined quantity	o Smartphone o Tablets o Phablets o Wearables Computing o Drones o Laptop o Semiconductor components	o Mobile phones o Tablet o Car parts o Inverter o Notebooks o Medicine o Wireless Speaker
Type of media	o Pictures o Description o Videos o Comments o Summary o...	o Pictures o Description o Videos o Comments	o Pictures o Description o Videos o Comments	o Product features o Parts list o Pictures o Videos o System-Block-Diagram o Production cost analysis	o Pictures o Videos o Functional assignment o Summary o Product information o Cost breakdown o Bill of materials o Comparative analysis
Usability of the portfolio	o Teardown Analysis o Functional analysis for circuit boards	o Teardown Analysis	o Teardown Analysis o Functional analysis	o Product specification description o Industry-specific teardown analysis o Component parts list incl. product number, manufacturer and function o System-Block-Diagram o Manufacturing Cost Breakdown o Cost Drivers	o Industry-specific teardown analysis o Component bill of material o Functional analysis o Benchmark analysis o Manufacturing Cost Breakdown
Authors/ users	o altruistically motivated users	o altruistically motivated users	o altruistically motivated users	o Research and technology analysts	o Research and technology analysts
Costs	o Free of charge without registration	o Free of charge without registration	o Free of charge without registration	o No data	o License fee of $50,000 to $60,000

Table 4.2 Advantages and restrictions of online media for the product development process via reverse engineering

	Advantages		Restrictions
A.1	Open content media provides a free source for the reverse engineering process	R.1	Altruistic authors define Open Content media offerings
A.2	Information via online media is available quickly and at any time	R.2	Analysis in the context of open content media must be questioned with regard to quality
A.3	By means of comment function, the open content media are a source for understanding customer needs	R.3	Commercial media is a costly source for the reverse engineering process
A.4	Source for potential determination within the framework of strategic product planning	R.4	No source of disruptive innovation in the same business area
A.5	Support of task clarification in the process of product development	R.5	Online media do not offer haptic perception
A.6	Support for early product conception (frontloading) based on established solution approaches (rule of ten)		
A.7	Possible time saving by means of reduction of iterations		
A.8	Existing solutions have a certain quality to show, as they have established themselves on the market		

"snowball principle" with the consequence of an avalanche growth of contributions. Some users use this platform to disassemble products in front of the camera and publish their knowledge about the products. In addition, companies and retailers also use YouTube to offer customers a free service, such as repair instructions, in order to draw attention to their core competence. Therefore, the medium YouTube also provides a source for a reverse engineering process.

Contrary to the open content media, on which altruistically motivated authors publish their knowledge free of charge, the business interest of the service providers of the commercial media is profit oriented. I.e., product analyses are made available to the licensees for a license fee. The market leaders for commercial teardown analyses are "Tech Insight" and "IHS Markit".

The following section presents the results of the studies on the media mentioned above (see Table 4.1) and discusses the advantages and restrictions (see Table 4.2). The quantitative analysis of the portfolio is intended to show, to some extent, for which product segments the respective medium can be an added value. By examining the diversity, the different means of communication with which the authors publish their experience are presented. Furthermore, the potential usability of the portfolios in terms of added value for the reverse engineering process is mentioned. Finally, the author types and the costs of the respective online media are listed (see Figs. 4.3 and 4.4).

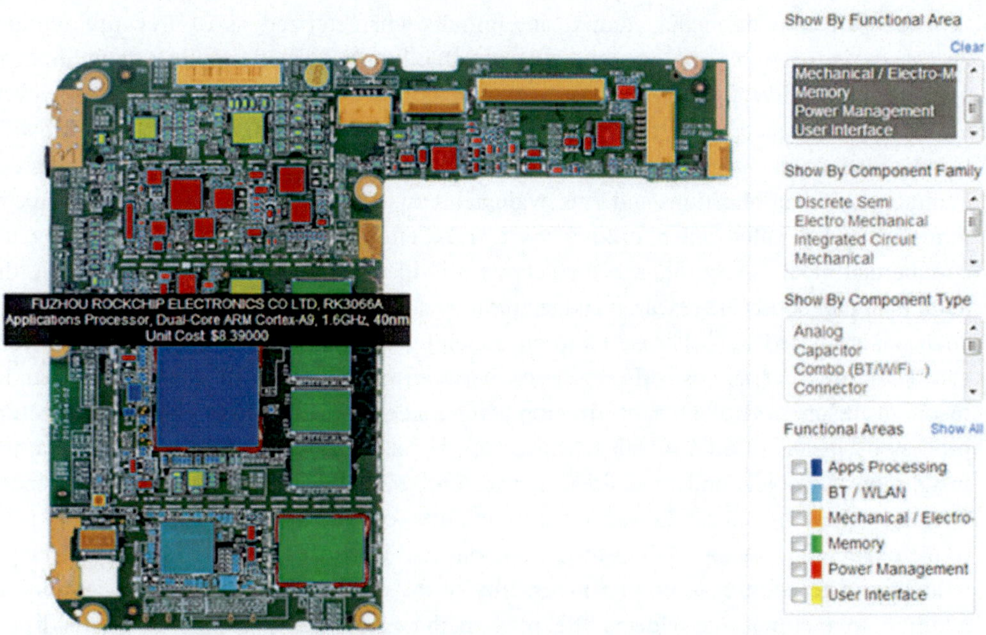

Fig. 4.3 Exemplary photo teardown analysis (IHS Markit 2018)

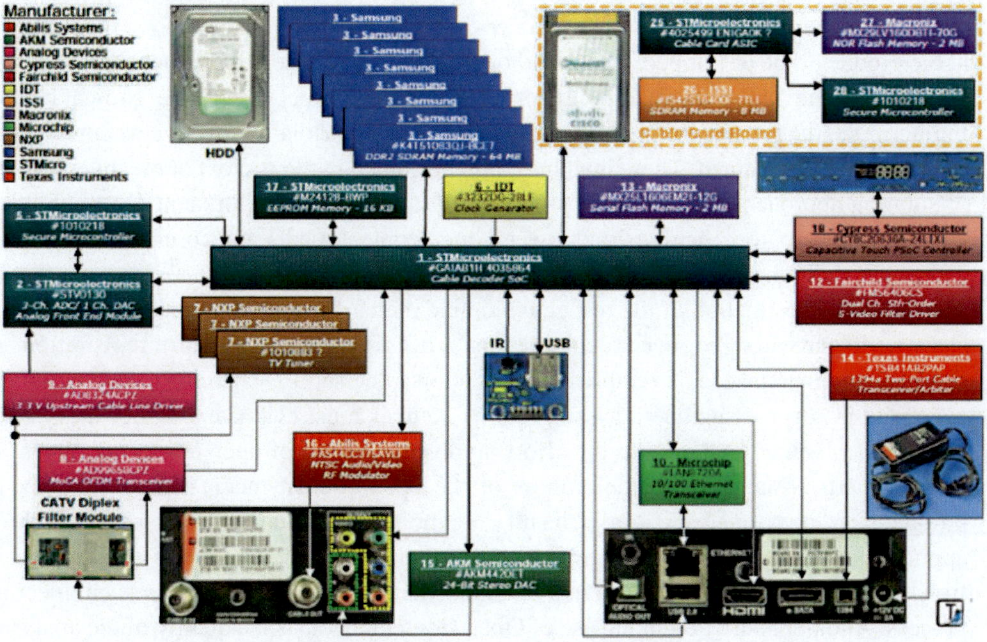

Fig. 4.4 Exemplary system block diagram (TechInsights 2018)

The open content media studied are initially characterized as of free use without necessary registration. This offer is made possible by altruistically motivated authors. As a result, the quality of the publication is mostly not based on trained expertise. However, the quality assurance of the publications is practiced by means of the so-called "crowd". For the media, which function according to the Wikipedia principles, all Internet users can comment on the publications and thus evaluate them. A more professional online source in terms of added value when used for a reverse engineering process is offered by the commercial media. According to the service providers Tech Insight and IHS Markit, the authors of the reports are research and technology analysts. Accordingly, the corresponding analyses are based on expertise and consequently, high-quality, structure-oriented work can be assumed. This cost differentiation between open content media and commercial media results in more diverse information being made available for the reverse engineering process by means of paid media. Through open content media, the developer gets a first insight into the object under consideration via images, videos, descriptions and comments. Certainly, these media can be sufficient to identify some components and individual parts of the existing product as well as their functions and interfaces. However, this informative value correlates significantly with the quality of the reporting by the altruistic authors. In addition to pictures and videos, the paid media also offer a component parts list, a functional assignment of the product components via a system block diagram and a manufacturing cost breakdown including identification of the cost drivers.

As described in the previous Sect. 4.1, the combination of the reverse engineering method and the product development process can make an early and effective contribution to the product to be developed. These and other advantages based on online media in terms of added value when used for a reverse engineering process will be shown below. Furthermore, the restrictions and disadvantages of online media for the reverse engineering process will be explained. These insights could be gained in the context of the investigation of solution approaches of reference products for a hardware development for a condition monitoring. Here, too, the division between open content media and commercial media is necessary, since these have different platform specifications.

First, the investigation of the respective online media revealed that open content media are a free source for the reverse engineering process and do not require registration. In contrast, commercial media require a subscription to be set up. Therefore, each company interested in reverse engineering using online sources must evaluate whether the service offers added value in relation to the effort involved. The use of open content platforms is free of charge. Accordingly, the authors of the open content media define the offer of product teardown analyses. As a result, the portfolio includes preferably household appliances and electronic devices. In conclusion, only a few industries can directly benefit from the Open Content media in terms of added value when used for a reverse engineering process. Another restriction in the use of Open Content media is the quality of the analyses. Since this information platform functions according to the Wikipedia principle, any user can upload a contribution. In addition to professional information, this platform also contains all kinds of technical nonsense. In contrast, the added value of online media with regard to the reverse engineering process is the information that can be accessed quickly and at any time. These information bases can result in an early and effective contribution to the product development process. The effective contribution of solution approaches is based on the fact that, for example, the requirements or the design solutions are derived from reference products. These reference products are mostly very successful products from companies that have years of experience in developing consumer products. I.e. these products have successfully passed through a consumer quality filter. On the other hand, the reference requirements and the reference design solutions of the investigated objects have already gone through several iteration loops and consequently a lot of expertise has been incorporated. Therefore, the references can be adapted as state of the art into the product to be developed with new framework conditions. The established method of "front-loading" based on the "Rule of Ten" advocates the early integration of references based on existing successful products. Thus, considering the availability of references already from the beginning of the product development process changes the understanding of the early phase of product development (Albers et al. 2017).

The disadvantage of this approach is that the adaptation of references only leads to incremental improvements in the respective products of the same business segment. As a result, disruption based on reverse engineering is rather unlikely. Another restriction of online media is that the information platforms do not offer any haptic perception. The information is only conveyed via images, videos and analysis displays. Compared to the conventional reverse engineering process, in which the products themselves are systematically disassembled, online media do not offer a direct experience, as required by LEWIN. In many use cases, however, teardown representations or the product specification description are already sufficient to support the product development process. Furthermore, the representations by means of online media offer a filter function in which a preliminary decision can be made by analyzing the corresponding media. This means that not every product that would be suitable for conventional reverse engineering has to be procured and disassembled. Instead, it is sufficient to find a relevant design solution via a picture and to explicitly examine this product using the conventional reverse engineering in order to obtain the haptic perception. Therefore, not every product in question has to be examined, but only the products that have been pre-selected via online media. This maximizes the effectiveness of the conventional reverse engineering process and minimizes the effort.

4.4 Recommendations and Practical Example of Use

Based on the combination of methods defined in Sect. 4.2 and the online media examined in Sect. 4.3, this chapter will explain a recommended course of action for dealing with online media as a basis for reverse engineering. This recommended course of action is then illustrated by a case study, which was documented as an example in the course of the previous analysis and application to the condition monitoring system. The reverse engineering process based on online media can be divided into four process steps. First, suitable

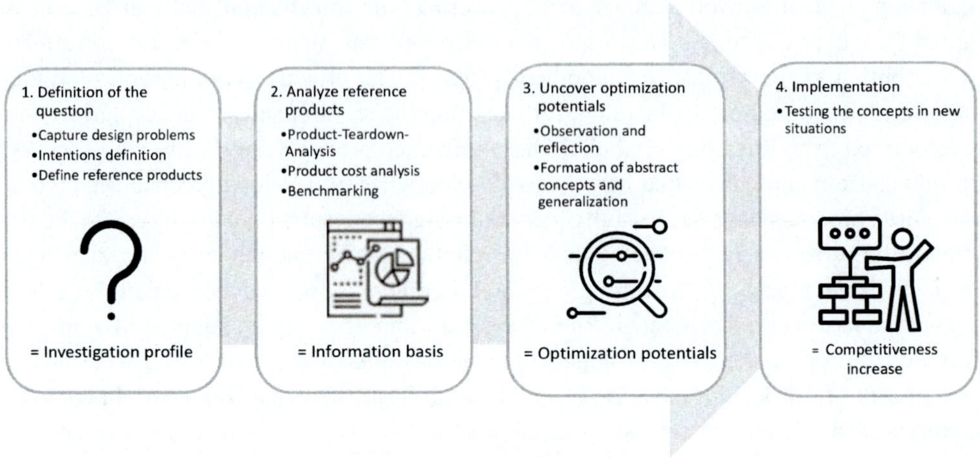

Fig. 4.5 Recommendation for action steps in dealing with online media as a basis for reverse engineering

questions must be defined. For this purpose, the definition of the questions from Fig. 4.2 have become established. In the second phase, the previously defined questions are to be examined. Consequently, depending on the selected online medium, the results are revealed as possible solutions. In the third phase the optimization potentials are to be uncovered by professional observation and reflection of the reference solution approaches. As a result, new concepts and generalizations can be formed. In the last phase, this gained expertise is to be tested iteratively in a new product environment (see Fig. 4.5).

For the advancement of the condition monitoring system PANDA I TIMESWIPE, the design problem was defined that the cooling system of the printed circuit board may not be sufficient. Thereby the heat dissipated to the environment through the housing via thermal emission and convection as well as direct heat conduction through the connections. Consequently, the intention was raised that there must be a higher heat dissipation from the CPU to the environment in order to prevent performance throttling or even disintegration (step 1). Therefore, the question in the first process step was defined: How are the permissible values of the CPU temperatures maintained in existing products using an active cooling management? The investigation profile was completed by the hypothesis that laptops offer an efficient solution due to the increasing performance of the electronic components and the small available installation space (step 2). By analyzing these reference products, it was possible, to obtain the information that heat pipes are primarily used in efficient cooling systems. The heat pipe is connected at one end to the warm side and at the other end to a cooling side in conjunction with a fan (see Figs. 4.6 and 4.7).

The pictures shows that the attachment of the heat pipe differs between the reference products. The connection of the heat pipe with the hot side is based on a friction-locked screw connection. However, the contact surfaces of the Nintendo Switch, for example, are connected with a thermal paste (see Figs. 4.8 and 4.9).

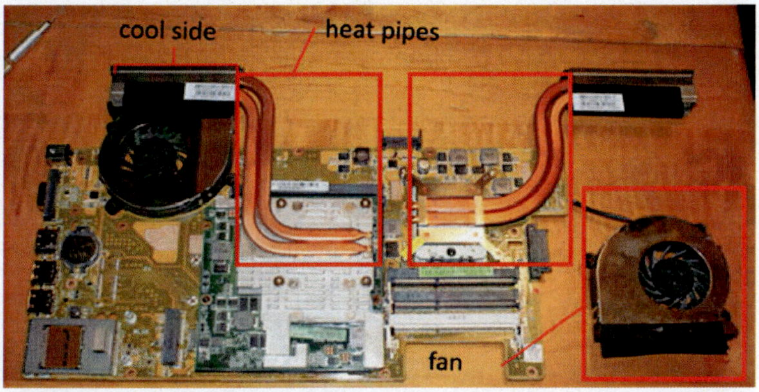

Fig. 4.6 Active cooling management from the Asus G73 as a reference system (iFixit 2021)

Fig. 4.7 X-ray view of the active cooling management from the Nintendo Switch as a reference system (iFixit 2021)

These information bases and corresponding reflection enabled generalizations to be formed and consequently optimization potentials to be uncovered. This is because heat pipes transport about 100–1000 times higher amounts of heat than a corresponding pipe made of solid copper. This is due to the physical principle that energy is absorbed during evaporation and released again during condensation. Thus, heat pipes are filled with a working medium that evaporates on the hot side and condenses on the cool side. Condensation on the cool side is promoted by a fan. The condensate is returned to the hot side by capillary forces and the cycle starts again. The capillary forces depend on the design of the heat pipe. Geometry and position influence the transport speed of the medium and thus the cooling capacity. Bending radii, diameter of the heat pipe and installation position must be taken into account. In addition, a heat-conducting paste can be used to improve the thermal

Fig. 4.8 Heat pipes connection on CPU from Asus G73 with friction-locked screw connection as a reference system (iFixit 2021)

Fig. 4.9 Heat pipe connection on CPU from Nintendo Switch with thermal paste as a reference system (iFixit 2021)

conductivity between the CPU and the heat pipe, for example (step 3). By this procedure existing reference product knowledge was compiled. By means of iterative testing of these concepts in new situations, this knowledge was implemented in the new product environment (step 4).

In conclusion, the following statement can be made based on the investigation of solution approaches using online media: The analysis of online media concerning the added value for the product development process using reverse engineering can offer a high degree of effectiveness and benefit with relatively little effort.

References

Albers A, Rapp S, Birk C, Bursac N (2017) Die Frühe Phase der PGE—Produktgenerationsentwicklung. In: Stuttgarter Symposium für Produktentwicklung

Albers A et al (2019) The reference system in the model of PGE: Proposing a generalized description of reference products and their interrelations. In: Wartzack S, Schleich B (eds) Proceedings of the 22nd international conference on engineering design (ICED19), 5–8 Aug 2019, vol 1. The Design Society, Delft, The Netherlands, pp 1693–1702

Badke-Schaub P, Frankenberg E (2003) Management Kritischer Situationen. Springer, Hamburg

Chickowsky EJ, Cross II JH (1990) Reverse engineering and design recovery, pp 13–17

Gruber H (1999) Erfahrung als Grundlage kompetenten Handels. Bern

iFixit (2021) iFixit. [Online] Available at: https://de.ifixit.com

IHS Markit (2018) Competitive benchmarking tool [Online]. Available at: https://technology.ihs.com

Martin N, Lessmann S, Voß S (2008) Crowdsourcing: Systematisierung praktischer. Institut für Wirtschaftsinformatik Universität Hamburg

Matthiesen S (2021) Gestaltung—Prozess und Methoden. In: Bender B, Gericke K (eds) Pahl/Beitz Konstruktionslehre. Methoden und Anwendung erfolgreicher Produktentwicklung, 9 Auflage 2019. Bd. 22. Springer, Berlin, pp 397–465

Meboldt M, Matthiesen S, Lohmeyer Q (2012) The dilemma of managing iterations in time-to-market development processes. In: Heisig P, Clarkson J (eds) DS 72: Modelling and management of engineering processes. University of Cambridge, Cambridge, UK, pp 127–140

Otto KN, Wood KL (2001) Product design; techniques in reverse engineering and new product development. Prentice Hall, Upper Saddle River, NJ

TechInsights (2018) [Online] Available at: http://www.techinsights.com/

Verein Deutscher Ingenieure (2019) VDI 2221—Entwicklung technischer Produkte und Systeme—Modell der Produktentwicklung. VDI 2221, Düsseldorf

Weber C, Husung S (2016) Solution patterns—their role in innovation, practice and education. In: International design conference—design 2016

Part II

Methods for Specific Systems and Products

Improving Products by Combining Usability and Emotions

Tina Buker, Jörg Miehling, and Sandro Wartzack

Abstract

User-friendly and successful products usually emerge when pure functionality meets good usability as well as an emotional and aesthetic design. Modern theoretical approaches consider all of those three aspects as equally important. In practice, however, they are often still treated independently from one another. Thus, neither positive nor negative interdependencies can be taken into account. This chapter tries to close this gap by examining the relationship of usability and emotions and introducing the idea of dual user integration that aims for combining both aspects. A concrete approach to balance both aspects focusing on emotional impressions and physical capacities of the user is presented to make dual user integration applicable. This proactive approach is called Application for Computer-Aided Design of Emotional impressions and Physical capacities (ACADE+P) and consists of three main steps: (1) user/product description, (2) product evaluation and (3) data-based derivation of quantitative recommendations for design improvements. In addition to the general framework, the specific workflow of the method including an analysis and synthesis phase is explained in detail.

5.1 Introduction

ISO 9241-11:2018 describes usability as an "extent to which a system, product or service can be used by specified users to achieve specified goals with effectiveness, efficiency and satisfaction in a specified context of use". Within this context, the definition of satisfaction

T. Buker · J. Miehling · S. Wartzack (✉)
Friedrich-Alexander-Universität Erlangen-Nürnberg (FAU), Engineering Design, Erlangen, Germany
e-mail: wartzack@mfk.fau.de

D. Krause, E. Heyden (eds.), *Design Methodology for Future Products*,
https://doi.org/10.1007/978-3-030-78368-6_5

has changed significantly over the past 30 years. When the standard was first published in 1988, satisfaction was mainly described by the absence of discomfort, whereas the revised version of 2018 explicitly mentions physical, cognitive and emotional reactions of the user in relation to the user's expectations. This trend of opening up to soft, affective factors like emotions, is evident in many different areas. Companies evolve from focusing on the pure capability of a product being used in an ergonomic context to designing overall positive user experiences (e.g. Seidel et al. 2005). Besides products, human factors are also increasingly recognised in human-centred work design and production systems (e.g. Gräßler et al. 2020b). The understanding in research has also changed. In addition to hierarchical models (e.g. Jordan's (2002) hierarchy of consumer needs), others nowadays consider functionality, attractiveness and usability as equally important and relate them to each other—like the A.C.T. model by van Gorp and Adams (2012) (see Fig. 5.1).

All three aspects contribute to a product's success. Functionality provides the basic usefulness of a product (van Gorp and Adams 2012) whereas usability is necessary for a technical system to be usable while it can also foster e.g. the learnability during product usage (cf. Gräßler et al. 2020a). Attractiveness improves the product's desirability and can also improve effective product usage (cf. Quinn and Tran 2010). There are different approaches in product development to address these three aspects. Meeting functional requirements is often a company's core competence and can be achieved by classic engineering design (cf. Bender and Gericke 2021). Usability and attractiveness, on the contrary, are more likely to be accomplished through applying user-centred design (UCD). UCD approaches generally aim for efficiently integrating the user into the development process to improve the overall perceived quality of a product. Thereby, most methods address the product's evaluation in late stages of the design process (Wallisch et al. 2019). The perceived quality of a product is influenced by many interdisciplinary factors like the different levels of experience in the usage of a product or the publicity of a brand (cf. e.g. Germann et al. 2020). Wolf et al. (2021) distinguish between reactive and proactive user integration within UCD. Reactive approaches, on the one hand, integrate the user personally, e.g. with focus groups or user tests. Proactive approaches, on the other hand, integrate user data in form of user, product and interaction models, e.g. by using digital human models for ergonomic assessment (e.g. Chaffin 2005; Wolf et al. 2020) or by using high level guidelines for enhancing positive emotions (e.g. Fenech and Borg 2007; Triberti et al. 2017). See Wallisch and Paetzold (2019) for a detailed overview of different methods of user involvement.

Difficulties are more likely to occur when product developers want to consider more than one aspect at a time. Methods that deal with correlations of functionality, attractiveness and usability are lacking as available and well-established methods often address only single aspects. Thus, neither positive nor negative interdependencies are taken into account. To close this gap, Schröppel et al. (2019b) introduced the need for a dual user integration that aims for combining usability and emotional aspects and providing quantitative recommendations for design improvements. This chapter focusses on the two aspects usability and emotions, while answering two main research questions:

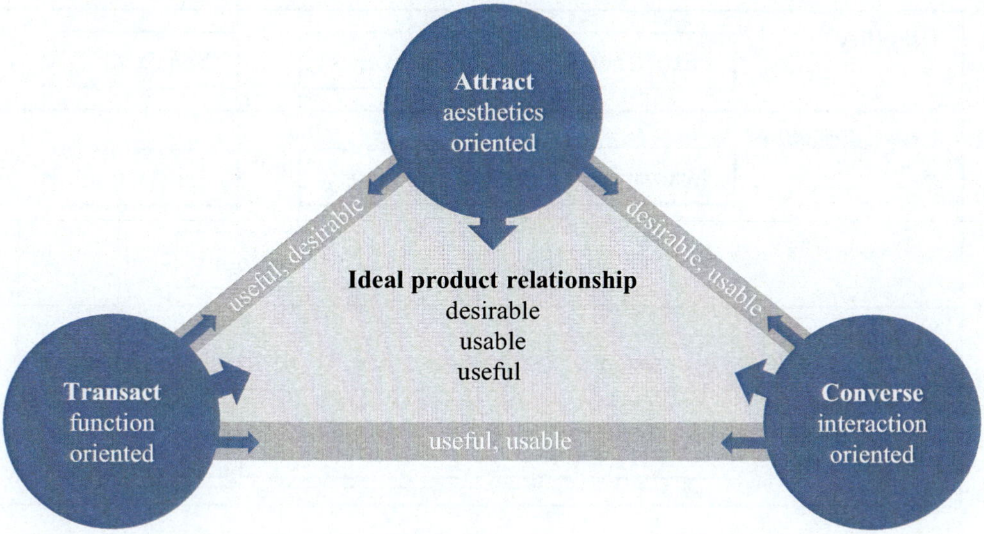

Fig. 5.1 A.C.T. model by van Gorp and Adams (2012)

- What is the relationship between usability and emotions?
- How can products be equally improved in terms of their physiological usability and an emotional appealing product design?

Based on those questions, Sect. 5.2 first deals with usability and emotions in product design in general, while Sect. 5.3 explains the idea of dual user integration. Afterwards, a novel Application for Computer-Aided Design of Emotional impressions and Physical capacities (ACADE+P) is introduced in Sect. 5.4 to derive quantitative recommendations for improving product design on a physiological and emotional level. Its potentials and limitations are discussed in Sect. 5.5.

5.2 Usability and Emotions in Product Design

Usability and emotions are both complex constructs without common, scientific definitions. While the ISO 9241-11:2018 describes usability as a combination of efficiency, effectiveness and satisfaction, other researchers include more aspects. Shackel (2009), for instance, describes effectiveness, learnability, flexibility and attitude as part of usability whereas Nielsen (2001) speaks of learnability, memorability, errors, efficiency and satisfaction. Those aspects are all very fuzzy and abstract making improving products in product design challenging, especially for unexperienced developers. A more practical solution is provided by van Welie et al. (1999). They introduced a layered model of usability for human-computer interaction (HCI) whereas the three usability aspects from ISO 9241-11:2018 (first layer) are divided into more concrete usage indicators (second layer) like learnability or memorability, which in turn are described by specific means (third

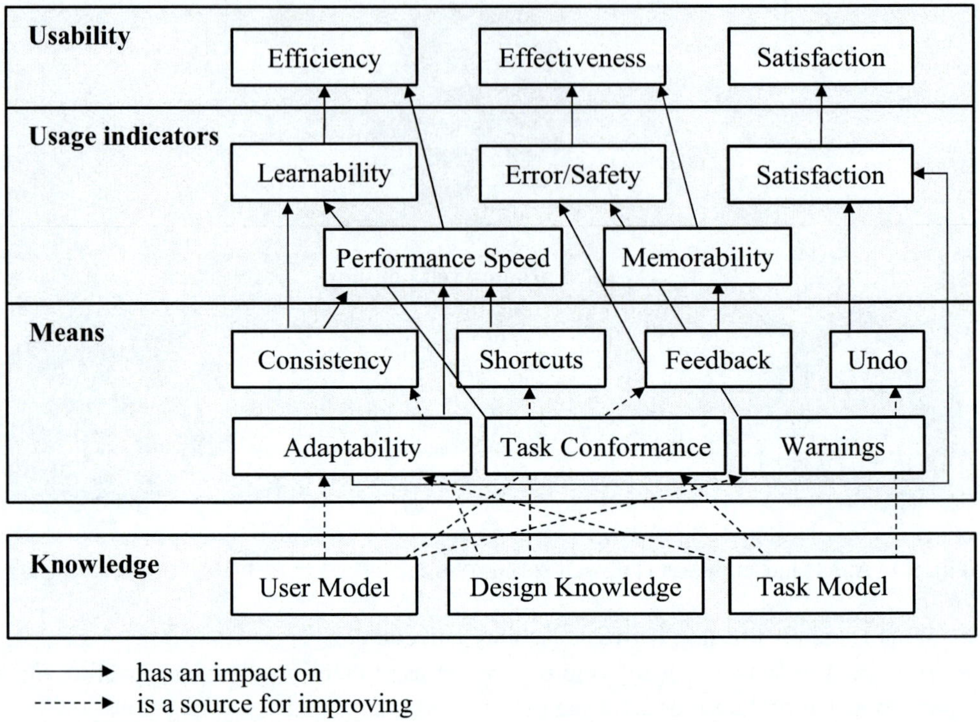

Fig. 5.2 Layered model of usability by van Welie et al. (1999)

layer) like adaptability or feedback (see Fig. 5.2). Those means can be adjusted by using knowledge (fourth layer) like user, product or task models. This layered model includes different models of usability and shows once more, that the understanding of the term usability is generally the same, but there is no common language.

Similar to a lack of common language for usability, there is no common understanding of emotions. Fehr and Russel (1984) get to the heart of the problem by stating that "everybody knows what an emotion is until to give a definition. Then, it seems, no one knows". Talking about emotions in the context of product design is also not just about emotions but rather emotional/affective design. Hereby, product perception is of great importance including the sensory perception of the stimulus, the human information processing of conscious perception and human experience, and the long-term impression of personal values, attitudes and experiences (Goldstein 2015; Goldstein and Brockmole 2017; Trommsdorff 2009). According to established models from Desmet (2002), Hekkert and Desmet (2002), Jordan (2002) and Hassenzahl (2018), the main aspects of emotional design can be reduced to appealingness, praiseworthiness, desirability and pleasure. Those aspects may be influenced by different means like aesthetics (van Gorp and Adams 2012), branding or sensory appeal (Fenech and Borg 2007).

As implied in Sect. 5.1, there are different methods for gathering and integrating relevant user data in the development process. Many of them focus on single aspects like usability or emotionally appealing product design without taking correlations or interdependencies into account. Surveys like AttrakDiff (Hassenzahl et al. 2003) or general approaches like Kansei Engineering (Nagamachi and Lokman 2011), for instance, contribute to improving a products attractiveness. Usability may be increased by using digital human modelling (e.g. Wolf et al. 2019), guidelines (e.g. Bongwald et al. 1995) or standards (e.g. DIN EN ISO 26800 2011). Integrated approaches that address usability and emotional design as equal contributors are hardly known in literature. Few available methods have often only an evaluative purpose (e.g. Puschmann et al. 2016) or focus on broad situational influences instead of providing specific design implications (e.g. Yannou et al. 2009). More pragmatic solutions can be found directly in the industry, where classical user tests or virtual car clinics are used, for example, to evaluate a vehicle exterior's sportiness (Hoermann and Schwalm 2015) or assess a vehicle's interior in terms of ergonomic suitability (Zimmermann 2008). Although these pragmatic solutions provide good results for single applications, the effort for planning and realisation is still very high. A huge challenge especially for small and medium-sized enterprises that have only limited financial or human resources. Depending on the scientific background of the approaches, superior interdependencies might not be considered. This poses the risk of drawing the wrong conclusions or negative effects occurring from supposedly good design adjustments. Redesigning the car door's handle might improve the grip but can also have negative effects on aesthetics. In order to avoid such problems, reduce the necessary effort for gathering user data and instead include superior correlations, a generic method that combines different views (e.g. usable and emotional design) seems reasonable. A structured and quantitative approach could also provide comparability and comprehensibility of subjective user data.

To provide such a generic and integrated method, the general correlations between usability and emotions need to be analysed first. From a purely theoretical point of view, their relationship has not yet been explored in sufficient depth. Instead, it is controversially discussed in the scientific community. Studies conducted in this context often focus on single aspects of emotional design like aesthetics. Usability itself is usually considered holistically, but due to the lack of a common definition, the studies are only partially comparable. Test objects are also mostly software tools (cf. Hassenzahl and Monk 2010) and rarely physical products. Table 5.1 gives an overview of selected studies examining physical products.

The studies presented confirm a general importance of emotional and usability factors (e.g. Jordan 1998) and provide first insights of the beneficial relationship between usability and emotions in different contexts (e.g. Quinn and Tran 2010; Trathen 2014). Even though the relationship has not yet been clarified in its theoretical entirety, useful implications can still be derived for product development. First of all, product design should take both usability and emotional design into account. Products can then be improved by adjusting specific influencing means. Hereby, the individuality of the user and the product has to be

Table 5.1 Selected studies examining the relationship between usability and emotional design

Source	Methods	Object	Key results
Creusen and Schoormans (2005)	Interviews (n = 142)	Telephone answering machines	1. Different roles of product appearance, a very personal choice what values most 2. Aesthetics and symbolic appearance salient 3. Appearance influence perceived ergonomic value for 1/3 of participants
Jordan (1998)	Interviews (n = 18)	Self-chosen products, 1× pleasurable, 1× displeasurable (e.g. computer, alarm clock, washing machine...)	1. Pleasure in product use involves more than usability alone 2. Features, usability, aesthetics, performance, reliability all influence pleasurable products
Kuijt-Evers et al. (2004)	Questionnaire (n = 50)	Hand tools (like screwdrivers, pliers and scrapers)	1. Descriptors of functionality are most related to comfort in using hand tools followed by descriptors of physical interaction 2. Descriptors of appearance become secondary in comfort in using hand tools
Quinn and Tran (2010)	Lab sessions, one participant at the same time (n = 106)	Mobile phones	1. Participants achieved more effective task performance using attractive versus unattractive phones, which supports the notion that "attractive things work better" than unattractive things 2. Suggestion that there are additional mechanisms underlying the relationship between attractiveness and perceived usability; further research needed
Seva et al. (2011)	Lab sessions and survey (n = 66)	Mobile phones	1. Product attributes related to form are relevant in eliciting intense affect and perception of usability in mobile phones especially those directly related to functionality and aesthetics

(continued)

Table 5.1 (continued)

Source	Methods	Object	Key results
			2. Some product attributes related to aesthetic perception of a product enhance apparent usability but not in general
Stavrakos and Ahmed-Kristensen (2013)	Lab session (n = 23)	Bluetooth headsets	1. Product attractiveness enhances perceived comfort during human-product interaction
Trathen (2014)	2× questionnaire (n = 49/37)	Cordless kettles, clock radios	1. Aesthetic/emotional factors have greater influence on liking or buying a product than usability factors

taken into account as Creusen and Schoormans (2005) stated that it is a personal choice which of the influencing factors of product appearance are considered most valuable. In the following, we introduce the general concept of dual user integration as a basis to improve products regarding emotional and usability factors.

5.3 Dual User Integration

Analogous to emotional and usability factors being mostly separated in the context of product design (see Sect. 5.2), both topics are equally separated in the scientific landscape. Affective Engineering (also referred to as Emotional Engineering) aims for improving physical product design in terms of its affective influences on the user (Schütte 2007). Human factors and ergonomics, on the contrary, mainly address the user-product interaction regarding individual needs, abilities and limitations (Karwowski 2005)—mainly focusing on physiological aspects. Dual user integration aims for linking both research areas and enables the development of a systematic approach to ensure an emotionally appealing and at the same time physically suitable product design for optimisation purposes (see Fig. 5.3) (Schröppel et al. 2019b). Design optimisation hereby means providing the product designer quantitative recommendations for design improvements.

Dual user integration has three contributing key elements: the user, the product and their interaction (see Fig. 5.3). Hereby, the interaction process is a complex feedback loop of perception and behaviour on a physiological and psychological level (Wartzack et al. 2019). It is either the user perceiving the product (user's perception) and interacting with it in order to fulfil a task (user's behaviour) or the product perceiving the user's behaviour (product's perception) and responding towards this behaviour in a specific way (product's behaviour).

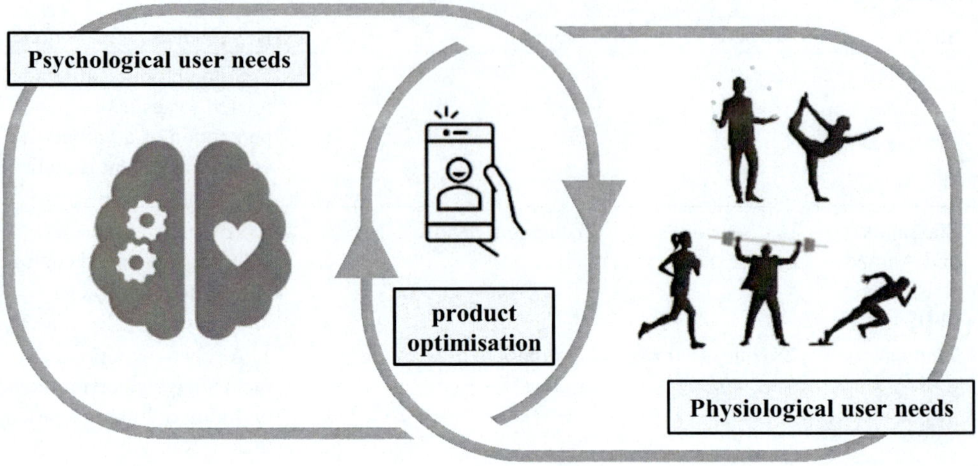

Fig. 5.3 General concept of dual user integration (according to Schröppel et al. 2019b)

The idea of dual user integration, i.e. the need to combine usability and emotional aspects as equally important, is simple, but still very challenging when trying to be applied in a specific approach. First of all, the large number of heterogeneous user needs is quite challenging. As shown in the model of user-product interaction (see Fig. 5.4), there are many different physical abilities and psychological factors of the user that might be necessary to consider in the context of product design. Given the current state of research, we are not able to provide a holistic model of the user for product design optimisation yet. It therefore appears reasonable to focus on these user characteristics that are relevant in the context of the occurring user-product interaction first. In terms of physical abilities, their relevance depends on the type of product and the occurring interactions. In the use case of a smartphone, for instance, the cognitive abilities of the users as well as the motor abilities of their hands are important to look at. Contrary, other products, like an e-scooter or a car cockpit, require motor abilities of the whole body. The product developer, therefore, has to decide anew for each use case which physiological abilities of the user are relevant and which are not. Furthermore, in order to develop a physically suitable product design for many users with different physical abilities, the product should require the lowest degree of a user's physical ability. Thus, a senior citizen is able to use the same product as a young adult. In terms of psychological aspects in the context of product development, product-personality-congruency proved to be useful (cf. Kett and Wartzack 2016). Hereby, a strong user-product attachment arises when the perceived quality of a product fits the personal values and attitudes of the user (see Fig. 5.5) (Sirgy 1982).

The second main challenge of applying dual user integration is data collection and processing. As one aim of dual user integration is to provide quantitative recommendations for design improvements, quantitative measurement techniques for physical user needs and personal attitudes is needed (Schröppel et al. 2019b). Furthermore, to be able to focus on

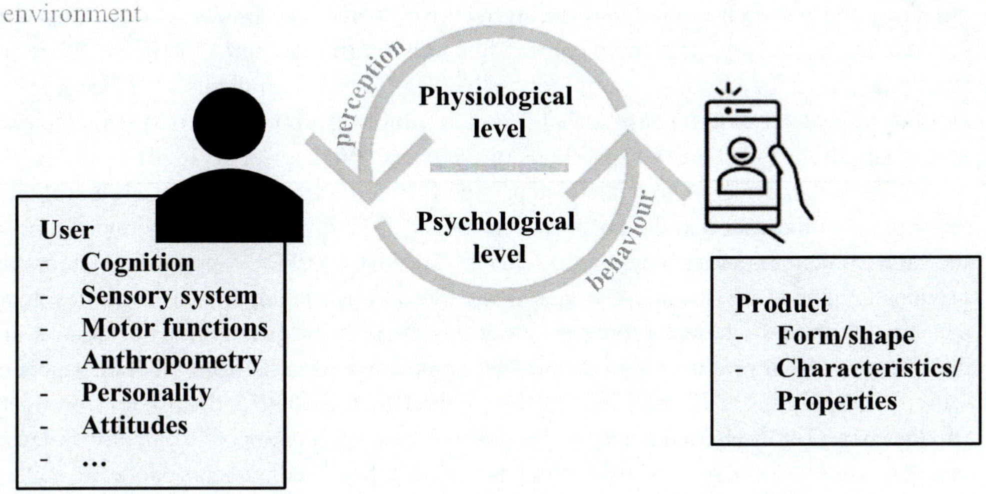

Fig. 5.4 Model of user-product interaction (according to Wartzack et al. 2019)

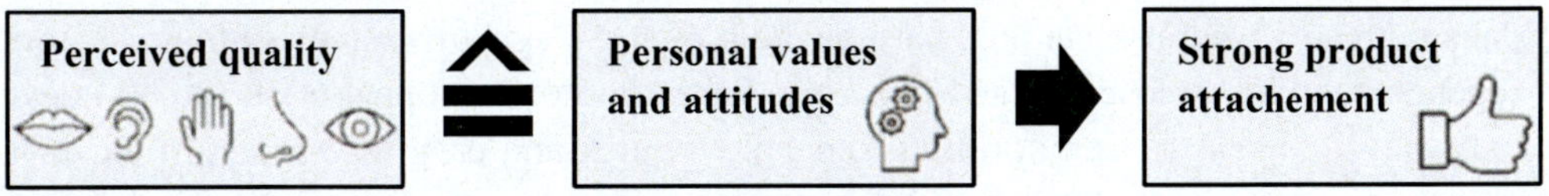

Fig. 5.5 Product-personality-congruency (according to Sirgy 1982)

those abilities and attitudes that are necessary in the specific use case scenario, a customisable measurement would be beneficial. The measurement techniques should also be comparable to each other to enable efficient data processing. A uniform data structure would not only support a combined and parallel processing but also shortens the training effort for the developer as user of the approach (Schröppel et al. 2019a). In the following, we introduce a novel approach applying the idea of dual user integration into a concrete application.

5.4 Application for Computer-Aided Design of Emotional Impressions and Physical Capacities (ACADE+P)

The Application for Computer-Aided Design of Emotional impressions and Physical capacities (ACADE+P) aims for improving products by increasing the fit between user and product on an emotional and usability level. ACADE+P focuses on the user's attitudes not only as important aspect of dual user integration but as part of the means of emotional design. Emotions usually arise as a direct response to our sensory perception, e.g. when using a product, making them rather spontaneous and situation-based (Kroeber-Riel et al.

2009). Attitudes, however, evolve for a longer period of time in response to perceived emotions and personal motivations during recurring perceptual processes (e.g. long-term product usage) making them more context-free and robust than pure emotions (Zöller and Wartzack 2017). In addition to attitudes, ACADE+P focuses on the user's physiology as ergonomic product design is the basis for good usability (Bubb et al. 2016). This includes motor, cognitive and sensory capabilities of the users (Wickens et al. 2004).

ACADE+P consists of four main steps (see Fig. 5.6), whereas first of all a sufficient description of the user and the product is necessary. The product is hereby understood as the sum of its properties adapting Weber's (2005) CPM/PDD approach. Herein, the product is defined by specific characteristics like radius or length which the developer can directly model whereas properties are a combination of different characteristics and describe the actual product's behaviour. The application benefits from this rather general view of properties, as it can therefore be applied in the context of different products. Product description is conducted by the product developer. Support to identify relevant properties in the context of user-product interaction is provided by Schröppel et al. (2020). User description considers physiological and emotional components. Zöller (2019) provides a compact instrument for measuring the user's attitudes with impression profiles (see Fig. 5.7a). These profiles consist of thirteen semantic differentials, i.e. opposing word pairs with an 11-point-scale (e.g. harmonic/dissonant; elegant/massive), in four categories (openness, style, atmosphere and prestige). Furthermore, Schröppel et al. (2019a) developed physiological capacity profiles to enable a quick and easy assessment of the user's physiological parameters (see Fig. 5.7b). Analogously to impressions profiling, physiological capacity profiling uses 35 semantic differentials with 11-point-scaling (e.g. distracted/attentive; slow/reactive) in the categories motor, sensory and cognitive abilities.

Both profiling is done via questionnaire with the respective target users as subjects. This data allows clustering of homogeneous user groups with similar attitudes and physical capacities, which in the end become the input for product optimisation. Hereby, profiles do not have to include the complete number of available semantic differentials. For a fast and target-oriented survey, it is advisable to focus on those differentials that are relevant according to the existing user-product interaction.

The second main step of ACADE+P is the assessment of different product variants in terms of its physiological requirements and emotional impressions. The product variants should ideally differ in the properties that have been classified as relevant in the context of product usage beforehand and will subsequently be considered in the product optimisation. Parametric CAD models, for instance, are suitable for this purpose as they can easily be adapted and are relatively inexpensive. Their visual appearance is also often very close to the original while rendered. Their assessment is conducted by using both profiling methods from step one. In order to create the impression profiles for the individual product variants, user surveys are carried out. For flexible and location-independent applicability, all user surveys in ACADE+P are online-based. Therefore, in the context of impression profiling, only a visual assessment can take place. However, the physiological requirements that the product demands from the user cannot be evaluated solely visually. Although it would be

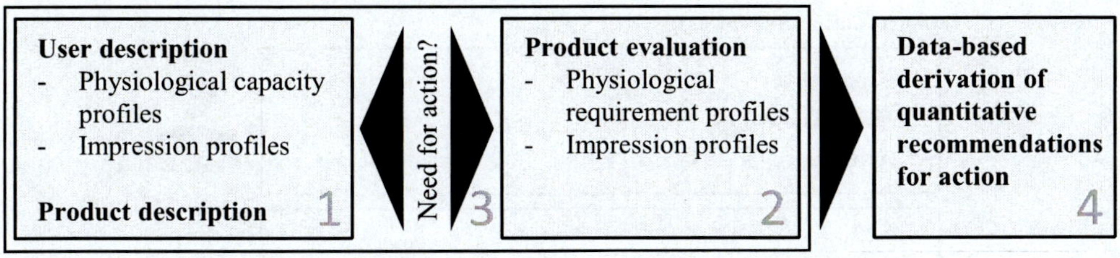

Fig. 5.6 General framework of ACADE+P

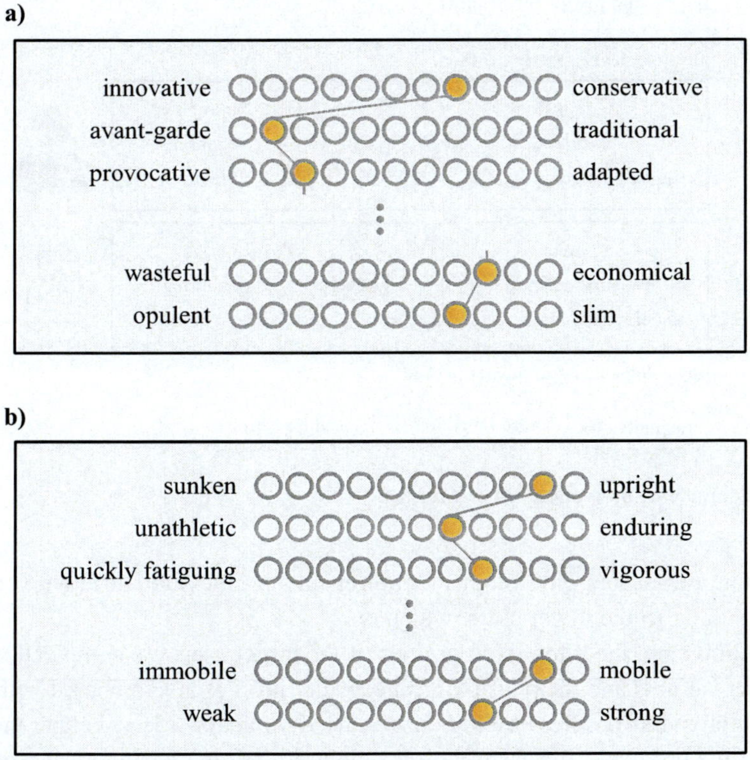

Fig. 5.7 Excerpt of (**a**) impression profile for user/product and (**b**) physiological capacity/requirements profile

possible to assess e.g. the necessary mobility of extremities, other aspects such as required strength are difficult to capture on basis of images. Physical requirements of the product should therefore be assessed by the product developer or usability/ergonomics experts. Nevertheless, this assessment is also documented via physiological capacity profiles which are then called physiological requirements profiles. On the basis of product profiles

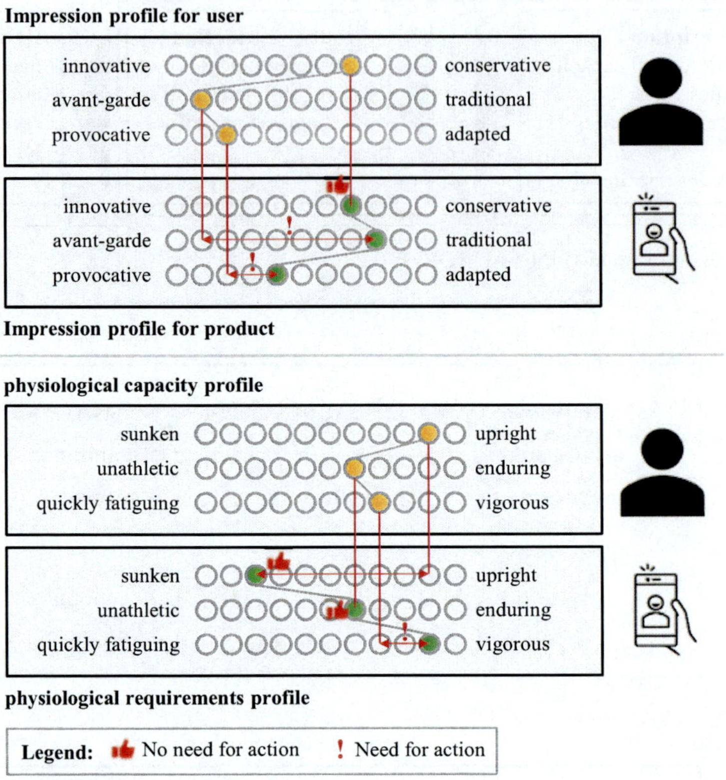

Fig. 5.8 Example of derivation of need for action

(impression profiles and physiological requirements profiles) the different variants can be clustered similar to the target users in step one.

In the third step, the impression profiles of the target users are compared with those of the product variants and the physiological capacity profiles are compared with the physiological requirements profiles (see Fig. 5.8). A need for action arises if either the impression profiles of the product differ from those of the users or the physiological requirements of the product exceed the physical capacity of the users.

After identifying a need for action, the product gets improved in the fourth and last step of the application. Therefore, different mathematical analyses are conducted using the data collected during user/product description (step 1) as well as product evaluation (step 2). Starting with the impression/physiological requirements profiles of the evaluated products, a classic correlation analysis reveals significant relations between the properties of the evaluated product variants and single elements of the profiles (i.e. single impressions or physical requirements). Afterwards, functional relations are modelled for the significant parings of property and profile element. For this purpose, regression analyses proved to be sufficient (Zöller et al. 2017). Figure 5.9 shows an example of a functional relation of the

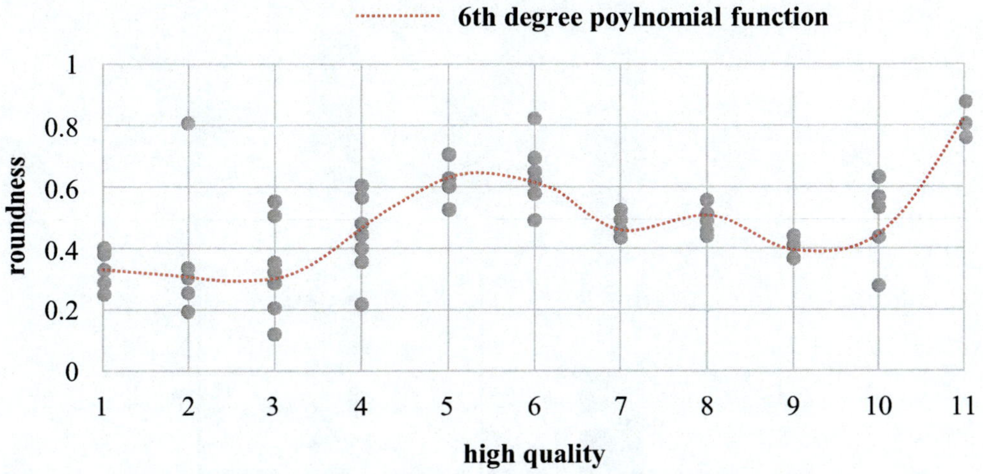

Fig. 5.9 Example of functional relation of the property roundness and the impression high quality

property roundness and the impression high quality modelled via a 6th degree polynomial regression function.

Product optimisation is finally conducted by redesigning/modifying product characteristics to match target property values. To calculate these target property values, the developer first needs the impression and physical capacity profiles of the target user. Clustering is beneficial in this context to identify target users with similar profiles. The quantitative data from the impression/physiological capacity profiles will then serve as input for the previously defined functional relations. Thus, target property values can be obtained for each of the significant parings of property and profile element. If there is more than one significant paring, they can be combined into a single target property value by using pareto optimisation. Weighting might be necessary if usability and emotional design is not rated equally important by the target users.

The presented four main steps can be transferred into a concrete workflow which is divided into analysis and synthesis. The dark boxes in Fig. 5.10 show the actual workflow, whereas the bright boxes on the left and right present the output of the individual activities.

5.5 Discussion

The idea of dual user integration is the combination of emotional and usability aspects during product design as equally important. In this context, ACADE+P provides a systematic approach to derivate quantitative recommendations for improving an emotionally appealing and at the same time physically suitable product design. Similar to other methods, ACADE+P has its potentials and limitations. The application aims to manage different heterogeneous user needs, which is one central challenge for dual user integration.

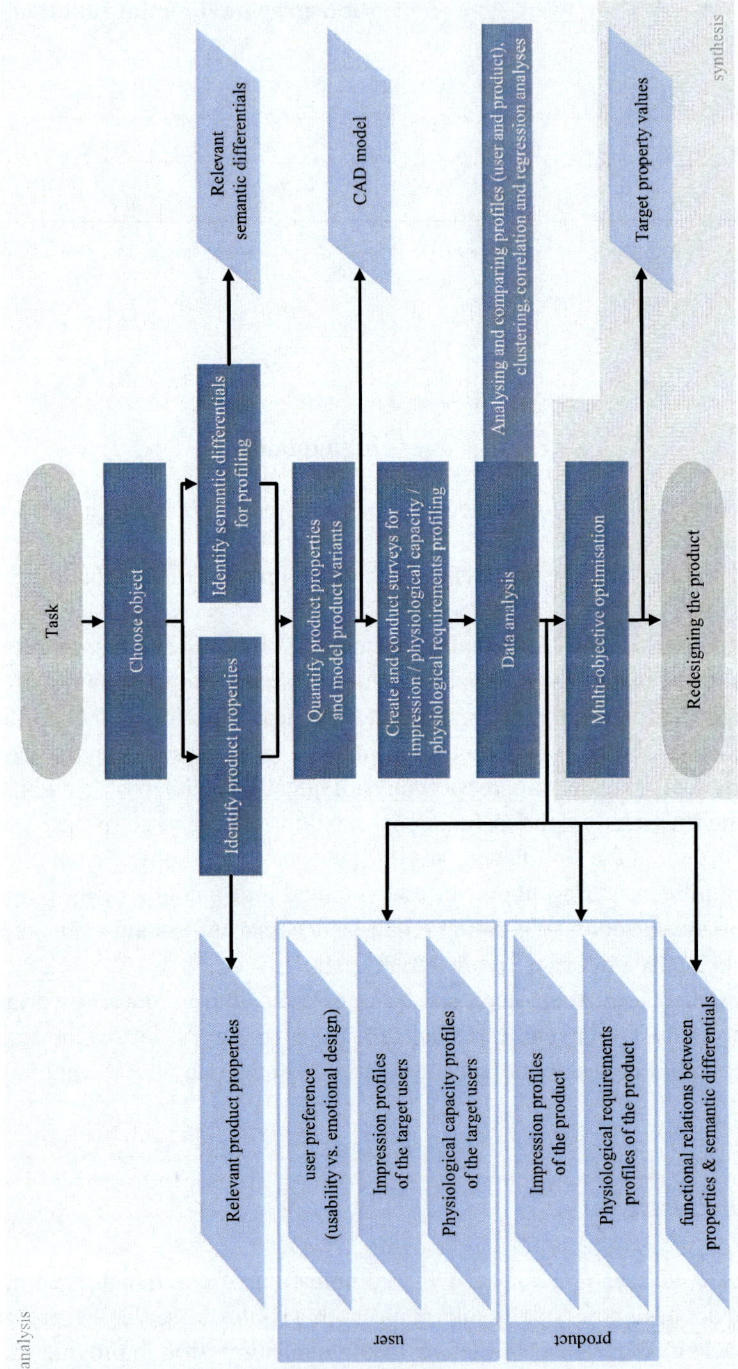

Fig. 5.10 Workflow of ACADE+P; dark boxes: single activities, bright boxes: output of activities

By using impression and physiological capacity/requirements profiling motor, sensory and cognitive abilities as well as personal attitudes of the user can be analysed. However, it should be acknowledged that this profiling techniques are a simplified assessment of physiological capacities and attitudes of the user and thus an approximation instead of a high-precision solution. Nevertheless, those profiles are very flexible and provide the same quantitative data structure which makes data analyses comparable and easy to conduct. This dataset, thus, enables the product developer to identify and analyse the target group fast and easy on an emotional and physiological level. Besides the user, products can also be assessed and compared to the task user via the different profiling methods. Instead of CAD models, rough drafts or first concepts can be evaluated. All these information provide a solid basis for design optimisation, i.e. redesigning/modifying product characteristics to match target property values. ACADE+P thus offers various possibilities for revealing information about the product and the user in early and late development phases and support product development with a moderate effort for proactive dual user integration.

5.6 Conclusion and Outlook

User-friendly and successful products usually emerge when pure functionality meets good usability as well as an emotional and aesthetic design. In this chapter, we focused on bringing together the two aspects usability and emotional design. By examining the relationship between these aspects we have seen that there is still a lack of common definitions and understanding. First studies indicate positive influences of emotionally appealing design on usability, but there is still a long way to go before the relationship between usability and emotional design is fully understood. Which role do the personal attitudes and values of the users play in this context? Is the relationship between these two aspects dependent on the product? Although we are not able to answer these questions validly yet, we can still offer basic methodical support for product development by applying dual user integration. With ACADE+P, an application was presented that not only evaluates users and products, but also identifies quantitative recommendations for improving product quality. However, there is still further research required as the presented method is not yet sufficiently evaluated. ACADE+P is currently under evaluation by conducting a use case of a smartphone for elderly. As we recommended the consideration of superior interdependencies of emotional and usable product design, the introduced approach provides a first basis to manage them in a structured manner. Still, in order to represent them more clearly, further research will focus in the analysis of various influencing factors on usability and emotional design and their correlations with each other. In parallel, we are going to improve ACADE+P's practicability by developing a software prototype.

Acknowledgments This work was supported by the German Research Foundation (DFG) under Grant WA 2913/32-1.

References

Bender B, Gericke K (eds) (2021) Pahl/Beitz Konstruktionslehre: Methoden und Anwendung erfolgreicher Produktentwicklung, 9th edn. Springer, Berlin

Bongwald O, Luttmann A, Laurig W (1995) Leitfaden für die Beurteilung von Hebe- und Tragetätigkeiten: Gesundheitsgefährdung, gesetzliche Regelungen, Messmethoden, Beurteilungskriterien und Beurteilungsverfahren. HVBG, Sankt Augustin

Bubb H, Popova-Dlugosch S, Breuninger J (2016) Ergonomische Produktgestaltung. In: Lindemann U (ed) Handbuch Produktentwicklung. Carl Hanser, München, pp 837–865

Chaffin DB (2005) Improving digital human modelling for proactive ergonomics in design. Ergonomics 48:478–491. https://doi.org/10.1080/00140130400029191

Creusen MEH, Schoormans JPL (2005) The different roles of product appearance in consumer choice. J Prod Innovat Manag 22:63–81

Desmet P (2002) Designing emotions. Delft University of Technology, Delft

DIN EN ISO 26800 (2011) Ergonomie—Genereller Ansatz, Prinzipien und Konzepte. DIN, Berlin

Fehr B, Russel JA (1984) Concept of emotion viewed from a prototype perspective. J Exp Psychol 113:464–486

Fenech OC, Borg JC (2007) Exploiting emotions for successful product design. In: Bocquet J-C (ed) DS 42: Proceedings of ICED 2007, the 16th international conference on engineering design, Paris, France, 28–31.07.2007, pp 1–12

Germann R, Kurth L, Matthiesen S (2020) Disruptive factors in the evaluation of perceived quality aspects—consideration of the brand influence. Int J Ind Ergon 76:102931. https://doi.org/10.1016/j.ergon.2020.102931

Goldstein EB (2015) Cognitive psychology: connecting mind, research, and everyday experience, 4th edn. Cengage, Boston, MA

Goldstein EB, Brockmole JR (2017) Sensation and perception, 10th edn. Cengage Learning, Boston, MA

Gräßler I, Roesmann D, Pottebaum J (2020a) Entwicklung eines Prüfstands für die Bewertung von kompetenzbildenden Assistenzsystemen in Cyber-physischen Produktionssystemen. In: Frühjahrskongress 2020—Digitaler Wandel, digitale Arbeit, digitaler Mensch? Gesellschaft für Arbeitswissenschaft e.V., Dortmund, B.6.4

Gräßler I, Roesmann D, Pottebaum J (2020b) Model based integration of human characteristics in production systems: a literature survey. In: 14th CIRP conference on intelligent computation in manufacturing engineering. Elsevier B.V, Amsterdam

Hassenzahl M (2018) The Thing and I: Understanding the relationship between user and product. In: Blythe M, Monk A (eds) Funology 2. Springer, Cham, pp 301–313

Hassenzahl M, Monk A (2010) The inference of perceived usability from beauty. Hum Comput Interact 25:235–260

Hassenzahl M, Burmester M, Koller F (2003) AttrakDiff: Ein Fragebogen zur Messung wahrgenommener hedonischer und pragmatischer Qualität. In: Szwillus G, Ziegler J (eds) Mensch & Computer 2003: Interaktion in Bewegung, vol 57. Teubner, Stuttgart, pp 187–196

Hekkert P, Desmet P (2002) The basis of product emotions. In: Jordan P, Green W (eds) Pleasure with products. CRC, London

Hoermann M, Schwalm M (2015) Evaluation of a vehicle exterior's sportiness under real vs. virtual conditions. In: Shumaker R, Lackey S (eds) Virtual, augmented and mixed reality, vol 9179. Springer, Cham, pp 470–479

ISO 9241-11:2018 Ergonomics of human-system interaction—Part 11: Usability: Definitions and concepts. ISO

Jordan PW (1998) An introduction to usability. CRC, London

Jordan PW (2002) Designing pleasurable products: an introduction to the new human factors. CRC, London

Karwowski W (2005) Ergonomics and human factors: the paradigms for science, engineering, design, technology and management of human-compatible systems. Ergonomics 48:436–463. https://doi.org/10.1080/00140130400029167

Kett SG, Wartzack S (2016) Considering emotional impressions in product design: taking on the challenges ahead. In: Krzywinski J, Linke M, Wölfel C (eds) Entwerfen, Entwickeln, Erleben 2016: Beiträge zum Industrial Design. TUD, Dresden, pp 215–232

Kroeber-Riel W, Weinberg P, Gröppel-Klein A (2009) Konsumentenverhalten, 9th edn. Vahlen, München

Kuijt-Evers LFM, Groenesteijn L, de Looze MP, Vink P (2004) Identifying factors of comfort in using hand tools. Appl Ergon 35:453–458. https://doi.org/10.1016/j.apergo.2004.04.001

Nagamachi M, Lokman AM (2011) Innovations of Kansei engineering. CRC, Boca Raton

Nielsen J (2001) Usability engineering. Kaufmann, Amsterdam

Puschmann P, Horlitz T, Wittstock V, Schütz A (2016) Risk analysis (assessment) using virtual reality technology—effects of subjective experience: an experimental study. Procedia CIRP 50:490–495. https://doi.org/10.1016/j.procir.2016.04.115

Quinn JM, Tran TQ (2010) Attractive phones don't have to work better: independent effects of attractiveness, effectiveness, and efficiency on perceived usability. In: Proceedings of the SIGCHI conference on human factors in computing systems. ACM, New York, pp 353–362

Schröppel T, Diepold T, Miehling J, Wartzack S (2019a) A concept for physiological user description in the context of dual user integration. Proc Int Conf Eng Des 1:3791–3800. https://doi.org/10.1017/dsi.2019.386

Schröppel T, Miehling J, Wartzack S (2019b) Roadmap für die Entwicklung einer Methodik zur dualen Nutzerintegration. In: Binz H, Bertsche B, Bauer W, Riedel O, Spath D, Roth D (eds) Stuttgarter Symposium für Produktentwicklung SSP 2019. Frauenhofer IAO, Stuttgart, pp 293–302

Schröppel T, Miehling J, Wartzack S (2020) How to identify relevant product properties in the context of user-product interaction? Procedia CIRP 91:615–620

Schütte S (2007) Towards a common approach in Kansei engineering: a proposed model. In: Marasek K, Sikorski M (eds) Proceedings of the conference: Interfejs użytkownika Kansei w praktyce. Wydawnictwo PJWSTK, Warsaw, pp 8–17

Seidel M, Loch CH, Chahil S (2005) Quo vadis, automotive industry? A vision of possible industry transformations. Eur Manag J 23:439–449. https://doi.org/10.1016/j.emj.2005.06.005

Seva RR, Gosiaco KGT, Santos MCED, Pangilinan DML (2011) Product design enhancement using apparent usability and affective quality. Appl Ergon 42:511–517. https://doi.org/10.1016/j.apergo.2010.09.009

Shackel B (2009) Usability – Context, framework, definition, design and evaluation. Interact Comput 21:339–346. https://doi.org/10.1016/j.intcom.2009.04.007

Sirgy MJ (1982) Self-concept in consumer behavior: a critical review. J Consum Res 9:287–300

Stavrakos KS, Ahmed-Kristensen S (2013) Investigating the role of aesthetics for interaction design. In: Lindemann U, Venkataraman S, Kim YS, Lee SW (eds) ICED 13, the 19th International conference on engineering design: 19th-22nd August 2013, Sungkyunkwan University (SKKU), Seoul, Korea. The Design Society, pp 557–566

Trathen S (2014) Aesthetics versus usability: what drives our product choices? In: Bohemia E, Eger A, Eggink W, Kovačević A, Parkinson B, Wits W (eds) Design education & human technology relations: Proceedings of the 16th international conference on engineering and product design education. Institution of Engineering Designers; The Design Society, Westbury, Wiltshire, Glasgow, pp 706–712

Triberti S, Chirico A, La Rocca G, Riva G (2017) Developing emotional design: emotions as cognitive processes and their role in the design of interactive technologies. Front Psychol 8:1773. https://doi.org/10.3389/fpsyg.2017.01773

Trommsdorff V (2009) Konsumentenverhalten, 9th edn. Kohlhammer-Edition Marketing. Kohlhammer, Stuttgart

van Gorp T, Adams E (2012) Design for emotion. Elsevier, Amsterdam

van Welie M, van der Veer GC, Eliëns APW (1999) Breaking down usability. In: Sasse MA, Johnson C (eds) Human computer interaction—Proceedings of interact 99 , 30th August–3rd September 1999, Edinburgh, Scotland. IOS, pp 613–620

Wallisch A, Paetzold K (2019) Designing user research designs for systematic user involvement in requirements engineering. In: Eriksson Y, Paetzold K (eds) Human behaviour in design: Proceedings of the 2nd SIG conference April 2019. Institut für Technische Produktentwicklung, Universität der Bundeswehr München, München, pp 51–62

Wallisch A, Sankowski O, Paetzold K, Krause D (2019) Overcoming fuzzy design practice: revealing potentials of user-centered design research and methodological concepts related to user involvement. In: 2019 IEEE international conference on engineering, technology and innovation (ICE/ITMC). IEEE, Piscataway, pp 1–9

Wartzack S, Schröppel T, Wolf A, Miehling J (2019) Roadmap to consider physiological and psychological aspects of user-product interactions in virtual product engineering. Proc Int Conf Eng Des 1:3989–3998. https://doi.org/10.1017/dsi.2019.406

Weber C (2005) CPM/PDD—an extended theoretical approach to modelling products and product development processes. In: Bley H, Jansen H, Krause F-L, Shpitalni M (eds) Proceedings of the 2nd German-Israeli symposium on advances in methods and systems for development of products and processes. Fraunhofer-IRB, Stuttgart, pp 159–179

Wickens CD, Lee JD, Liu Y, Gordon Becker SE (2004) An introduction to human factors engineering, 2nd edn. Pearson Prentice Hall, Upper Saddle River, NJ

Wolf A, Binder N, Miehling J, Wartzack S (2019) Towards virtual assessment of human factors: a concept for data driven prediction and analysis of physical user-product interactions. Proc Int Conf Eng Des 1:4029–4038. https://doi.org/10.1017/dsi.2019.410

Wolf A, Miehling J, Wartzack S (2020) Challenges in interaction modelling with digital human models—a systematic literature review of interaction modelling approaches. Ergonomics 63:1442–1458. https://doi.org/10.1080/00140139.2020.1786606

Wolf A, Buker T, Molz C, Scherb D, Wartzack S, Miehling J (2021) Proactive integration of user data into engineering processes - a generic approach and specific examples. In: Duffy VG, Lehto MR, Yih Y, Proctor RW (eds) Human-automation interaction: manufacturing, services and UX

Yannou B, Wang J, Rianantsoa N, Hoyle C, Drayer M, Chen W, Alizon F, Mathieu J-P (2009) Usage coverage model for choice modeling: principles. In: American Society of Mechanical Engineers, U.S. (ed) Proceedings of the ASME 2009 international design engineering technical conferences & computers and information in engineering conference: Volume 5: 35th Design automation conference, part A & part B. ASME, pp 407–418

Zimmermann P (2008) Virtual reality aided design. A survey of the use of VR in automotive industry. In: Talaba D, Amditis A (eds) Product engineering. Springer, Dordrecht, pp 277–296

Zöller SG (2019) Mapping individual subjective values to product design: an attitudes-based approach for user centered design. Dissertation, FAU University Press

Zöller SG, Wartzack S (2017) Considering users' emotions in product development processes and the need to design for attitudes. In: Fukuda S (ed) Emotional engineering, 5th edn. Springer, Cham, pp 69–97

Zöller SG, Schröppel T, Wartzack S (2017) Ein Vergleich von Datenanalysemethoden für eine Affective Engineering Methode. In: Krause D, Paetzold K, Wartzack S (eds) Design for X: Beiträge zum 28. DfX-Symposium Oktober 2017. TuTech, Hamburg, pp 287–298

Challenges in the Development of Biomechatronic Systems

Marc Neumann and Beate Bender

Abstract

This book chapter is dedicated to the methodological challenges in the development of biomechatronic systems, which require an integrative consideration of biological and mechatronic systems in the development process. There are no uniform process models, methodologies and tools that structure and support the individual development phases. Rather, it must be stated at the current time of the numerous research activities that appropriate methods and procedures must be conceived and designed for a given development project and goal in the form of a coevolution. This is particularly necessary in the early development phases of biomechatronic systems and thus in the system design. The variety of biomechatronic developments requires situationally adapted procedures as well as a frequently differing but goal-oriented application of analytical, experimental and numerical tools. This applies equally to the two thrusts in the development of biomechatronic systems, which are (bio)medical technology and bionics. Both directions are based on system theoretical approaches in system design, which enable a model-based and finally a simulation-based development of biomechatronic systems. The theoretical explanations of the chapter are exemplified by two current research projects. The development of a movement trainer to promote implant healing of hip end prostheses is addressed as a representative example of medical technology. As an example from bionics, the transfer of musculoskeletal lightweight design to technical applications is thematized.

M. Neumann (✉) · B. Bender
Ruhr-University Bochum, Product Development, Bochum, Germany
e-mail: neumann@lpe.rub.de; bender@lpe.rub.de

D. Krause, E. Heyden (eds.), *Design Methodology for Future Products*,
https://doi.org/10.1007/978-3-030-78368-6_6

6.1 Introduction

Mechatronic systems are based on the integrative interaction of mechanics, electronics and information processing. As an interdisciplinary field of science and technology, mechatronics dissolves the historically strict separation of the individual disciplines and treats them holistically (Czichos 2015). This results in technical systems that are not defined by the properties of their components but rather by their structural and functional integration and interaction. In their wholeness, mechatronic systems are more than the sum of their parts. The parts themselves are only significant from the point of view of the whole.

Biological systems are multilayered in the same way. The upright gait of a human being, for example, or the fine manipulation of objects with the hands are only made possible by the perfect interplay between the sensory-motor system of the brain and spinal cord as well as the receptors and the more than 650 muscles of the human body. For this purpose, billions of neurons enter into communication with each other, whereby complex chemical and electrical processes take place within milliseconds at thousands of cells.

It is the human organism in particular, that has always been a source of inspiration and a model for numerous technical innovations. For example, numerous research groups around the world are pursuing the goal of replicating human motorics in humanoid robots (Goswami and Vadakkepat 2019). The attempt to simulate the perceptive and cognitive abilities of humans in the same way is promoting developments in sensor technology and is driving innovations in the field of artificial intelligence (Wittpahl 2019).

Although biology and the human organism are a particular source of inspiration for new technical developments, they are also the target group and application area for new technologies. For example, the human body as a highly complex physiological system is susceptible to pathologies. These are deviations from physiological conditions of varying severity that can affect both individual organs and the organism as a whole. Not least in view of a steadily aging society, technical solutions are therefore needed to adequately counter such pathologies.

The field of mechatronics in particular is paving the way for completely new approaches in diagnostics, therapy and rehabilitation. Whereas until a few years ago prosthetic legs were purely mechanical constructions with severely limited functionality, mechatronic solutions today enable an amazingly natural gait pattern, allow alternating steps on stairs and walking backwards on uneven surfaces (Chui et al. 2019).

6.2 Biomechatronics

The field dedicated to the integrative consideration of mechatronic and biological systems is biomechatronics. This can be defined according to (Neugebauer 2019) as follows:

> *Biomechatronics refers to the development and improvement of mechatronic products and processes based on knowledge of the structure and mode of action of biological systems.*

The highly multidisciplinary research field differentiates two fundamentally opposing research directions. (Bio)medical technology focuses on the interaction between technical and biological systems in the development of mechatronic systems specifically for use on or in biological organisms (Kramme 2016). The research direction of bionics takes a contrasting perspective by using biological organisms as models for the development of technical systems (Wanieck 2019).

Although the orientations of the two main research areas are diametrically opposed per se, both strive to continuously expand the knowledge base on biological organisms for the benefit of sophisticated technical systems.

Traditional approaches to the development of technical products regularly reach their limits in the development of biomechatronic systems. For this reason, biomechatronics partly uses its own methods and tools, which are themselves still the subject of research to a large extent (Kuhl et al. 2020). The realization of biomechatronic systems thus often takes place in coevolution with the further development of the methodology. Analogous to conventional technical and mechatronic systems, the system design and the system modeling and simulation of the system to be developed contained therein are at the core (Gehrke 2005).

6.3 Concept Development of Biomechatronic Systems

As a reference model for the development of biomechatronic systems, the so-called V-model for mechatronic systems can be used in its main features and core (VDI 2206 2004). After all, the product to be developed represents a mechatronic product that is to be developed and/or improved by taking into account and integrating biological data, information and findings. The decisive difference to the known procedure is the mandatory extension of the reference model by a suitable approach for the integration of this knowledge, which is schematically visualized in Fig. 6.1.

Thus, Fig. 6.1 shows in its center the well-known V-model of mechatronics with the decisive process steps and required iteration loops. Starting from the requirements, the cross-domain system design is carried out, followed by the domain-specific rough and detailed designs. Verification and validation of the product takes place during system integration. In conventional mechatronic product development, the individual development steps are accompanied by modeling and model analysis adapted to the development task at hand.

However, Fig. 6.1 illustrates in particular that the development of biomechatronic systems requires adequate integration of the biological system into the development process. This is best done in the form of targeted modeling and analysis of the overall biomechatronic system throughout the individual development phases of the product to be developed.

Here, the early phases of development and consequently the system design of biomechatronic systems are of particular importance. In this phase, a cross-domain solution

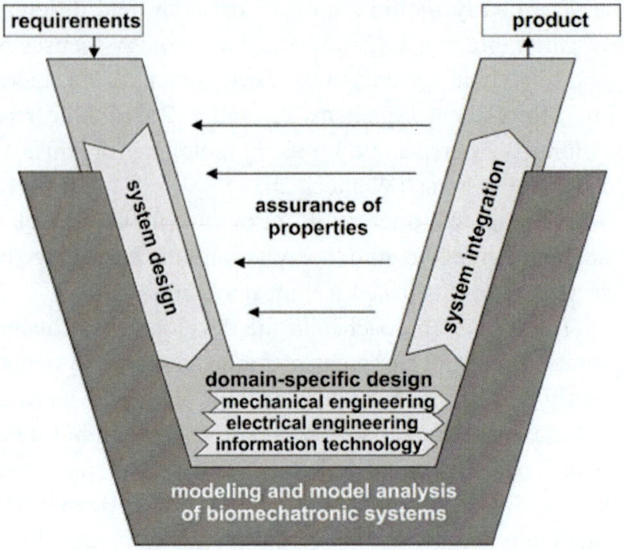

Fig. 6.1 Integration of the biological system in the development of biomechatronic systems

concept must be developed that takes into account all relevant functions and interactions (Janschek 2010). This is only possible with the help of system-theoretical approaches (Döring 2011). Thus, only a system-theoretical view and working and thinking in models allows a suitable handling of the complexity of mechatronic and biological systems as well as an early validation of the future product (Forsteneichner et al. 2018).

6.4 Modeling of Biomechatronic Systems

Sophisticated mechatronic systems and biological organisms are often so complex that they cannot be treated in their entirety by the engineering sciences. Only a system-theoretical view makes their complexity manageable and the systems accessible to a theoretical treatment (Huth and Vietor 2020). The basic principle of the system theoretical view is the structural decomposition of the system into manageable elements, which are joined to a whole by relational connections. Such a compound of system elements delimits itself from its environment by a system boundary. The relationship between the input and output variables of the system describes the system function (Bender and Göhlich 2020).

The structural design of mechatronic systems is illustrated in Fig. 6.2. According to this, mechatronic systems have a basic system that is usually mechanically dominated and whose behavior is influenced with the help of actuators. Sensors monitor the state of the basic system by recording functionally relevant measured variables and converting them into electrical measurement signals. Information processing generates electrical control signals from the measurement signals, which are used to address the system's actuators. Due to the feedback of state variables of the basic system, mechatronic systems are typically closed control loops. Here, physical variables such as force, speed and

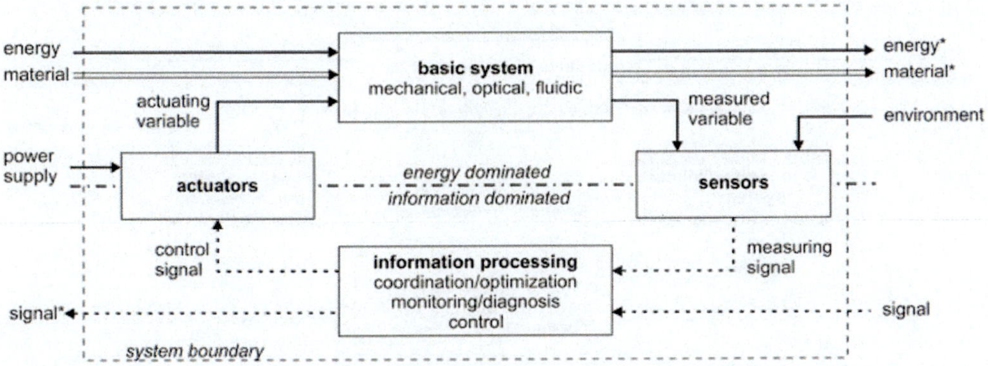

Fig. 6.2 Structural design of mechatronic systems

temperature determine the energy-dominated subsystems within the framework, of which a power conversion takes place, while physical variables in the information-dominated subsystems merely serve as carriers for information.

Despite its generally valid character, the basic model of mechatronic systems is only suitable to a limited extent for adequately describing biomechatronic systems. Due to the limited perspective of the model on the technical elements of the system, the interaction of the technical with the biological system, which is characteristic of biomechatronic systems, is largely excluded from the structural consideration. An adequate description of biomechatronic systems therefore requires a model extension that supplements the basic model of mechatronic systems with its biological counterpart.

In this respect, the human body shows astonishing parallels to the mechatronic system from a system-theoretical point of view. Thus, the passive musculoskeletal system, consisting of bones, joints, ligaments etc. can be seen as part of what constitutes the biological counterpart to the basic mechanical system in mechatronics. The passive musculoskeletal system is actuated by the active musculoskeletal system, including muscles and tendons, which thus can be understood as analogous to the actuators of the mechatronic system. The execution of controlled movements is made possible by the central nervous system (CNS), which can consequently be equated with the information-processing subsystems of mechatronic systems. Our sensory system, including for example the proprioceptors, establishes the relation to the sensory subsystems of mechatronics.

Although the analogies presented take a highly simplistic view of the human organism, it is clear from the explanations that the human organism can be described structurally in a similar way to a mechatronic system from a systems theory perspective. In this sense, the basic structure of mechatronic systems is transferred to the human body in Fig. 6.3.

The structure of mechatronic systems and the structure of the human body derived by analogy can be extended at will by appropriate physical and information technology couplings. Hierarchical as well as non-hierarchical relations between the subsystems can be used for a model extension. In this way, mechatronic systems can be linked to other

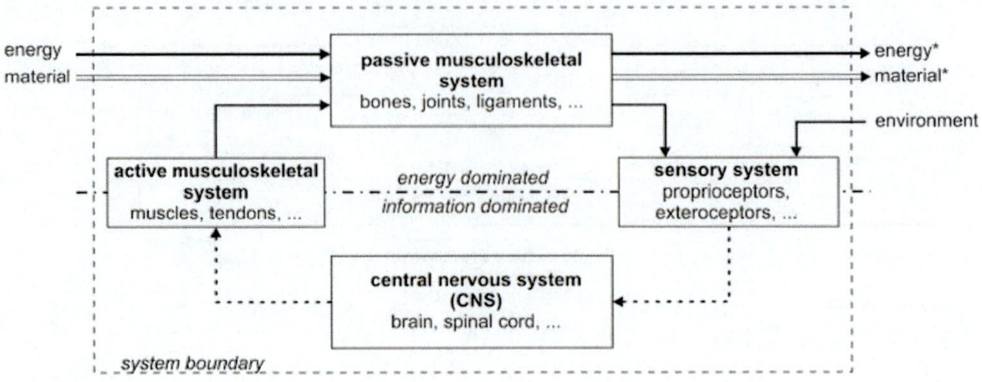

Fig. 6.3 Structural design of biological systems

mechatronic systems or resolved down to the component level. In the same way, the modeling of biological systems is in principle possible down to the cell level.

6.4.1 Structural Modeling of Biomechatronic Systems

Figure 6.4 shows an example of how the mechatronic and biological systems can be linked to form a biomechatronic system and what forms such a link can take. The figure shows a training device that can be used to treat motor impairments of the hand after a stroke. The basis of neurological rehabilitation is cortical plasticity, which describes the ability of the human brain to develop modified organizational structures in response to functional and morphological changes. This occurs in a process analogous to learning in a healthy person. Through repeated execution of physiological movements, neural connections are established and strengthened, and the muscle contraction patterns underlying the movement are consolidated. Learning success is highly dependent on training intensity, frequency and volume.

The training device shown reproduces the physiological movement of the hand by means of substitute kinematics. The training device moves the hand, while the hand in turn influences the behavior of the training device. The hand and the training device thus enter into a kinematic-kinetic coupling relationship.

In addition to such a mechanical coupling between the mechatronic and the biological system, further coupling relationships can exist in sophisticated biomechatronic systems. This becomes clear when looking at the design variants of the training device (I–IV) outlined in Fig. 6.4. Whereas in passive training the movement is dictated by the training device alone and is imposed on the hand completely from the outside (I), in active training the training device must recognize the intention to move and merely support the patient in completing the movement (II–IV).

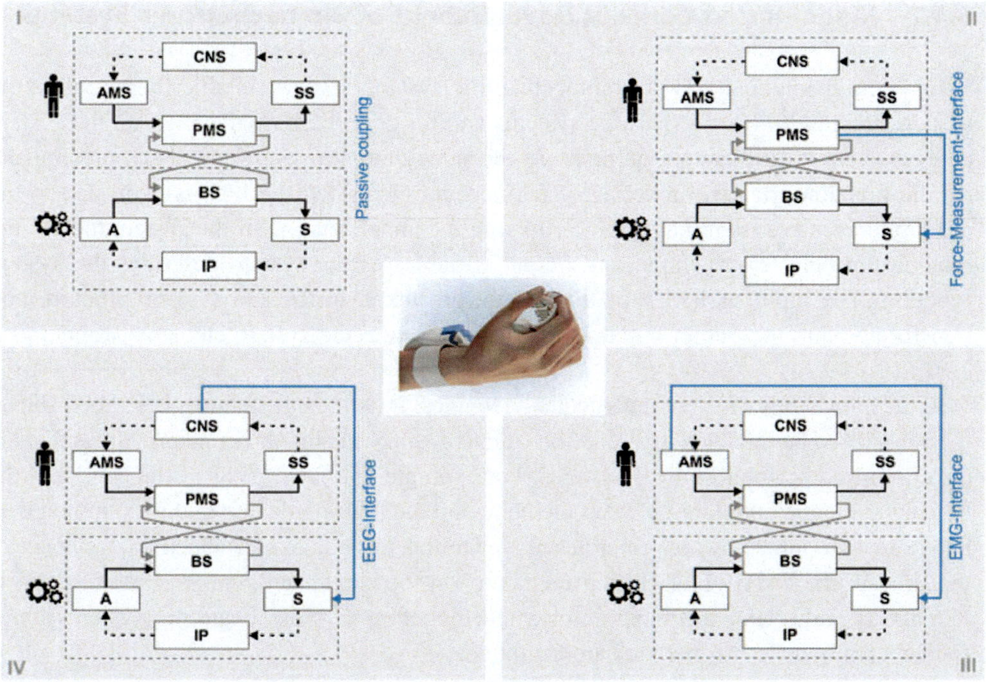

Fig. 6.4 Conceptual system design of a mechatronic training device for hand rehabilitation (*CNS* central nervous system, *SS* sensory system, *PMS* passive musculoskeletal system, *AMS* active musculoskeletal system, *BS* basic system, *S* sensors, *IP* information processing, *A* actuators)

In order to be able to realize such an active training with the presented training device, which has been proven to have a positive influence on the success of the therapy, another system interface to the human being is required. If there is residual motor function in the affected part of the body, this can, for example, be a force measurement sensor or a switch on the training device. If, on the other hand, the motor disorder is so pronounced that voluntary movement of the fingers is not possible at all, the patient's intention to move can, if necessary, be registered by means of electromyography (EMG) (III) or electroencephalography (EEG) (IV). Here, electrodes applied to the skin are used to measure muscle action currents in the case of EMG and brain currents in the case of EEG.

In the context of the system design of a biomechatronic system, different concepts for the product to be developed can obviously be considered, which manifest themselves, for example, in differently pronounced interfaces between technical and biological system (Reischl 2006). While the realization of force measurement on the training device can be largely detached from the biological system, the coupling of the systems via an EMG or EEG interface inevitably requires a much more comprehensive understanding of the biological system.

6.4.2 Model-Based Concept Development of Biomechatronic Systems

The system-theoretical view of biomechatronic systems is thus suitable for revealing the system structure of biomechatronic systems and for concentrating the view on the system areas relevant for the system purpose. At the same time, however, it is also a prerequisite for the function-oriented modeling of biomechatronic systems, which is dedicated to the description of the system behavior. Following the understanding of the system theory, the relationship between the input and output variables of the system describes the system function. The input variables of the system are taken up by the system structure and transformed into the output variables by the interconnected system elements (Lindemann 2016).

If the human body is now regarded as a technical system in the manner described above, it can basically be modeled in a function-oriented manner using the same principles that are also used to describe technical systems. For example, the movement of the hand and the kinematic-kinetic coupling between the hand and the training device in the example given above are determined by the principles of Newtonian mechanics. The electrical influence of the skin on the EMG or EEG measurements, on the other hand, can be described using Kirchhoff's rules. For the function-oriented modeling of biomechatronic systems, three fundamentally different methods are available:

- Theoretical modeling: Theoretical or physical modeling describes the real system on the basis of scientific laws. The subject of the analysis are the internal mechanisms of action of the system, the mathematical representation of which is based on scientific laws.
- Experimental model building: If it is not possible to derive the model by physical model building, the behavior of the real system is approximated or identified by observations and experiments. Here, the input and output signals of the system or the system elements are measured and evaluated. The internal mechanisms of action themselves do not have to be known. The description is done by external observation alone.
- (Semi-)Empirical Modeling: Since it is often difficult to adequately describe the system behavior by one method alone, especially for complex systems, it is often useful to use theoretical and experimental modeling in combination (Üreten et al. 2020). The model structure is then determined, for example, by theoretical model building, while the model parameters are assessed experimentally.

However, it is also obvious that the human body can be interpreted as a highly complex variant of a technical system. For example, the human body is characterized by inhomogeneous, anisotropic composite tissues whose properties are further affected by age, gender and pathological factors. When describing the interrelationships, therefore, highly simplifying assumptions must necessarily be made, which means that the modeling of biomechatronic systems generally has only limited validity.

6.4.3 Simulation-Based Concept Development of Biomechatronic Systems

Starting from the modeling of biomechatronic systems, system development is usually necessarily supported by simulations, which lead to a better understanding of the system. Through simulations it is not only possible to make statements about the system behavior even before the system is physically complete. It is also possible to infer internal system states that would otherwise remain hidden from analysis.

Model-based, simulation-supported system design in the context of mechatronic development can now be regarded as established. Commercial and freely available tools are available for this task and are continuously being further developed. For example, the software tools Matlab/Simulink and Dymola can be mentioned here, which primarily enable interdisciplinary modeling and simulation of technical, dynamic systems. The tools that can be used are mostly based on object-oriented modeling, which allows the construction of hierarchical models, the use of model components from model libraries and the reuse of own models. In this way, any technical system can be modeled by coupling individual system elements, which in turn represent, for example, mechanical, electrical or hydraulic system components.

Tools for modeling and simulating biological systems are also highly mature. Here, for example, the tools AnyBody Modeling System (Damsgaard et al. 2006) and ArtiSynth (Lloyd et al. 2012) can be mentioned. These systems have been specially developed for modeling biological systems, so that special boundary conditions are taken into account that cannot be used as a basis for technical systems. This can be illustrated by the example of the tool AnyBody Modeling System.

The numerical tool was developed specifically with the aim of determining muscle and joint forces in the course of the interaction of the human body with its environment. For this purpose, the musculoskeletal system is interpreted as a multibody system to which multibody dynamics methods can be applied. The bones, modeled as rigid bodies, are connected by abstracted joints and ligaments and driven by simplified muscle-tendon units. A restriction of the degrees of freedom of the individual segments enables time-dependent kinematics. Corresponding motion data can be obtained, for example, using motion capture techniques, and taken into account in the numerical calculation. Furthermore, external forces can be imposed on the models as time-dependent boundary conditions. For the computations of muscle forces inverse dynamic routines are used. Since the musculature of the human body is highly redundant and the mechanisms of muscle recruitment by the central nervous system are not fully understood, additional assumptions must be made for the calculation of muscle forces. Therefore, to solve this redundancy problem, AnyBody uses the minimization of the total muscle activity as a target criterion in addition to the kinematic and anatomical constraints.

6.5 Procedure in Modeling and Simulation of Biomechatronic Systems

In the following, the modeling and simulation of biomechatronic systems will be further deepened on the basis of two current research projects at the Chair of Product Development at the Ruhr-University Bochum. The examples given not only show the potential of model- and simulation-based system design, but also illustrate the fact, that a given biomechatronic development task is always accompanied by the parallel coevolution of its own, suitable methodology for solving the development task.

6.5.1 Medical Technology Lead Example: Development of a Movement Trainer to Promote Implant Healing of Hip Endoprostheses

With approximately 250,000 implantations per year, hip endoprosthetics (hip TEP) is one of the most frequently performed operations in Germany. The mobilization of the patient after the operation is decisive for success, since the ingrowth of the prosthesis is only stimulated by mechanical loads. The subject of current research is the investigation of the effects of bicycle ergometer training on the biomechanics of the bone-implant union. The vision is to develop a movement trainer that for the first time specifically addresses the mechanobiological rules of implant healing and thereby shortens treatment times, causes faster mobilization of the patient and ensures greater long-term stability of the implant.

The healing of cementless hip TEP takes place through primary (contact osteogenesis) and secondary bone formation processes (distance osteogenesis) (Gradinger and Gollwitzer 2006). Both processes are stimulated by mechanical loads, which lead to a gradual densification of the bone mass, resulting in a firm bond between the implant and the bone. Mobilization of the patient is mandatory, since it is only through movement that the mechanical stimuli are induced that trigger the bone remodeling processes (Pancanti et al. 2003). The intensity, number and timing of these stimuli are decisive for the build-up of bone substance. Incorrect movements, on the other hand, can permanently disrupt the ingrowth of the implant. In the worst case, aseptic loosening of the prosthesis occurs, necessitating hip revision.

Loosening of the prosthesis can be caused both by excessive loads, which lead to shearing of the primarily formed bone connections and by insufficient loads, which result in immobilization osteoporosis. Thus, although implantation creates the basis for the patient's convalescence, the decisive factor for the success of the hip implant is rehabilitation (Jöllenbeck and Schönle 2005).

Rehabilitation measures focus on strengthening the hip muscles, improving mobility and everyday motor skills, and strengthening the cardiovascular system, which has been weakened as a result of immobility (Güth et al. 2004). Common measures include bicycle ergometer training, but its biomechanical effects on implant stability are largely unknown to date. Particularly for patients who have to maintain partial weight-bearing, there is a risk of permanently disturbing implant stability by excessive loads. On the other hand, ergometer training, in contrast to other therapy methods, offers the possibility of dosing loads specifically via the training intensity in order to comply with partial load specifications.

Clinical studies of the long-term behavior of hip TEPs are time-consuming, not always ethically defensible, and inevitably limited in terms of the number of factors that can be studied. Bone stresses and muscle forces cannot be determined in vivo, or only to a limited extent (Bergmann et al. 1993).

By contrast, digital methods can be used to simulate the implant stability during the bicycle ergometer training without direct measurements (Sauerhoff 2020). The procedure for a simulation-based evaluation of the implant stability is shown in Fig. 6.5. As part of the overall procedure, the multibody simulations serve to determine the hip joint load and the finite element simulations to analyze the stresses on the bone-implant composite. The anatomical structures of the femur are synthesized by anatomical modelling.

For the multibody simulations, a modelling approach is used that takes into account all muscle and joint forces acting on the femur in order to generate physiological load cases for the assessment of implant stability. The software tool AnyBody Modelling System (AMS) is used for this purpose. In AMS, a parametric, musculoskeletal human model is coupled with the model of a bicycle ergometer by constraints, respecting the machine geometry, the training performance and the anthropometry of the trainee. Inverse dynamics calculations are then used to determine the muscle and joint forces that occur during the interaction of the two systems. The effects of different training intensities as well as different geometric parameters of the bicycle ergometer (for example saddle height, saddle position and pedal length) on the hip joint load are assessed by parameter variations.

The load cases calculated in this way are then used as boundary conditions for finite element (FE) simulations of the femur with a virtually implanted prosthesis. The morphology of the femur is derived through anatomical modelling based on CT images, where the material properties of bone are assessed by segmentation and gray value analysis (Fleischmann et al. 2021). The previously calculated muscle and joint forces are applied to the generated femur implant model. Through this, the implant stability can be determined under the influence of the different training intensities. To evaluate the implant stability, the micro-movements between bone and implant on the one hand and the compressive stresses between bone and implant on the other hand are assessed. In this context, micromovements <40 μm and compressive stresses between -0.25 and 20 MPa are defined as the physiological range that has sufficient implant stability.

Figure 6.6 shows exemplary results of the simulation-based analyses. The left side shows the resulting hip joint force over the crank angle at different powers and a cadence of 60 rpm, which was determined using the musculoskeletal human model. The right side of the figure gives an example of the maximum micromovements at the bone-implant junction assessed using FE analyses.

The coupled use of the simulation methods for the biological system makes it possible to analyze the effects of the training performance, the body posture and the kinematics of the training device on the resulting hip joint force in detail. Parameter constellations can be

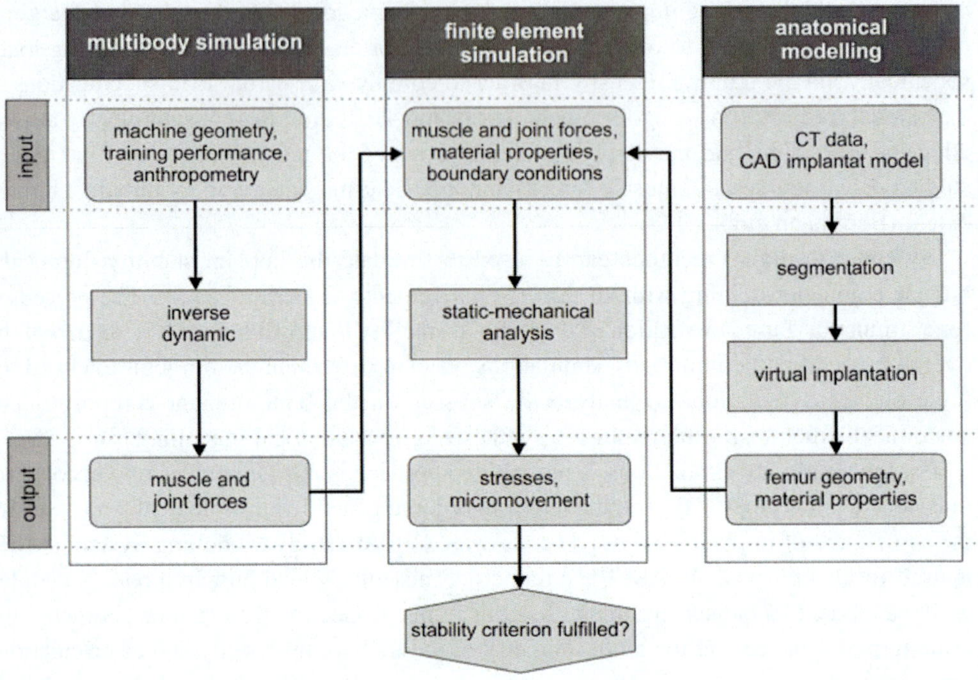

Fig. 6.5 Schematic representation of the procedure for simulation-based evaluation of implant stability (Sauerhoff 2020)

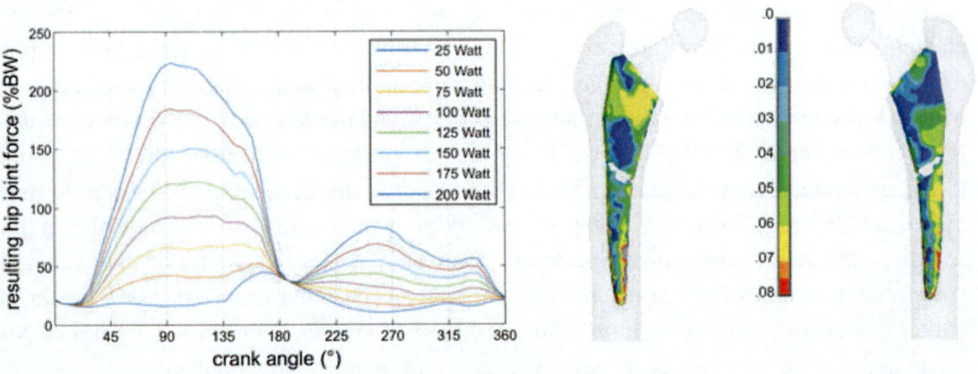

Fig. 6.6 Resulting hip joint force via crank angle (left) and occurring micromovements under axial load (right) (Sauerhoff 2020)

assessed that keep the mechanical stresses and relative movements at the bone-implant junction within physiological limits and induce the optimal stimuli for healing. These data and findings from the described modeling and analysis are taken up in the system design for a new ergometer trainer that addresses the mechanobiological rules of implant healing and respects the requirements of early postoperative rehabilitation. Based on the knowledge gained from the models of the biological system, the concept envisages that, in addition to the power, the crank torque can also be specified on the training device. A force measurement system is provided for the control of the pedal forces.

6.5.2 Bionics Lead Example: Transfer of Musculoskeletal Lightweight Design to Technical Applications

In the course of current efforts to develop energy- and resource-efficient products, technical lightweight construction plays a key role. The aim of technical lightweight construction is to save mass by exploiting the structural load-bearing capacity of a structure without sacrificing the stiffness and function. A proven approach is to adopt lightweight principles from nature (Nachtigall and Wisser 2013). Biological systems are subject to a natural selection pressure in which systems with low mass and minimal energy requirements have an advantage. Biological light-weight construction can thus be an important source of ideas for technical systems. One source of inspiration is the musculoskeletal system of humans, which is based on the coordinated interaction of a whole series of lightweight principles. In addition to the hierarchical structure of bones at the micro level, three light-weight principles are important at the macro level. These are the functional adaption through remodeling of bone mass along main stress trajectories, the active and passive tension chording of the bones by muscles and ligaments and the bending-minimized motion control of the extremities by the sensory-motor system.

While the aforementioned principles interact through overarching control and optimization strategies in the human body, they have so far only been used isolated in technical applications. The potential of a systemic integration of the principles mentioned remains unused with only some exceptions in the areas of, for example, structural optimization (Glamsch et al. 2019; Mattheck 1997; VDI 6224-3 2017) or finite element structure synthesis (Witzel and Preuschoft 2005; Gößling 2010). Therefore, the aim of current research work is to transfer the lightweight construction principles of the human body in their integrative interaction to technical applications (Bartz 2019; Bartz et al. 2019). Due to the similarity with the extremities of the human body, open kinematic chains are considered, as they exist in technology, for example, in articulated arm robots or coupling gears. The research work is based on the following two central starting points shown in Fig. 6.7:

- While classical systems in technical applications are usually kinematically unambiguously actuated via joint motors, the movement of human extremities is achieved through the interaction of a multitude of muscles. Only kinematic overdetermination enables the biological system to perform movements simultaneously while minimizing external bending loads. For the transfer of biological lightweight construction to technical applications, the muscular redundancy of the human body is therefore adopted in the actuator concept of the bioinspired system (left side of Fig. 6.7).

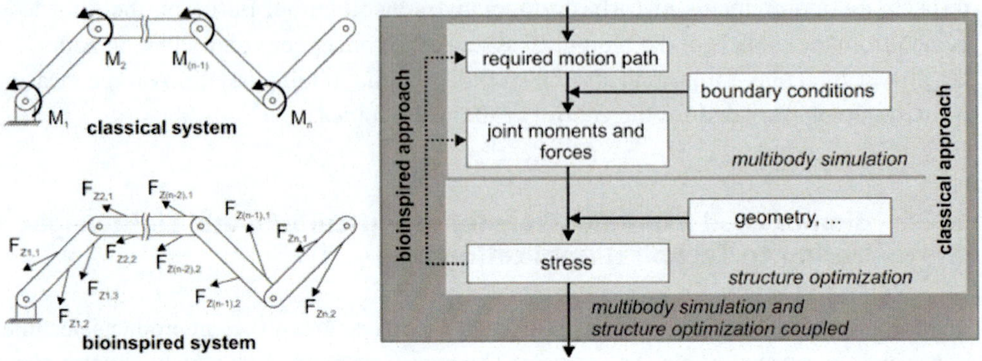

Fig. 6.7 Mechanical models of classical and bioinspired systems (left) and extension of the classical to a bioinspired approach (right) (Bartz et al. 2019)

- The design of open kinematic chains in technical applications is carried out sequentially according to the classical approach. Based on the required motion path, joint moments and joint forces are determined within the classical approach using multibody simulations. Based on this, the geometry of the system is then synthesized through structural optimization and strength verification is provided. In biological systems, however, the adaptation of structure and muscle forces as well as movement path planning takes place synchronously. The transfer of biological lightweight principles thus requires an iterative procedure that feeds back the results of the structural optimization into the multibody simulations (right side of Fig. 6.7).

The principle is applied to the example of an articulated arm robot with two joints and two arm segments shown in Fig. 6.8. The joint motors usually used in the classical system are replaced by two pairs of tension belts within the bioinspired system as shown. In each case, one pair of tension belts engages at the end of the segment, the other at the dynamic centre of gravity of the segment. The restriction to two pairs of tension belts is made on the one hand to keep the complexity of the drive concept within limits. On the other hand, it could be shown in previous theoretical investigations that already two tension belts significantly improve the bending moment curve. The type of tension chord thus imitates the muscles of the biological model without, however, completely adopting the muscle arrangement.

Once the actuator concept including redundancies for motion generation has been defined, the optimal movement path is calculated by a multibody simulation. In contrast to classical approaches to motion control, the bioinspired approach aims to reduce the bending moments in the system. By means of inverse dynamic calculations, the optimal interaction of the forces is determined, which minimizes the bending stress as best as possible. Both the calculation of the optimal motion path and the inverse-dynamic calculation of the forces are carried out in Matlab with the aid of multicriteria optimization using

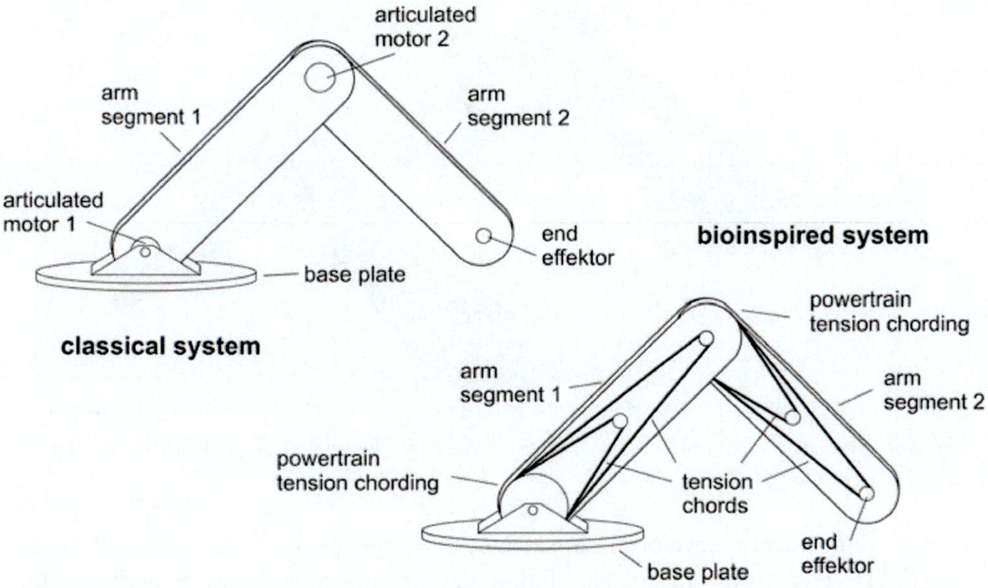

Fig. 6.8 Designs of the articulated arm robots following the classical (left) and the bioinspired approach (right) (Bartz et al. 2019)

the MatlabOptimisationToolbox. The modelling for the multibody simulation is carried out according to Bernoulli's beam theory.

Based on the previously determined load cases, a topology optimization is carried out using the finite element software ANSYS Workbench. Here, the minimization of the mass with maximization of the stiffness is defined as the target value. As a special feature of the bioinspired approach, the results of the topology optimization are fed back into the multibody simulations. This takes into account the fact that the changed mass influences the bending moment distribution and thus a new motion path and a changed set of tension chord forces become necessary. As the calculations converge, a strength verification is provided by finite element analyses. If the strength is not given, both the geometry is revised and the principle of the tension chord is modified. The post-processing of the geometry is finally done using Space Claim and SolidWorks.

Figure 6.9 shows the results of the study by contrasting a system optimized according to the classical approach with one synthesized according to the bioinspired approach. As can be seen from the plot, the classical system has a structure that is optimized for bending stresses. This is achieved by positioning the material far away from the beam axis to increase the moment of resistance against bending. Due to the maximum bending moment, more material remains near the joints to ensure bending and shear stiffness. The bioinspired system deviates completely from the classical system. The shear force absorbing structures in the upper and lower edge areas of the segments are omitted due to the minimized bending load. Due to the normal force acting mainly in the beam axis, a full load-bearing

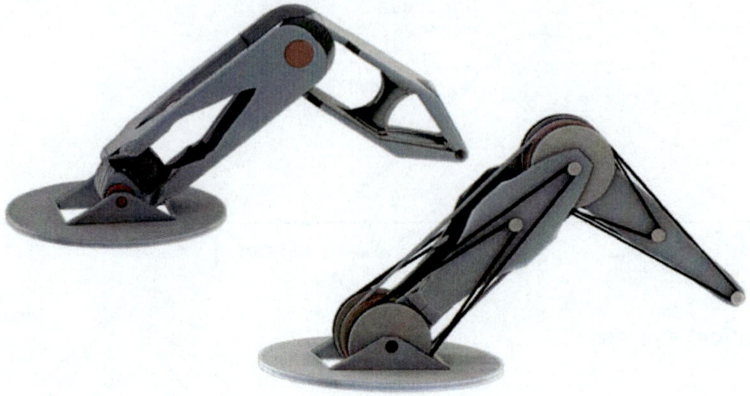

Fig. 6.9 Synthesized articulated arm robots derived by classical (left) and bioinspired approach (right) (Bartz et al. 2019)

compression structure develops in the area near the beam axis. In the middle of the beam material is saved, as the tension chording forces are applied at the edges of the segments. Compared to the classic system, which is optimized in terms of bending, mass can be saved in the bioinspired system, which is optimized in terms of compression.

6.6 Summary and Conclusion

Biomechatronics represents a field dedicated to the integrative consideration of mechatronic and biological systems. The objective is the development and improvement of mechatronic products and processes on the basis of knowledge about the structure and mode of action of biological systems. In this context, a basic distinction can be made between the two research fields of (bio)medical engineering and bionics. Although the V-model for mechatronic systems can be used as a reference model for the development of biomechatronic systems, it must be extended to include a suitable approach for integrating the available biological knowledge. This is best done in the early development phases and in particular in the system design, modeling and analysis of the biomechatronic system. Here, the system-theoretical view and the working and thinking in models play a superordinate role in order to finally enable a suitable handling of the complexity of mechatronic, biological as well as biomechatronic systems. An adequate description of biomechatronic systems thus requires a model extension that supplements the basic model of mechatronic systems with its biological counterpart.

In the simplest case, approaches and tools from mechatronics and biology cannot simply be used for the system design of biomechatronic systems. Rather, biomechatronics must make use of its own methods and tools as well as their integration, which, however, are themselves to a large extent still the subject of research. The realization of biomechatronic systems in the sense of the application of knowledge thus often takes place in coevolution with the further development of the methodology as a prerequisite for the acquisition of knowledge.

This is illustrated in the present chapter by two examples from medical technology and bionics. For example, the development of a motion trainer to promote implant healing of hip end prostheses requires, on the one hand, a special two-stage modeling and simulation approach to analyze and evaluate in detail the effects of training performance, posture, and kinematics of the training device on the resulting hip joint force and thus the healing process. The transfer of lightweight principles of the human body in their integrative interaction to technical applications on the other hand requires a completely novel approach and innovative procedure to extend the classical process steps in the design of lightweight principles by their biological inspiration.

Finally, the current research work represents a significant contribution in the further development of biomechatronics as well as the usable methods and tools. With the subsequent work and projects, it must be considered which unifications and standardizations in the system design of biomechatronic systems can be sustainably achieved in order to be able to provide engineers and scientists with a goal-oriented guide for the development of these systems.

References

Bartz M (2019) Transfer des muskuloskelettalen Leichtbaus auf offene kinematische Ketten. Doctoral thesis, Ruhr-University Bochum

Bartz M, Uttich E, Bender B (2019) Transfer of lightweight design principles from the musculoskeletal system to an engineering context. Design science. Cambridge University Press, Cambridge

Bender B, Göhlich D (eds) (2020) Dubbel Taschenbuch für den Maschinenbau 1: Grundlagen und Tabellen. Springer Vieweg, Berlin

Bergmann G, Graichen F, Rohlmann A (1993) Hip joint loading during walking and running, measured in two patients. J Biomech 26(8):969–990

Chui K, Jorge M, Yen S-C, Lusardi M (2019) Orthotics and prosthetics in rehabilitation. Mosby US

Czichos H (2015) Mechatronik – Grundlagen und Anwendungen technischer Systeme. Springer Vieweg, Berlin

Damsgaard M, Rasmussen J, Christensen ST, Surma E, de Zee M (2006) Analysis of musculoskeletal systems in the AnyBody modeling system. Simul Model Pract Theory 14(8):1100–1111

Döring D (2011) Eine kurze Einführung in die Systemtheorie. Vieweg + Teubner, Herausgeber

Fleischmann C, Scherb D, Leher I, Wolf A, Miehling J, Wartzack S, Sesselmann S (2021) Segmentation of musculotendinous structures of the hip from 3D imaging for patient-specific individualization of biomechanical simulations. In: Proceedings of the 3rd international conference on human interaction and emerging technologies: future applications

Forsteneichner C, Paetzold K, Metschkoll M (2018) Continous integration of model validation into product development. Proceedings of the design society: design conference

Gehrke M (2005) Entwurf mechatronischer Systeme auf Basis von Funktionshierarchien und Systemstrukturen. Doctoral thesis, University Paderborn

Glamsch J, Deese K, Rieg F (2019) Methods for increased efficiency of FEM-based topology optimization. Int J Simul Model 18(3):453–463

Gößling R (2010) Biologische Vorteile und mechanische Probleme der langen Schnauze am Beispiel Plateosaurus. Doctoral thesis, Ruhr-University Bochum

Goswami A, Vadakkepat P (2019) Humanoid robotics: a reference. Springer, Netherlands

Gradinger R, Gollwitzer H (2006) Ossäre Integration. Springer Medizin Verlag, Heidelberg. Available via http://site.ebrary.com/lib/alltitles/docDetail.action?docID=10145315

Güth V, Kramme WG, Naeve F, Rödig S, Schönle C (2004) Rehabilitation. In: Schönle C (ed) Praxiswissen Halte- und Bewegungsorgane. Thieme, Stuttgart

Huth T, Vietor T (2020) Systems Engineering in der Produktentwicklung: Verständnis, Theorie und Praxis aus ingenieurwissenschaftlicher Sicht. Gr Interakt Org 51:125–130

Janschek K (2010) Systementwurf mechatronischer Systeme. Springer, Berlin

Jöllenbeck T, Schönle C (2005) Die Teilbelastung in der orthopädisch-traumatologischen Rehabilitation zwischen Vorgabe und Realisierung. Homo Reparandus. Abstracts 9. Dortmunder Unfallchirurgie- und Orthopädie-Tagung

Kramme R (2016) Medizintechnik – Verfahren, Systeme, Informationsverarbeitung. Springer, Berlin

Kuhl J, Sankowski O, Krause D (2020) Investigation on methods and characteristics in medical device development. In: Proceedings of the design society: design conference. Cambridge University Press, Cambridge

Lindemann U (ed) (2016) Handbuch Produktentwicklung. Carl Hanser Verlag, München

Lloyd JE, Stavness I, Fels S (2012) ArtiSynth: a fast interactive biomechanical modeling toolkit combining multibody and finite element simulation. In: Payan Y (ed) Soft tissue biomechanical modeling for computer assisted surgery. Springer, Berlin

Mattheck C (1997) Design in der Natur. Rombach GmbH + Co Verlagshaus KG, Freiburg

Nachtigall W, Wisser A (2013) Bionik in Beispielen: 250 illustrierte Ansätze. Springer Spektrum, Wiesbaden

Neugebauer R (ed) (2019) Biologische transformation. Springer Vieweg, Berlin

Pancanti A, Bernakiewicz M, Viceconti M (2003) The primary stability of a cementless stem varies between subjects as much as between activities. J Biomech 36(6):777–785

Reischl M (2006) Ein Verfahren zum automatischen Entwurf von Mensch-Maschine-Schnittstellen am Beispiel myoelektrischer Handprothesen. Doctoral thesis, University Karlsruhe

Sauerhoff M (2020) Simulationsbasierte analyse der Stabilität von Hüftimplantaten in der frühen postoperativen rehabilitation. Doctoral thesis, Ruhr-University Bochum

Üreten S, Eisenmann M, Nelius T, Garrelts E, Krause D, Matthiesen S (2020) Current challenges and solution approaches in empirical engineering design research. In: Proceedings of the design society: design conference, vol 1. Cambridge University Press, Cambridge, pp 61–70

VDI-Gesellschaft (2017) Bionik – Bionische Strukturoptimierung im Rahmen eines ganzheitlichen Produktentstehungsprozesses. Beuth-Verlag, Berlin

VDI-Gesellschaft Produkt- und Prozessgestaltung (2004) VDI 2206 – Entwicklungsmethodik für mechatronische Systeme. Beuth-Verlag, Berlin

Wanieck K (2019) Bionik für technische Produkte und Innovation. Springer Spektrum, Wiesbaden

Wittpahl V (ed) (2019) Künstliche Intelligenz – Technologien, Anwendung. Gesellschaft, Springer Vieweg

Witzel U, Preuschoft H (2005) Finite-element model construction for the virtual synthesis of the skulls in vertebrates: case study of diplodocus. Anat Rec Part A, Discov Mol Cell Evol Biol 283 (2):391–401

Design Methodologies for Sustainable Mobility Systems

Dietmar Göhlich, Anne Magdalene Syré, Michel Joop van der Schoor, Dominic Jefferies, Alexander Grahle, and Ludger Heide

Abstract

The balance between the environment, traffic and freedom of mobility is one of the great challenges of our time. During the last decades, vehicles have become significantly more efficient, however, motorized transport still causes severe negative environmental impacts through the emission of greenhouse gases (GHG), air pollutants and noise, as well as land use and resource consumption. Sustainable mobility has been subject to a large number of research projects on the individual technologies and innovations have been carried out and published. However, most studies only consider individual modes of transport and not the entire transport system. Furthermore, the aspect of sustainability is often limited to environmental issues or GHG emissions. We discuss sustainability in a more holistic approach which integrates environmental, economic and social sustainability and we provide design methodologies and exemplary applications to support the transformation towards sustainable mobility in the future. This should be of interest to both, engineering design researchers and practitioners from the automotive industry as well as fleet operators. The applicability of our methodologies is shown with examples which are taken from current research at TU Berlin as well as solutions which have already been validated in operational application.

D. Göhlich (✉) · A. M. Syré · M. J. van der Schoor · D. Jefferies · A. Grahle · L. Heide
Technische Universität Berlin, Institute of Machine Design and System Technology, Berlin, Germany
e-mail: sekretariat@mpm.tu-berlin.de

D. Krause, E. Heyden (eds.), *Design Methodology for Future Products*,
https://doi.org/10.1007/978-3-030-78368-6_7

7.1 Introduction

Mobility of people and transport of goods are key factors for economic exchange, employment, prosperity and personal freedom in our society. At the same time, however, the current transport system is characterized by heavy environmental and societal burdens. A transformation of individual vehicles and, indeed, the entire mobility system is needed to maintain mobility of people and goods in accordance with environmental targets. With the 2030 Climate Target Plan, the EU greenhouse gas (GHG) emissions should be reduced to 55% below 1990 levels by 2030.

Today, road transport accounts for a fifth of the EU's GHG emissions. This corresponds to an emissions increase by over a quarter since 1990 (European Commission 2020). It is evident that new solutions for sustainable road transport have to be developed and deployed in order to reverse this trend. Although a modal shift towards public transport and bicycles is desirable and already taking place in many cities, motorized transport will continue to play an important role. It is therefore essential to increase its sustainability. In this work, we point out several approaches to do so.

Sustainable mobility has been subject to a large number of research projects. Examples from the vast body of literature on this topic are (Brown et al. 2020; Eckhardt et al. 2020). However, most studies only consider individual modes of transport and not the entire transport system. Furthermore, the aspect of sustainability is often limited to environmental issues such as GHG emissions. In this work, we discuss sustainability in a more holistic way. Abele et al. define Ecodesign as "... the holistic ecological, economic and technical optimization of products taking into account their entire life cycle" (Abele et al. 2008). This is supported by Birkhofer, who also emphasizes the consideration of the entire life cycle (Birkhofer 2011). In this study, we complement this definition with the social dimension of sustainability. We begin in Sect. 7.2 by introducing the three dimensions of sustainability and providing a general overview of qualitative and quantitative methods to evaluate sustainability. In Sect. 7.3, we present an example of integrating social sustainability in product design: The construction of an urban service robot for automated waste collection from litter bins. We then move on to the overall transport system in Sect. 7.4. Here, we illustrate methods to tackle the ecological and economic dimensions of sustainability in the context of urban transport systems: We present a comprehensive life-cycle assessment (LCA) method for motorized individual transport as well as design and total cost of ownership (TCO) evaluation methods for urban freight transport and urban bus fleets. Application of these methods is illustrated using scenarios in which the current fleet of fossil-fuel based vehicles is replaced by battery-electric vehicles (BEV). Section 7.5 concludes our work.

7.2 Design for Sustainability

The German word for sustainability—"Nachhaltigkeit"—has its origins in eighteenth-century forestry research. It referred to the science of determining the amount of continuous logging that can be achieved without depleting the available resources. Since the industrial revolution, humanity's ability to change its environment has only grown and many resources that appeared "infinite" to our predecessors must be managed sustainably. This protection of natural resources must be seen in the context of maintaining, and improving, a functioning and just society, as shown, for example, in the UN Sustainable Development Goals (United Nations 2015). Particularly in developed countries, digitalization and ever-increasing automation make addressing this social sustainability dimension more and more important. Finally, the economic dimension must also be considered, since a functioning economy is also necessary for a healthy society and societal costs (externalities) of change should always be taken into account. Hence, sustainability can be divided into three core categories (the "three pillars" or "triple bottom line") (Elkington 1998):

- Environmental: A product's environmental impact must be compatible with the long-term existence of earth's biosphere
- Social: A product's impact on society must improve the overall well-being
- Economic: A product must make economic sense in order for it to exist

Today, these dimensions are often treated separately, or as conflicting with each other. They can be modeled as spheres (Fig. 7.1a). However, the model with three concentric spheres postulates that for sustainability to be achieved, an economy cannot exceed its society's capacity and both society and economy are limited by the available ecological resources (McKenzie 2004), represented by the spheres embedded within each other as shown in Fig. 7.1b).

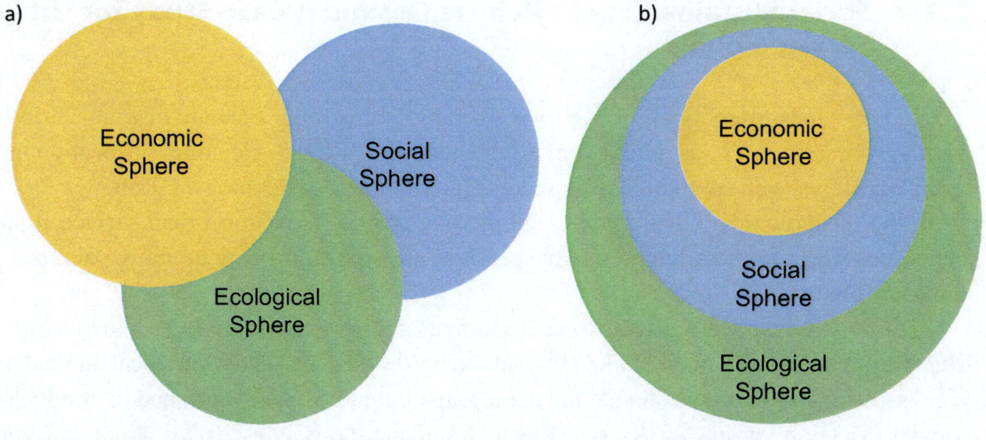

Fig. 7.1 The three dimensions of sustainability. (**a**) Interlaced model, (**b**) Embedded model

In sustainable Engineering Design, different variants of a technological product need to be evaluated and compared in each sustainability category. The interdisciplinary development of complex systems requires methodologies which support the understanding of the system, its environment and its requirements (Hentze and Graessler 2015). Several methods exist, each applicable to different categories and with varying levels of detail. (VDI 3780, 2000) offers a general approach for evaluating the consequences of a given technology. It highlights the need to clarify the relationships between different design goals and societal or personal values, highlighting potential conflicts and emphasizing the need for participation of all stakeholders. For evaluating the ecological impact, an LCA according to (DIN EN ISO 14040, 2021) is commonly used. This methodology works by defining a given system boundary, creating an "inventory" of ecological impacts and then assessing their effect on a given endpoint (i. e. human health). For the development of mobility systems, an LCA can be conducted either of individual vehicles or transportation systems as a whole, with a tradeoff between a broad scope and high accuracy. An approach to introduce a life cycle development framework in the design process is given in (Cudok et al. 2017).

Life-Cycle Costing (LCC) is a collection of techniques for calculating the monetary costs incurred throughout the lifetime of a given product. For transportation systems, TCO is commonly used to decide between different system solutions. These methods usually do not include externalities, which are costs created by the product but paid for (or gained) by society as a whole (Estevan et al. 2017).

While LCA and TCO are standardized methodologies, methods to assess the social impact are still being developed. The authors of (United Nations 2009) recommend an approach similar to an ecological LCA: Identifying all social effects of a given product along its life cycle and then assessing their positive or negative contributions to different aspects of human well-being.

7.3 Social Sustainability in Vehicle Design: A Case Study for Urban Service Robots

The industry for professional service robots is rapidly growing. They are making entries in various domains such as public environments, professional cleaning, inspection and maintenance as well as the medical sector (International Federation of Robotics 2020). With the technological advancements, numerous projects envision urban service robots supporting cities in providing municipal services and opting for new mobility solutions in the near future.

In the logistics and mobility sector, autonomous deliveries in urban areas like Starship in Hamburg, Germany (Brandt et al. 2018) are being tested. The field of medical and elderly care, growing in importance due to the demographic trend, is being enriched by robots like Stevie (McGinn et al. 2019) or Pepper (Pandey and Gelin 2018). Another much investigated area is municipal waste management, as the project "European Coordination

Hub for Open Robotics Development" (Grau Saldes et al. 2017) suggests. In the city of Berlin, the sanitation department (BSR) supports initiatives such as the case study SWEEP (Schneider and Lindau 2020), comprising a human-machine cooperation for street cleaning, or the project MURMEL in which a fully autonomous service robot is being developed (MPM TU Berlin 2020).

All these initiatives foster the development and progress of urban service robots and build upon the idea to integrate robots into our everyday life. As a logical consequence, robots will increasingly enter public spaces or institutions leading to an inevitable interaction between human and machine. As a result, the design process has to consider not only the functional requirements, but equally the problems induced by the robots' impact on society. To showcase applicable methods for both issues, the design and concept phase of the MURMEL prototype are described in this section.

7.3.1 Vehicle Design

The project aims to improve the process of emptying litter bins in an urban environment. Its goal is to eliminate GHG emissions in the process (ecological aspect) and to increase efficiency and, therefore, quality of service, through automation (economic aspect), whilst having a beneficial impact on society (social aspect). A functional prototype (shown in Fig. 7.2) is being designed to prove the feasibility of the concept and to evaluate its social and environmental impact. As the ecological aspect is discussed in Sect. 7.4.1, this chapter focuses on the social aspect and the product design.

MURMEL's modular product architecture (MPA) is designed to serve multiple purposes. A basic platform with interchangeable service modules, also seen in a similar concept (Barckmann and Jahn 2020), can cover numerous use cases and provides the flexibility to react to changes in requirements. One aspect that may change in the future is the infrastructure of cities itself, e.g., the redistribution of public space to facilitate

Fig. 7.2 CAD model of the urban service robot MURMEL

sustainable modes of travel. Another one is the jurisdiction which is currently in its infancy for urban service robots. Therefore, it is necessary to identify parts of the product that are probably not affected by those changes and are therefore considered robust, and parts of the product that are likely to undergo changes, considered as flexible, as Greve and Krause point out for a future robust MPA design (Greve and Krause 2018).

Following the general model of product design (VDI 2221-1, 2019), a clarification of the task and problems had to be executed at first. For MURMEL, this was achieved by investigating the process of emptying litter bins and interviewing workers at the sanitation department (BSR). At the same time, the robot's environment was examined to determine the type of locomotion required. The emphasis of this activity is placed on the requirements engineering, as Bender and Gericke (2021) stress the importance of gathering, specifying, analyzing and structuring requirements in a first iteration.

They serve as the input for a structured planning and management of our development process, in which the requirements have to be reconsidered, sharpened and evolved, as Göhlich and Fay describe (Göhlich and Fay 2021). By the use of methods like the morphological analysis (Bender and Gericke 2021), solution concepts are found, evaluated and finally shaped for prototyping. The agile product engineering is accompanied by continuous validation as proposed by (Albers et al. 2017).

7.3.2 Integrating and Evaluating Social Sustainability in the Design Process

During the design process, virtual prototypes and simulation models (e.g. energy consumption, material flow) help understand the system as a whole and evaluate its impact on the three dimensions of sustainability. A social life cycle assessment looks at each life phase of a product and defines possible and hazardous impacts as well as benefits which have to be considered (Commission of the European Union 2015). As one of the first urban service robots, MURMEL represents a technology that is completely new to the public urban environment. Hence, we developed a guideline for the evaluation of social impact focusing on the use phase (Kohl et al. 2020b).

The guideline, shown in Fig. 7.3, proposes several indicators of social sustainability organized in four main areas. Each of these indicators has to be reviewed for the given automation initiative and assigned to one of the four columns "not needed (benefit at hand)", "not applicable", "applicable" or "applicable and urgent" according to the probable outcome. The last two, "applicable" and "applicable and urgent", imply a negative effect on social sustainability and require the use of the columns "Short Suggestion" and "Evaluation Method". The first contains a recommendation to counteract negative effects and the latter advocates a way to evaluate the measures taken once the automation is put into place. To illustrate the applicability of this method, we implemented the guideline in the design process of an urban service robot as shown in the next section.

Suggestions | Evaluation Methods | Benefits | Applicability

Indicators for Social Sustainability

Indicators of Social Sustainability		Short Suggestion	Evaluation Method	The Guidelines are: not needed (benefit at hand)	not applicable	applicable	applicable and urgent
					for automations that		
Quantitative effects on work	Loss of Employment	Transforming instead of Replacing	Long-term census of positions and working hours		create new tasks, replaces dangerous or ecologically unsustainable tasks.	replace existing tasks and reduces need for human labour	replace/reduce skilled work
	Repositioning	Including instead of Transferring	Long-term census of work migration, consulting, including affected workers		create new tasks, replaces dangerous or ecologically unsustainable tasks.	shift demand of human work from one set of tasks to another	replace/reduce skilled work and affect people with long-time experience
⋮	⋮	⋮	⋮	⋮	⋮	⋮	⋮

Fig. 7.3 Guideline for social sustainability in automation initiatives (Kohl et al. 2020a)

By applying the guideline to the service robot in the project MURMEL, we evaluated its impact on social sustainability. As an auxiliary tool and visual feedback, we suggest the use of a radar plot with 10 axes, one for each indicator. If an indicator qualifies as "not needed", the graph remains in the center. The inner first perimeter corresponds to the category "not applicable", the second to "applicable" and the outer perimeter to "applicable and urgent". As a result, we obtain a plane whose size is dependent on the estimated impact of the evaluated automation initiative. Since the outer perimeters correspond to negative effects, the size of the plane reflects the risk of an overall negative effect a product will have on social sustainability and hence, the necessity of appropriate countermeasures as described in the previous section. This is showcased for the project MURMEL and depicted in Fig. 7.4.

The obtained information helps to form a new set of requirements that will complement the initial list. The various objectives are used as additional contextual factors as defined in (VDI 2221-1, 2019). Some of the objectives catering to social sustainability are represented by qualitative indicators. In a similar framework van Haaster et al. point out that operationalizing such indicators poses a challenge (van Haaster et al. 2017) and advise against a quantification of qualitative indicators. The methods for an evaluation should always be cooperative and inclusive for all stakeholders. To evaluate the impact an automation initiative may have, we suggest social audits, gathering information and

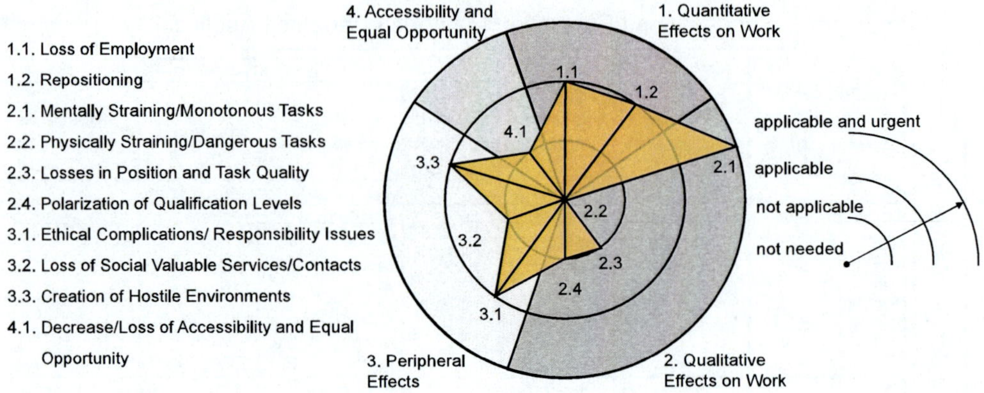

Fig. 7.4 Risk estimation for negative effects on social sustainability applied to MURMEL (based on (Kohl et al. 2020b))

feedback from employees and affected groups (Kohl et al. 2020b). Also, the method of real-world laboratories was tested in Germany and just recently evaluated (Bergmann et al. 2021), promising an evaluation under real conditions including all relevant stakeholders. Furthermore, forecasting methods like scenario writing can help to make assumptions of future conditions and deduct corresponding requirements. For an extensive assessment, a combination of multiple methods is recommended as well as several iterations of them.

For MURMEL, the technology assessment can be described as problem-induced and innovative, i.e., the assessment of a not yet existing technology at an early stage of the design process (VDI 3780, 2000). At this point in the project, the use of the aforementioned methods is reflected in a set of requirements and objectives. For example, functional requirements to assure the locomotion in an urban area, as well as requirements for the appearance and a social trajectory planning (Du et al. 2019) for the objectives "social acceptance" and "safety in a public environment". Regarding quantitative and qualitative factors on work, MURMEL is supposed to supplement the workforce by taking on physically straining tasks in a cooperative way. Fulfilling additional services like gathering data about air quality can foster the objective of improving the overall quality of life.

7.4 Design of Sustainable Mobility Systems

Sustainability in the transport sector is a much-discussed topic. Especially in urban areas, the adverse effects of pollutants and noise on health, safety and quality of life combine with an ever-increasing scarcity of space. Many solutions to these issues are being developed and some are already on the roads. Replacing conventional powertrains with locally emission-free powertrains, namely battery electric and fuel cell electric, can solve the local emission problem and reduce the impact of noise. Shared mobility on demand, especially with autonomous vehicles, could make mobility cheaper, more flexible and

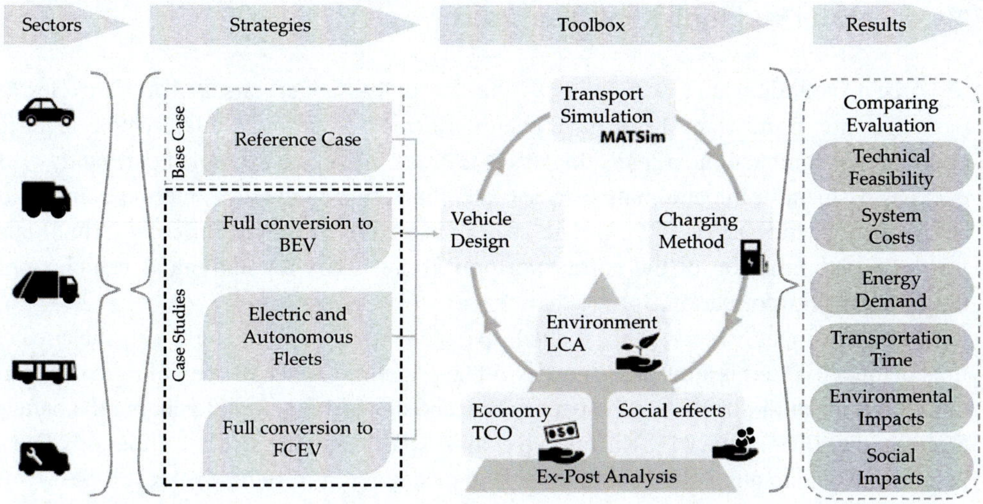

Fig. 7.5 Methodology design for sustainable mobility systems (Göhlich et al. 2021)

safer while reducing the number of vehicles needed, freeing up space in cities (Bischoff et al. 2019). But which is the right combination of technologies to reach sustainability goals in urban transport? And how does the transition to these technologies affect the environment, economy and society? There is a lack of holistic research approaches that attempt to answer these questions at the system level.

Therefore, we developed a new methodology to derive and analyze strategies for a fully decarbonized urban transport system (Göhlich et al. 2021). It combines vehicle design, a large-scale agent-based transport simulation, TCO, and LCA for a complete urban region. The approach evaluates technical feasibility, system cost, energy demand, transportation time and sustainability-related impacts of various decarbonization strategies which are applied to all segments of motorized urban traffic. The methodology follows the approach depicted in Fig. 7.5.

We subdivide urban road transport activity into five segments. Then we develop and analyze three different decarbonization strategies. We first consider a complete conversion of all segments from conventional propulsion technology to battery electric vehicles. In a second strategy, we assume the replacement of privately-owned vehicles with a fleet of shared electric autonomous vehicles. A third strategy investigates other zero emission vehicle technologies, e. g. using fuel cell technology. Each strategy is compared with the status quo of each segment.

The methodology can be applied to arbitrary regions and transport systems and is capable of analyzing various strategies to improve road transport. In addition to BEV, other propulsion technologies such as fuel cell electric or hybrid vehicles can also be considered using the methodology. For this study we choose the metropolitan region of Berlin-Brandenburg to analyze the aforementioned strategies. For further details, see (Göhlich et al. 2021).

7.4.1 Motorized Individual Transport

Motorized individual transport refers to the use of passenger cars and motorcycles for passenger transport. Many LCA have been performed on an individual vehicle level for both internal combustion engine vehicles (ICEV) and BEV one example is the study "Life cycle assessment in the automative sector: a comparative analysis of internal combustion engine and electric cars" (Del Pero et al. 2018). However, to evaluate different technologies for the decarbonization of the entire transport sector, LCA for individual vehicles only provide limited information.

There is a great variety in the vehicle parameters. The parameters with the greatest impact on GHG emissions are, among others, lifetime mileage, vehicle size, vehicle consumption, and—for BEV—the battery size and the grid mix used to charge the battery. Some studies (Dér et al. 2018; Kawamoto et al. 2019) present results for a variation of these parameters, which allows the comparison and evaluation of different vehicle types. However, the results cannot reflect changes at the transport system level including modal shift or new policies like speed limits or drive-through bans. Agent-based transport simulation makes these changes measurable.

Therefore, an LCA, which combines various parameters in a set of vehicles and uses an agent-based transport simulation to obtain detailed information on the use phase, is one solution to evaluate the decarbonization of the transport sector or its segments (Syré et al. 2020). By using simulation results, an LCA can already be performed and show potential environmental impacts, before the actual product is produced and real-life data is available. The basic approach is shown in Fig. 7.6.

The method includes an LCA according to (DIN EN ISO 14040, 2021) and the agent-based transport simulation MATSim (Horni et al. 2020). In the case presented here, the

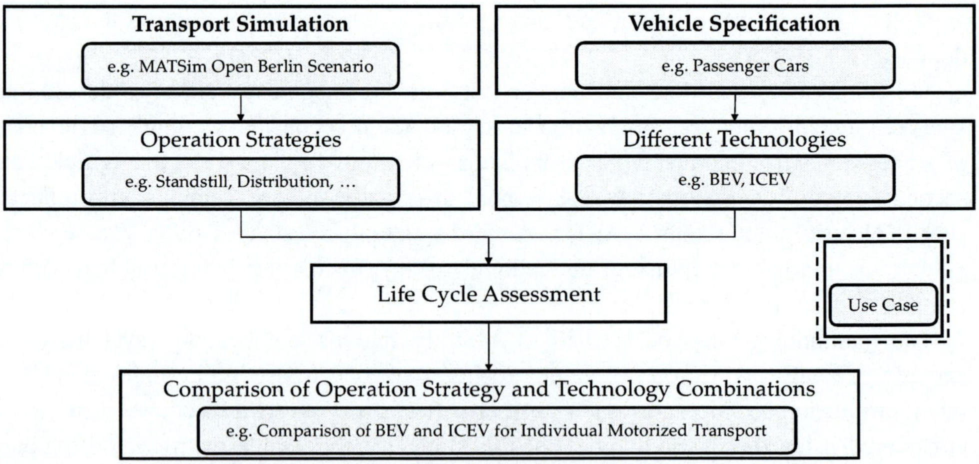

Fig. 7.6 Life cycle assessment with agent-based transport simulation (based on (Syré et al. 2020))

MATSim OpenBerlin scenario for the metropolitan region of Berlin-Brandenburg (Ziemke et al. 2019) was used to assess motorized individual transport.

MATSim uses one generic vehicle that is insufficient to reflect the variety of vehicle segments in private transportation. Therefore, we implemented three vehicle classes for BEV and ICEV in the post-processing of the simulation results: small, medium, and large (compare (Syré et al. 2020)). For ICEV, we distinguish diesel- and gasoline-fueled vehicles. The emissions from the production and the End-of-Life of the vehicles are computed with data from a literature review and common data sets (e.g. greet or ecoinvent). Moreover, the data sets deliver the emissions from the fuel and electricity supply chains and fuel combustion. The use phase emissions of BEV strongly depend on the grid mix used for charging the vehicle. Therefore, sensitivity analyses are performed to evaluate different grid mixes. The road sections in MATSim have different attributes like permitted speeds and possible capacity; this allows the vehicles' consumption to be calculated for individual road sections, which might strongly differ from the average consumption. We define three different consumption values for different speeds (compare (Syré et al. 2020)). As the MATSim scenarios cover only one average, synthetic day (Horni et al. 2020), we extrapolate the single simulation day to a whole vehicle lifetime.

The functional unit in this LCA is one kilometer driven within the transport system. This differs from other studies: single-vehicle LCAs mostly display one kilometer driven with a specific vehicle. The functional unit here represents the mixture of the vehicles according to the vehicle distribution used—in this case, the current vehicle distribution in Germany (compare (Syré et al. 2020)). Here, the results for several impact categories are displayed in Fig. 7.7. Detailed results on the share of the respective life cycle phases and the effects of increasing and decreasing lifetime mileages are presented in (Syré et al. 2020).

Our key findings (Fig. 7.7) show that the motorized individual transport with BEV offers benefits in the impact categories global warming and photochemical ozone formation

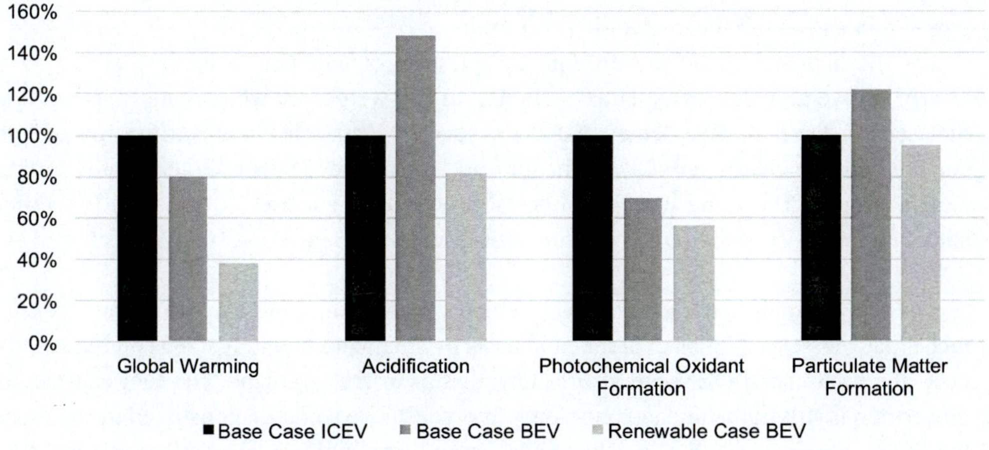

Fig. 7.7 Results life cycle assessment transport system (based on (Syré et al. 2020))

potential. The motorized individual transport with ICEV shows advantages for acidification and particulate matter formation. The renewable case shows additional BEV advantages for particulate matter formation potential and the life-cycle GHG emissions of the motorized individual transport segment can be reduced by 62%. The remaining 38% are mainly caused during the production phase of the vehicles. In some of these categories, emissions from BEV are dominated by the production and EoL phases, while emissions from ICEV are dominated by the use phase. Therefore, with increased lifetime mileages BEV will reach the break-even point against ICEV.

The presented case study shows that including agent-based transport simulation delivers sufficient results for the use phase of transport systems, as required for LCA. It enables the analysis of other versions of the Open Berlin Scenario and can therefore deliver results for changes on a transport system level. In contrast to the use of real-world data, future scenarios can be analyzed. The analysis of other transport sectors, like waste collection or freight traffic, can be performed with the respective vehicle data. Other databases and/or traffic simulations are usable with a certain effort.

7.4.2 Urban Freight Transport

Commercial vehicles are responsible for about 35% of the GHG emissions caused by road traffic (Bundesministerium für Umwelt, Naturschutz und nukleare Sicherheit 2020). Therefore, alternative powertrains hold the potential to significantly improve the sustainability of urban transport. Two completely different powertrains are being discussed: The electric powertrain and the conventional powertrain with synthetic (CO_2 neutral) fuels. However, since the latter cannot solve the local problems of particulate matter, NO_x and noise and thus does not improve living conditions, it is not considered further. The electric powertrain is again divided into two variants, which differ in their energy storage: the battery-electric powertrain and the fuel cell-electric powertrain.

For the battery electric powertrain, the question of technical feasibility seems to be clearly answerable for commercial vehicles in all weight classes. Companies such as Steetscooter have already demonstrated for several years that the electrification of light commercial vehicles is possible in medium to large scales (Streetscooter GmbH Elektro Nutzfahrzeuge 2021). In this segment, established OEM such as Volkswagen (e-Crafter) and Daimler (e-Vito) are now following suit (Mercedes-Benz AG 2021; Volkswagen AG 2021).

For a long time, heavy commercial vehicles with battery electric powertrains were a niche that was served on an experimental basis by small and medium-sized enterprises. But now, the established OEM are also following suit in this segment. The fuel cell electric powertrain is still immature and only few prototypes have so far demonstrated the technical feasibility. Compared to battery-electric drivetrains, these offer two clear advantages: short refueling times and very long ranges without restricting the payload. However, they also have a clear disadvantage: the well-to-wheel (WTW) efficiency of electricity-based hydrogen is about three times worse than the WTW efficiency of BEV.

This means that in order to achieve a real reduction in CO_2 emissions in the urban transport sector using existing power generation with a significant share of carbon based electricity, technical solutions must be found to deploy BEV. This requires a battery design tailored to the application and specific charging strategies.

To achieve this, we apply the approach described above to two significant sub-segments of urban freight transport. Since the urban transport sector includes applications with different range and energy requirements, we will look at two extreme sub-sectors. First, the supply of food retailing. Although this sub-sector does not have any special requirements in terms of kilometer-related energy demand, it does have a very wide range of distances to be covered. The second sub-segment is municipal waste collection. Here, the range requirements are relatively low, but a high proportion of the energy required is used by the secondary aggregates for loading and pressing the waste.

For the food retailing scenario, we are using a MATSim model by (Schröder and Liedtke 2014) with modifications by (Martins-Turner et al. 2020). This model represents the supply of the largest supermarket chains in Berlin operating 1057 stores which are supplied from 17 distribution centers. Some of the distribution centers are close to the city, while others are several hundred kilometers away. There are three different categories of goods: fresh, frozen and dry, which are delivered separately. The planning of the delivery tours is performed with the open source algorithm jsprit (jsprit 2018) which optimizes for total operating cost. In each depot, several sizes of trucks are available. 7.5 t, 18 t, 26 t and 40 t.

As part of the zeroCUTS project, a completely new transport simulation model is developed for urban household waste collection. The model is based on publicly available information. First, the amount of waste generated in Berlin is obtained from the disposal statistics and distributed evenly among all residents. Using data on population density, the amount of garbage to be collected per street can be determined. Finally, it is considered that densely populated areas are served twice a week and less densely populated areas only once. Now the tour planning algorithm jsprit is used to generate cost optimized tours for this task. In Berlin, only one class of waste collection vehicles with a total permissible weight of 26 t is used.

Representative prototypes or production vehicles are researched for each vehicle class in both segments in order to obtain the most reliable information possible on fuel consumption, prices and chassis weight. Our market analysis shows that the BEV chassis are relatively consistently about 60% more expensive than comparable ICEV across all vehicle classes. The battery price, which is for commercial vehicles on average 600 €/kWh at system level, must be added to the chassis price. Subsequently, vehicles with different battery sizes are specified for each segment and weight class. Larger batteries extend the range, but also reduce the payload and drive up the price. By simulating different battery sizes with the corresponding payloads and costs, a better estimation of a correctly dimensioned battery for the application is possible.

The planned tours together with the vehicle parameters are simulated in MATSim to obtain the distances driven by all vehicles and the associated energy consumption. The results indicate that waste collection is already feasible with a relatively small battery size of 155 kWh. Although slightly more vehicles are used than for vehicles with a larger battery, this option results in the lowest overall total operational cost. Recharging during the single-shift operation of 8 working hours per weekday is not required. It is therefore possible to resort to low-cost slow charging at the depot.

In food delivery, even with the largest battery dimensions, complete electrification is not possible without intermediate charging. Without rapid recharging during stops to pick up new goods at the depot, 56% of the trips can be electrified with the largest possible battery. Therefore, we conducted another study to determine adequate charging strategies. If recharging at the depot is enabled with 600 kW, 90% of the tours can be driven battery-electrically. A 100% electrification of this segment requires the provision of 27 additional fast charging stations with 600 kW at strategic points within the urban area. More results on the charging study can be found in (Miranda Jahn et al. 2021).

Figure 7.8 shows the changes in costs and WTW emissions due to a switch to BEV compared to the ICEV base case using the emission factor of the electricity mix in Germany from 2018. Both cases result in an increase in TCO which is greater for food delivery due to larger batteries. In addition, the waste collection vehicles have a very high potential for CO_2 savings using BEVs due to their specific driving profile with a lot of stop and go driving. The complete studies including all detailed results can be found in (Ewert et al. 2020; Martins-Turner et al. 2020).

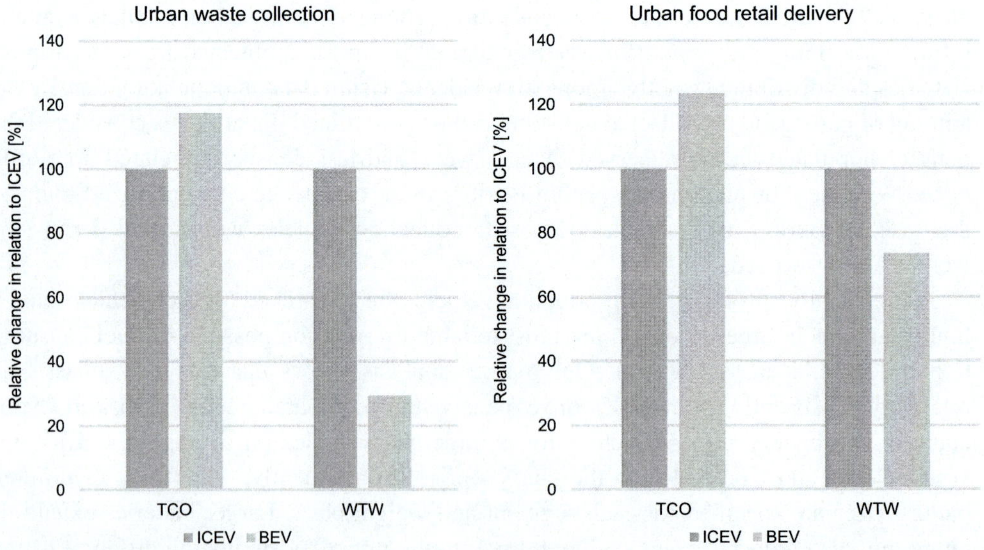

Fig. 7.8 Results well-to-wheel and total cost of ownership for urban waste collection and urban food retail delivery (data based on (Ewert et al. 2020; Martins-Turner et al. 2020))

For now, we have not conducted a full LCA. This will be one of the next steps in the analysis of this traffic segment.

7.4.3 Electric Bus Systems

As in the case of freight traffic, city bus operations are characterized by daily *vehicle schedules* that are currently tailored towards conventional vehicles. The schedules are designed to satisfy the *timetable*—a list of passenger trips—with a minimum vehicle demand. Electric buses, however, are limited in range such that existing vehicle schedules often cannot be covered, as we show in (Jefferies and Göhlich 2020) for a real-world metropolitan bus network. Thus, direct substitution of conventional buses for electric buses is not always a possibility.

Electric buses are commonly charged at the depot (depot charging) or during dwell periods at bus stops, usually at terminal stops (opportunity charging) (UITP 2018), see Fig. 7.9. While the latter alleviates the range limitation, it gives rise to new constraints in the form of charging time required at terminal stops. We demonstrate in (Jefferies and Göhlich 2020) that the layover times in the existing schedules of a metropolitan bus network are often not sufficient for stable electric bus operation as they are frequently diminished by delays.

The design of a fully electric bus system thus requires a re-scheduling of vehicle operations, i.e. solving the vehicle scheduling problem (VSP) under range and/or charging time constraints. In the case of opportunity charging, it is also desirable to determine cost-optimized charging point locations, especially for large bus networks. Furthermore, to perform a TCO and LCA evaluation of competing system concepts—e.g. different charging technologies or different vehicle configurations—a simulator is required that determines, among other quantities, the required fleet size and its energy demand.

(a) (b)

Fig. 7.9 (**a**) Depot charging (credit: BVG/Kevin Doan); (**b**) Opportunity charging (credit: Photothek)

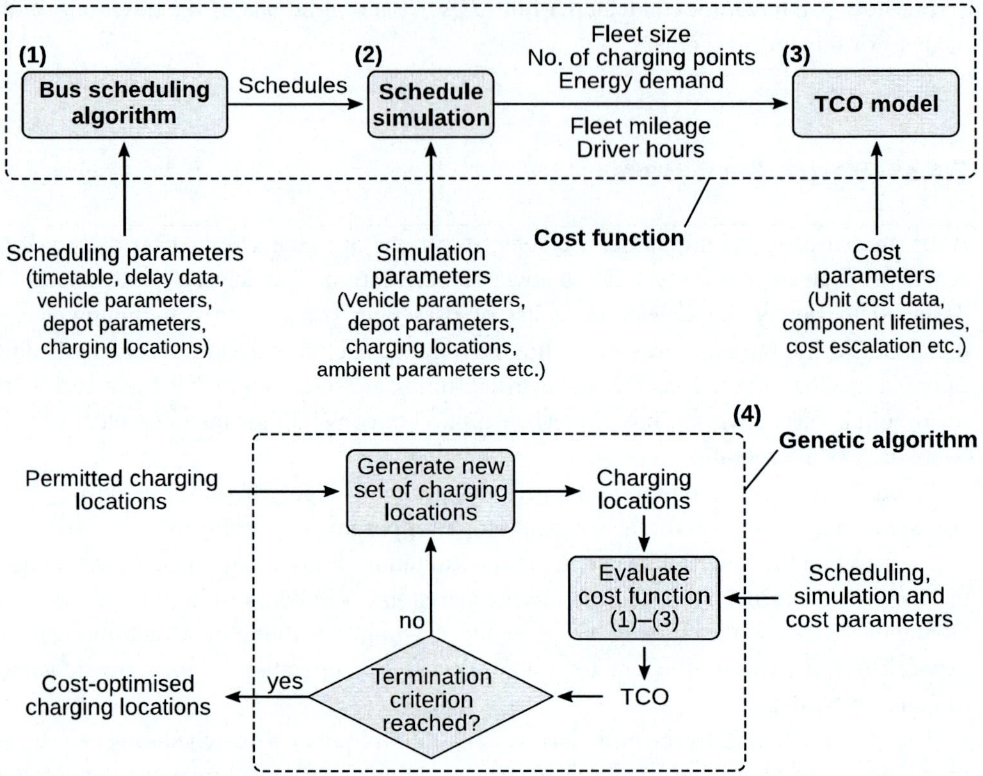

Fig. 7.10 Electric bus scheduling, simulation, TCO calculation and charging location optimization

Our integrated methodology enables the design and TCO evaluation of electric bus systems. It comprises four core components: (1) A scheduling algorithm to plan vehicle schedules suitable for electric buses, (2) a combined fleet and depot simulation model, and (3) a TCO model. A genetic algorithm (4) is wrapped around these three components to enable cost-optimized charging station placement. LCA calculation is currently not yet implemented, but the data for inventory analysis of the bus system, required as an input for LCA analysis, is determined. It is therefore possible to evaluate electric bus systems with regard to economic and ecological sustainability. Figure 7.10 gives an overview of the methodology.

The bus scheduling algorithm (1) can plan schedules for depot charging and opportunity charging at terminal stops. It is constructed as a greedy algorithm, enabling very fast solving of the VSP, although an optimal solution with minimum vehicle demand is not always found. Efforts are currently underway to improve the algorithm in this regard. The simulation model (2) enables detailed bus system simulation based on a set of vehicle schedules. It is implemented as an object-oriented, discrete-event based model, and contains a representation of vehicles, depots and charging facilities. Several traction models are available for the vehicles. Also, a model of the vehicle HVAC system is included in order to determine HVAC consumption. Battery charging can take place at charging facilities within the route network and at the depot, such that the entire operational day of all vehicles is simulated. This is a major improvement over other methods (see (Jefferies and Göhlich 2020)) for a comprehensive review) and enables an accurate determination of the required bus fleet size and driver hours for different system configurations.

A TCO model (3) determines the system cost for the simulation case. It employs dynamic costing and considers all relevant cost elements (investments for vehicles, batteries and charging infrastructure as well as operational cost for energy, drivers, maintenance etc.).

The genetic algorithm (4) determines TCO-optimized locations for opportunity charging stations. Contrary to other works, not only the required number of charging stations is reflected in the optimization, but also the effects of the choice of charging locations on fleet size and required driver hours.

As an example for the application of our methodology, Fig. 7.11 displays the results of a TCO comparison of several electric bus system configurations: Two depot charging variants with a range of 120 and 300 km, and two opportunity charging variants with a charging power of 300 and 450 kW. A diesel scenario was included as a reference. A set of vehicle schedules adapted to the vehicle parameters was constructed for each scenario; also, for the opportunity charging scenarios, TCO-optimized locations for the charging stations were determined. The study was carried out for a real-world bus network of 39 lines. For a detailed account of the parameters, the reader is referred to the original source (Jefferies and Göhlich 2020).

As public transport generally relies on public funding, it is important to assess the total system cost that transport operators—often commissioned or owned by local governments—will face by transitioning to electric buses. Therefore, contrary to other

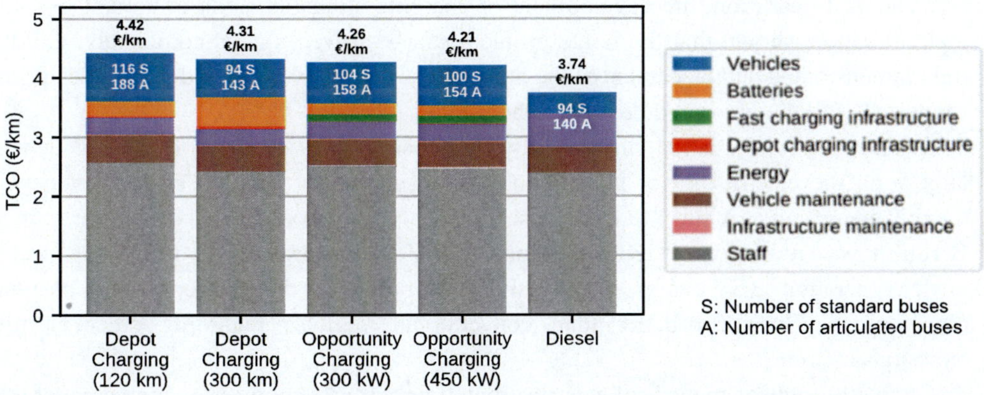

Fig. 7.11 TCO results in € per km of revenue service for different electric bus system variants (data from (Jefferies and Göhlich 2020))

sources, we fully include the driver wages in our analysis. Depending on the choice of electric bus technology, an increase in TCO of 13% to 18% is observed compared to diesel buses.

The 120 km depot charging scenario incurs the highest TCO because it requires the largest fleet and the highest number of empty trips. (The latter is reflected in the driver cost.) Increasing the range to 300 km reduces fleet size and driver cost, however, the vehicles become more expensive due to the larger batteries, leading only to a moderate TCO reduction of ≈3%. The opportunity charging scenarios enable a slight further cost reduction due to the smaller batteries required; however, no savings in staff cost can be gained compared to depot charging because of the additional charging time required at terminal stops. Generally, due to the modest TCO differences between the individual electric bus scenarios (at most, 5%), we argue that the selection of the "best" electric bus technology should consider more factors than TCO, e.g., operational concerns or lifecycle impact.

7.5 Summary and Outlook

When planning and developing innovative mobility systems, a holistic view of sustainability—i.e., including the ecological, social and economic dimensions of sustainability—is imperative from our point of view. Our methodology takes all three dimensions into account. However, the integration of social sustainability in this comprehensive approach needs substantial further work.

We applied our methodology to analyze a fully decarbonized urban transport system for a complete urban region. The decarbonization can be realized using different zero emission technologies. Our research clearly shows that the greatest barriers are the high cost of the battery and the limited range of the vehicles. Accordingly, a successful establishment of BEV requires, above all, a specific design of the batteries, tailored to the respective application. In addition, the development of charging strategies is an important research topic. It can be shown that the range problem can be solved almost completely with the right dimensioning and placement of the infrastructure. Both, vehicle and system design are supported with the presented design methodology.

Regarding environmental sustainability, the benefits of a conversion to BEV depend largely on the electric power generation. Our results show positive effects, even with today's energy mix: A full life cycle assessment of the motorized individual transport in Berlin shows a reduction of GHG emissions by 20%. In the case of fully renewable electric power generation, BEV can make a decisive contribution for the decarbonization of urban transport, they perform better than conventional vehicles regarding other harmful emissions as well.

Our well-to-wheel analysis of electric urban freight trucks shows even higher benefits in terms of GHG emissions- For f the full life cycle of urban freight vehicles similar results can be expected, however, our in-depth analysis is still pending.

To assess the economic sustainability of various decarbonization strategies, we calculate the total cost of ownership (TCO) for several different electric vehicle technologies. In this work, we analyzed the examples of urban freight trucks and urban buses. In both cases, electrification requires a combination of system design and re-scheduling or re-routing of vehicle operations to accommodate the range limitations of electric vehicles. Again, our findings can support the decision making to find the appropriate systems technology for given applications. For electric freight trucks, we determined an increase in TCO of 18–24% compared to conventional vehicles; electric buses currently incur a 13–18% higher cost than diesel buses. It can be expected, however, that electric trucks and buses become economically advantageous in the future (Göhlich et al. 2013).

Currently, there are no established methods for the evaluation of social sustainability in the context of mobility systems. With the upcoming of automated systems, however, this topic becomes particularly important. Therefore, we developed a guideline to assess social sustainability in the use phase of automation initiatives in general and applied it to the case study of the service robot MURMEL. Through the application of the guideline, the level of automation and the manner of human-machine cooperation were identified as two central factors for the social sustainability of service robots. Another fundamental insight is the importance of social acceptance. We learned that looking into (and working with) other disciplines like product design, computer science or social science can create solutions out of the usual engineering scope (e.g. body language and interaction design, social trajectory planning). This first example already showed us that in order to identify the relevant goals of social sustainability in the design of urban transport systems, there is a need for a more inclusive process taking into account the requirements of all stakeholders. Since this aspect of sustainability still seems mostly neglected in the predominant product development methods, we plan to advance methods like the introduced guideline and establish their use in the design community.

In this work, we are concerned with *motorized* transport in the city. We are fully aware that it would be very much in the interest of sustainability if more people switched to public transport, cycling or walking. However, the consideration of modal shifts is a highly complex sociological problem which is out of the scope of this work.

In the near future we will complete our work with the analysis of fully autonomous fleets of BEV and a conversion to hydrogen powered vehicles. We will then also include the dimension of social sustainability of urban transport systems in our comparison.

We have shown how different sustainability categories can be evaluated using methodical approaches based on proven engineering design principles. We are able to analyze the impact of current technologies and make predictions about the impact of future technologies. This approach can be used as a basis for decision-making in planning and development in the transport sector. However, as discussed at the beginning, a fully sustainable solution without any downsides is not possible and, rather, trade-offs must always be made between the individual dimensions of sustainability and other requirements. Our task as scientists and engineers is to provide the necessary factual basis for these societal and political decision-making processes.

Acknowledgements The authors gratefully acknowledge the financial support of several institutions. Parts of this work have been supported by the Federal Ministry of Education and Research of Germany through funding of Research Campus Mobility2Grid (www.mobility2grid.de) and the Federal Ministry of Transportation and Digital Infrastructure. The project 'MURMEL' is funded within the Berlin Program for Sustainable Development (BENE) sponsored by the European Regional Development Fund and the Senate Department for Environment, Transport and Climate Protection Berlin. The Project "zeroCUTS" is funded by the Deutsche Forschungsgemeinschaft (German Research Foundation).

References

Abele E, Anderl R, Birkhofer H, Rüttinger B (2008) EcoDesign. Springer, Berlin, Heidelberg

Albers A, Behrendt M, Klingler S, Reiß N, Bursac N (2017) Agile product engineering through continuous validation in PGE – product generation engineering. Design Sci 3:E5. https://doi.org/10.1017/dsj.2017.5

Barckmann J, Jahn M (2020) EDAG CITYBOT: Ganzheitliches Ökokonzept für die Stadt mit Zukunft. https://www.edag-citybot.de/. Accessed 8 January 2021

Bender B, Gericke K (eds) (2021) Pahl/Beitz Konstruktionslehre: Methoden und Anwendung erfolgreicher Produktentwicklung. Springer, Berlin, Heidelberg

Bergmann M, Schäpke N, Marg O, Stelzer F, Lang DJ, Bossert M, Gantert M, Häußler E, Marquardt E, Piontek FM, Potthast T, Rhodius R, Rudolph M, Ruddat M, Seebacher A, Sußmann N (2021) Transdisciplinary sustainability research in real-world labs: success factors and methods for change. Sustain Sci. https://doi.org/10.1007/s11625-020-00886-8

Birkhofer H (2011) The future of design methodology. Springer, London

Bischoff J, Führer K, Maciejewski M (2019) Impact assessment of autonomous DRT systems. Transp Res Procedia 41:440–446. https://doi.org/10.1016/j.trpro.2019.09.074

Brandt C, Böker B, Bullinger A, Conrads M, Duisberg A, Stahl-Rolf S (2018) Fallstudie: Delivery Robot Hamburg für KEP-Zustellung. https://www.bmwi.de/Redaktion/DE/Downloads/C-D/delivery-robot-hamburg.pdf?__blob=publicationFile&v=4. Accessed 20 December 2020

Brown AL, Fleming KL, Safford HR (2020) Prospects for a highly electric road transportation sector in the USA. Curr Sustain Renew Energy Rep 7:84–93. https://doi.org/10.1007/s40518-020-00155-3

Bundesministerium für Umwelt, Naturschutz und nukleare Sicherheit (2020) Klimaschutz in Zahlen. https://www.bmu.de/publikation/klimaschutz-in-zahlen-2020/. Accessed 8 February 2021

Commission of the European Union (2015) Social life cycle assessment: state of the art and challenges for product policy support. Publications Office, Luxembourg

Cudok A, Huth T, Inkermann D, Vietor T (2017) Life cycle development – a closer look at strategies and challenges for integrated life cycle planning and upgrading of complex systems. In: Proceedings of the 21st international conference on engineering design (ICED17), vol 5. Vancouver, Canada

Del Pero F, Delogu M, Pierini M (2018) Life cycle assessment in the automotive sector: a comparative case study of internal combustion engine (ICE) and electric car. Procedia Structural Integrity 12:521–537. https://doi.org/10.1016/j.prostr.2018.11.066

Dér A, Erkisi-Arici S, Stachura M, Cerdas F, Böhme S, Herrmann C (2018) Life cycle assessment of electric vehicles in Fleet applications. In: Herrmann C, Mennenga MS, Böhme S (eds) Fleets go green. Springer, Cham, pp 61–80

DIN EN ISO 14040 (2021) Umweltmanagement – Ökobilanz. Deutsches Institut für Normung

Du Y, Hetherington NJ, Oon CL, Chan WP, Quintero CP, Croft E, van der Machiel Loos HF (2019) Group surfing: a pedestrian-based approach to sidewalk robot navigation. 2019 international conference on robotics and automation (ICRA): 6518–6524. https://doi.org/10.1109/ICRA.2019.8793608

Eckhardt J, Lauhkonen A, Aapaoja A (2020) Impact assessment of rural PPP MaaS pilots. Eur Transp Res Rev 12:49. https://doi.org/10.1186/s12544-020-00443-5

Elkington J (1998) Cannibals with forks: the triple bottom line of 21st century business. New Society Publishers, Gabriola Island, CT

Estevan H, Schaefer B, Adell A (2017) Life cycle costing: state of the art report. https://iclei-europe.org/publications-tools/?c=search&uid=JVTB5WYD. Accessed 19 July 2021

European Commission (2020) Communication COM/2020/562: stepping up Europe's 2030 climate ambition investing in a climate-neutral future for the benefit of our people. https://knowledge4policy.ec.europa.eu/publication/communication-com2020562-stepping-europe%E2%80%99s-2030-climate-ambition-investing-climate_en. Accessed 3 February 2021

Ewert R, Grahle A, Martins-Turner K, Syré A, Nagel K, Göhlich D (2020) Electrifi-cation of urban waste collection: introducing a simulation-based method-ology for feasibility, impact and cost analysis. https://doi.org/10.14279/depositonce-10314. Accessed 19 July 2021

Göhlich D, Fay T-A (2021) Arbeiten mit Anforderungen: Requirements Management. In: Bender B, Gericke K (eds) Pahl/Beitz Konstruktionslehre: Methoden und Anwendung erfolgreicher Produktentwicklung. Springer, Berlin Heidelberg, pp 211–229

Göhlich D, Spangenberg F, Kunith A (2013) Stochastic total cost of ownership forecasting for innovative urban transport systems. In: IEEE international conference on industrial engineering and engineering management (IEEM), 2013: 10–13 Dec. 2013, Bangkok, Thailand. IEEE, Piscataway, NJ, pp 838–842

Göhlich D, Nagel K, Syré AM, Grahle A, Martins-Turner K, Ewert R, Miranda Jahn R, Jefferies D (2021) Integrated approach for the assessment of strategies for the Decarbonization of urban traffic. Sustainability 13:839. https://doi.org/10.3390/su13020839

Grau Saldes A, Bolea Monte Y, Sanfeliu Cortés A, Puig-Pey Clavería AM (2017) ECHORD++ Robotics in a public economy. Economic and social development: 24th International scientific conference on economic and social development, pp. 40–49

Greve E, Krause D (2018) An assessment of methods to support the design of future robust modular product architectures. In: Proceedings of the DESIGN 2018 15th international design conference. The Design Society, Glasgow, pp 335–346

Hentze J, Graessler I (2015) Lebenszyklusgerechte Umfeldmodellierung. In: DFX 2015: proceedings of the 26th symposium Design for X, Herrsching, Germany, pp. 245–254

Horni A, Nagel K, Axhausen KW (2020) The multi-agent transport simulation MATSim. http://ci.matsim.org:8080/job/MATSim-Book/ws/partOne-latest.pdf. Accessed 17 August 2020

International Federation of Robotics (2020) World robotics: service robots 2020. https://ifr.org/ifr-press-releases/news/service-robots-record-sales-worldwide-up-32. Accessed 30 December 2020

Jefferies D, Göhlich D (2020) A comprehensive TCO evaluation method for electric bus systems based on discrete-event simulation including bus scheduling and charging infrastructure optimisation. WEVJ 11:56. https://doi.org/10.3390/wevj11030056

jsprit (2018) jsprit. Accessed 02 dez 2018

Kawamoto R, Mochizuki H, Moriguchi Y, Nakano T, Motohashi M, Sakai Y, Inaba A (2019) Estimation of CO2 emissions of internal combustion engine vehicle and battery electric vehicle using LCA. Sustainability 11:2690. https://doi.org/10.3390/su11092690

Kohl JL, van der Schoor MJ, Syré AM (2020a) Guideline for social sustainability in automation initiatives. https://www.mpm.tu-berlin.de/fileadmin/fg89/PDFs/Forschung/MURMEL/Guideline.pdf. Accessed 24 January 2021

Kohl JL, van der Schoor MJ, Syré AM, Göhlich D (2020b) Social sustainability in the development of service robots. Proc Des Soc Des Conf 1:1949–1958. https://doi.org/10.1017/dsd.2020.59

Martins-Turner K, Grahle A, Nagel K, Göhlich D (2020) Electrification of urban freight transport – a case study of the food retailing industry. Procedia Comput Sci 170:757–763. https://doi.org/10.1016/j.procs.2020.03.159

McGinn C, Bourke E, Murtagh A, Donovan C, Cullinan MF (2019) Meeting Stevie: perceptions of a socially assistive robot by residents and staff in a long-term care facility. 14th ACM/IEEE international conference on human-robot interaction (HRI)

McKenzie S (2004) Social sustainability: towards some definitions. https://rgio.org/kraisusqb.pdf. Accessed 19 July 2021

Mercedes-Benz AG (2021) eVito Kastenwagen | Mercedes-Benz Transporter. https://www.mercedes-benz.de/vans/de/vito/e-vito-panel-van. Accessed 3 February 2021

Miranda Jahn R, Syré A, Grahle A, Martins-Turner K, Göhlich D (2021) Methodology for determining charging strategies for freight traffic vehicles based on traffic simulation results. Procedia Comput Sci 184:656–661. https://doi.org/10.1016/j.procs.2021.03.082

MPM TU Berlin (2020) Methoden der Produktentwicklung und Mechatronik: MURMEL. https://www.mpm.tu-berlin.de/menue/forschung/projekte/murmel/. Accessed 8 January 2021

Pandey AK, Gelin R (2018) A mass-produced sociable humanoid robot: pepper: the first machine of its kind. IEEE Robot Automat Mag 25:40–48. https://doi.org/10.1109/MRA.2018.2833157

Schneider J, Lindau V (2020) Casestudy Sweep. https://www.studio-hint.com/sweep. Accessed 2 January 2021

Schröder S, Liedtke G (2014) Modeling and analyzing the effects of differentiated urban freight measures - a case study of the food retailing industry. Annual Meeting Preprint. https://trid.trb.org/view/1289723. Accessed 19 July 2021

Streetscooter GmbH Elektro Nutzfahrzeuge (2021) Home | Streetscooter GmbH Elektro Nutzfahrzeuge. https://www.streetscooter.com/de/. Accessed 3 February 2021

Syré AM, Heining F, Göhlich D (2020) Method for a multi-vehicle, simulation-based life cycle assessment and application to Berlin's motorized individual transport. Sustainability 12:7302. https://doi.org/10.3390/su12187302

UITP (2018) ZeEUS eBus Report #2: An updated overview of electric buses in Europe. http://zeeus.eu/uploads/publications/documents/zeeus-report2017-2018-final.pdf. Accessed 21 August 2020

United Nations (2009) Guidelines for social life cycle assessment of products: social and socio-economic LCA guidelines complementing environmental LCA and life cycle costing, contributing to the full assessment of goods and services within the context of sustainable development. UNEP, Paris

United Nations (2015) Transforming our world: the 2030 agenda for sustainable development. https://sustainabledevelopment.un.org/content/documents/21252030%20Agenda%20for%20Sustainable%20Development%20web.pdf. Accessed 19 July 2021

van Haaster B, Ciroth A, Fontes J, Wood R, Ramirez A (2017) Development of a methodological framework for social life-cycle assessment of novel technologies. Int J Life Cycle Assess 22:423–440

VDI 2221-1 (2019) Entwicklung technischer Produkte und Systeme.. Verein Deutscher Ingenieure

VDI 3780 (2000) Technikbewertung. Verein Deutscher Ingenieure

Volkswagen AG (2021) Der e-Crafter | Volkswagen Nutzfahrzeuge. https://www.volkswagen-nutzfahrzeuge.de/de/modelle/e-crafter.html. Accessed 3 February 2021

Ziemke D, Kaddoura I, Nagel K (2019) The MATSim open Berlin scenario: a multimodal agent-based transport simulation scenario based on synthetic demand modeling and open data. Procedia Comput Sci 151:870–877. https://doi.org/10.1016/j.procs.2019.04.120

Methods for In Situ Sensor Integration

Maximilian Hausmann, Eckhard Kirchner, Gunnar Vorwerk-Handing, and Peter Welzbacher

Abstract

A novel research topic to accelerate the digitalisation of mechanical engineering is the integration of measuring functions directly into components close to the process at in situ positions. Often it is neither obvious where to measure nor which measurand is suitable to fulfil the required measuring function. Measuring concepts differ in terms of their measuring location, the used physical effects as well as required system and material properties. A methodically supported identification of potential measurands is addressed in this contribution. Therefore, physical effect catalogues are used to establish a connection between different physical quantities. Potential measurands are contrasted by uncertainty regarding the dependencies of the underlying measuring concepts from environmental and boundary conditions. The identification and consideration of uncertainty is mandatory for a reasoned decision for a specific measuring concept. This includes the development of measures to reduce uncertainty based on Robust Design strategies. A methodical approach for the systematic identification and consideration of uncertainty following the Uncertainty Mode and Effects Analysis (UMEA) is described. The chapter concludes by introducing the concept of Sensing Machine Elements (SME) as promising approach for in situ sensor integration in (existing) technical systems, which combine a reduction of additional installation space and uncertainty.

M. Hausmann · E. Kirchner (✉) · G. Vorwerk-Handing · P. Welzbacher
Technical University of Darmstadt, Institute for Product Development and Machine Elements (pmd), Darmstadt, Germany
e-mail: kirchner@pmd.tu-darmstadt.de

D. Krause, E. Heyden (eds.), *Design Methodology for Future Products*,
https://doi.org/10.1007/978-3-030-78368-6_8

8.1 Introduction

The progressive digitalisation of products and processes in the context of Industry 4.0 offers a significant potential for innovations, in particular in the field of condition monitoring and predictive maintenance (Anderl and Fleischer 2016; Hirsch-Kreinsen et al. 2019). However, this process requires large amounts of reliable data regarding relevant process and state variables of the technical systems under consideration (Dörr et al. 2019; Martin et al. 2018). Current research mostly addresses the processing, the use and security aspects of such large amounts of data (Akerkar 2019). In most cases, however, the data required for a large-scale digitalisation is not available in sufficient quantity or quality, but is often assumed to be available (Naeini and Prindle 2018). In order to obtain this required data, measuring functions must be integrated into technical systems. Existing standardised sensor solutions, such as measuring flanges for torque measurement, are not practical to realise these functions in many applications due to their influence on the technical function of the system as well as its system behaviour and the required installation space (Martin et al. 2018).

For a simple and accelerated digitalisation of future and in particular existing technical systems, flexible sensor solutions are required, that can be integrated into technical systems with reasonable technical and economic effort (Vorwerk-Handing et al. 2018). In this context, sensor integration close to the process under investigation is particularly suitable in order to be able to directly measure a quantity of interest with a reduced influence of disturbance factors (Vogel et al. 2018).

The measurement locations that can be realised in a technical system can be classified into in situ and ex situ locations. In case of an in situ measurement (Latin "*in the original location*"), the quantity of interest is measured at its point of origin, whereas in the case of an ex situ measurement (Latin "*outside the original location*"), the quantity of interest is measured outside its point of origin. According to Hausmann et al., the ex situ measurement locations can be further subdivided into process-close, process-distinct and off-process. Figure 8.1 illustrates the classification of measurement locations using the torque measurement in an electric drive train as an example. The difference between all these measurement locations is referred to as distance. In addition to a spatial dimension, the distance primarily describes the complexity of the model-based relation between the quantity of interest to be determined and the actual measured quantity at the measurement location (Hausmann et al. 2021).

The presented classification of potential measurement locations in a technical system alone is not sufficient to select a suitable sensor concept in practice. This requires the definition of the measurement task to be realised in order to derive therefore suitable quantities of interest. The methodological and systematic identification of these quantities and the associated measurement locations is presented in the following section. Subsequently, the uncertainty associated with the identified measurands, respectively their measuring concepts, must be identified and considered in order to be able to make an early and reasoned decision for a specific measurand and measuring concept. For this

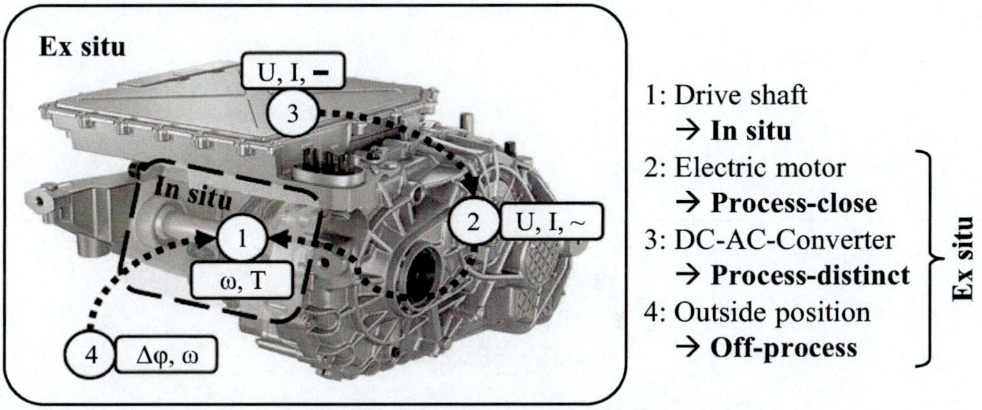

Fig. 8.1 Classification of measurement locations using an electric drive train as an example (Hausmann et al. 2021, based on Vorwerk-Handing et al. 2020a, picture on courtesy of Valeo Siemens eAutomotive Germany GmbH)

purpose, a methodological approach is presented in the third section of this chapter. The chapter concludes with a brief introduction of the concept of Sensing Machine Elements (SME), which are a flexible and promising (retrofit) solution for the integration of measuring functions into (existing) technical systems and thus enable in situ measurements in technical systems. The continuous development of these SME requires comprehensive methods, as presented in this chapter.

8.2 Identification of Potential Measurands and Measuring Locations

In order to support the integration of measuring functions, a methodology for identifying suitable measurands at potential measurement locations in existing technical systems or systems under development is introduced. Depending on the horizon, the scope of technical systems ranges from entire (production) plants to individual machines to assemblies and their components, e.g. in form of standardised machine elements.

An important step towards sensor integration into a technical system is the determination of potential measurands including a preliminary step of determining potential measuring locations (Fleischer et al. 2018; Löpelt et al. 2019; Zeller 1995). In order to avoid the unfounded restriction in the search for solutions through pre-fixation, the importance of a solution-neutral discussion of different potential measurands is emphasised by Fleischer et al. (2018). This is based on the fact that both, direct and indirect, measurements have to be considered for the determination of relevant process or state variables of a technical system (Fleischer et al. 2018). In the following, a consistent distinction is therefore made between the process or state variable of interest, as the target of the measurement and its interpretation, as well as the actual measured variable, as the input variable of a sensor.

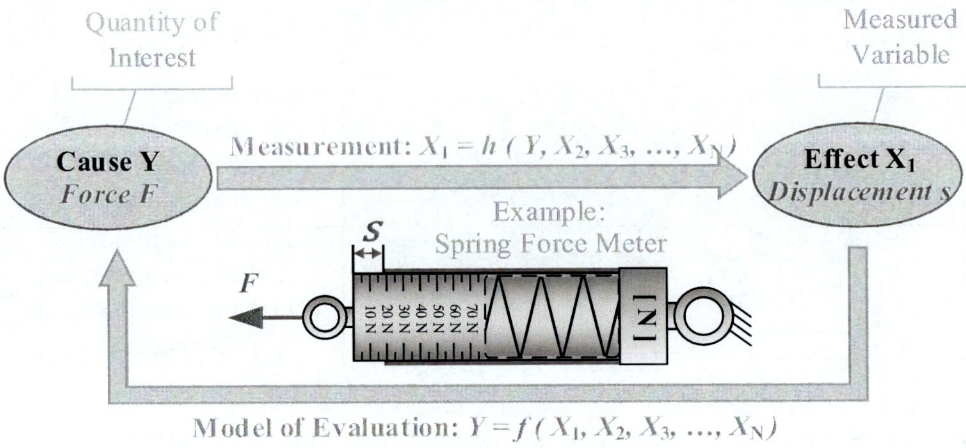

Fig. 8.2 Differentiation between cause and effect–Development of a model of evaluation (based on Tränkler and Reindl 2015 and Weckenmann et al. 2006)

Since a sensor changes a physical quantity into an electric signal, a measurement can also be seen as recording the effect X_1 of a cause Y on the sensor. The aim is to deduce from the measured effect X_1 to the cause Y with the model of evaluation (cf. Fig. 8.2). The therefore required connection between cause and effect is established on the basis of physical effects. Influences and imperfections that are not taken into account while establishing this connection (e.g. of the measured object, the environment or the measuring device) have an effect on the measuring chain and cause (measurement) uncertainty.

Defining the System Boundary and Structuring the System

Before potential measurands can be systematically identified, the scope of consideration must be defined in terms of a system boundary. This step is system-specific and determined by specific requirements as well as boundary conditions of the sensor integration. In addition to the technical requirements and boundary conditions, financial or approval-related aspects can also play a significant role.

After defining the system to be considered by defining a system boundary, it is structured for the methodologically supported identification of potential measurands. The structuring of the technical system takes place starting from the variable of interest along the occurring flow variables.

This approach is based on multipole modelling. Technical systems with a multidisciplinary character are modelled uniformly using concentrated network elements as well as general principles of energy conservation. The connection of the discrete network elements to an overall model of the technical system is carried out via poles on the basis of interconnection laws (*Kirchhoff's laws*). The basis of multipole-based modelling is the exchange of energy between network elements by means of a pair of conjugated

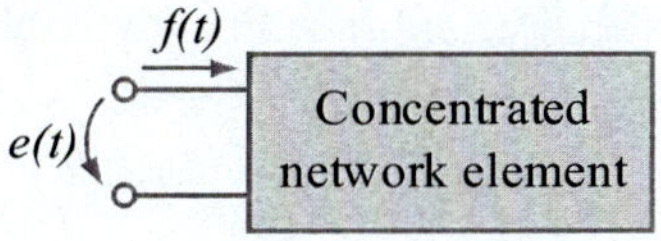

Fig. 8.3 Concentrated network element as basis for multipole-based modelling (Janschek 2009)

generalised energy or network variables (cf. Fig. 8.3). These are defined as generalised *e*ffort *e* and generalised *flow* *f* (Janschek 2009; MacFarlane 1967; Wellstead 1979).

The generalised effort and flow variables are classified according to the main domains of classical physics in mechanics, electricity and magnetism as well as thermodynamics. An overview of the generalised effort and flow variables used in the different domains is given in Table 8.1.[1] For a detailed presentation of the fundamentals and contexts of the approach, see Vorwerk-Handing (2021).

A structure according to the presented approach has two essential advantages with regard to the identification of potential measurands in the context of technical systems (Vorwerk-Handing 2021):

- Since the energy exchange between discrete network elements within the model can be described with flow and effort variables, energy flows and hence the conversions and changes of a variable of interest in the system can be modelled.
  - It is possible to structure the system gradually along nodes using flow variables and to model and consider it sequentially (Vorwerk-Handing et al. 2018). The term "node" goes back to the consideration of the electrical current in electrical networks according to *Kirchhoff's laws* and can be transferred analogously to other flow variables (MacFarlane 1967). In mechanics, for example, this corresponds to the balancing of forces on a free-cut element (cf. Fig. 8.4).
  - Via the effort variables, relationships between different discretely modelled elements in a system can be depicted along meshes (MacFarlane 1967). In electrical networks (cf. Fig. 8.4), this corresponds, for example, to the consideration of all partial voltages in the circuit of a mesh according to *Kirchhoff's laws.*
- In addition, due to this kind of modelling, it is possible to draw a direct conclusion on the metrological properties of the considered variables.
  - Exactly one spatial point is necessary to determine a flow variable *f*. Examples are the force *F* or the current *I*.
  - For the determination of an effort variable *e*, two spatial points are necessary. Examples are the velocity *v* or the voltage *U*.

[1]In literature, contrary classifications are also described under the assumption of a different point of view. Further information can be found in Janschek (2009) and Wellstead (1979).

Table 8.1 Overview of the generalised effort and flow variables classified according to the main domains of classical physics (cf. Vorwerk-Handing 2021)

Domain of classical physics	Generalised flow variable f	Generalised effort variable e
Mechanics	Force F	Velocity v
	Torque T	Angular velocity ω
	Mass flow rate $\dot{m}$	Gravitational potential Φ
	Volume flow rate $\dot{V}$	Pressure p
Electricity and magnetism	Electric current I	Electric voltage U
	Magnetic flux Φ_B	Magnetic voltage U_B
Thermodynamics	Entropy current $\dot{S}$	Temperature T
	Mass flow $\dot{m}$	Chemical potential μ

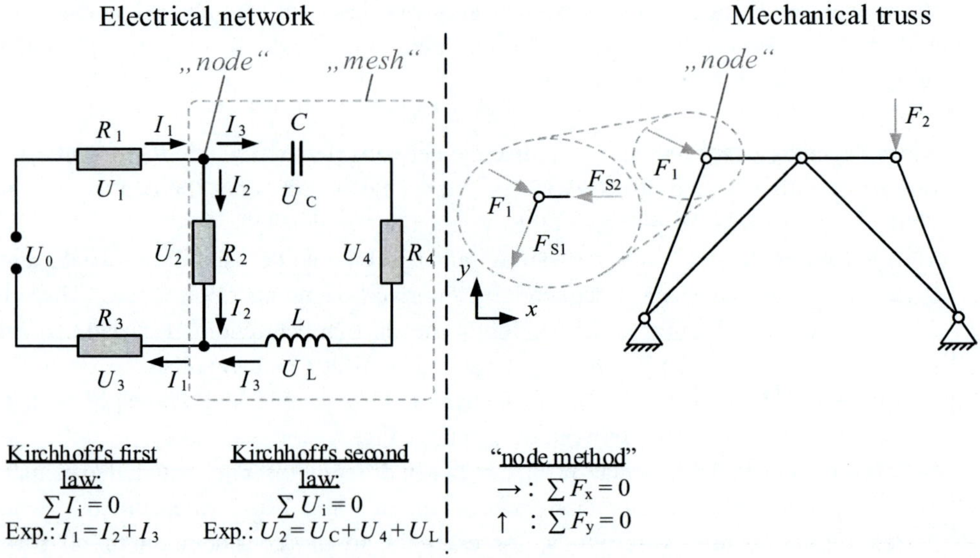

Fig. 8.4 *Kirchhoff's laws* in an electrical network (left) and "node method" on a free-cut mechanical truss (right)

Identification of Potential Measurands

In a technical system, the quantity of interest initially occurs at a specific location and is then transformed or changed within the system (cf. Fig. 8.5). The approach to identify systematically potential measurands starting from the quantity of interest is based on the consideration of cause-effect relationships. Physical effects thereby establish a relationship between two physical quantities. In conventional product development approaches, the relationship between the input and output variables of a technical system is considered in particular. In this context, physical catalogue systems enable the linkage of a desired effect with potentially applicable causes via known physical relationships (cf. Fig. 8.5). The basic idea of establishing a connection between physical quantities using physical effect

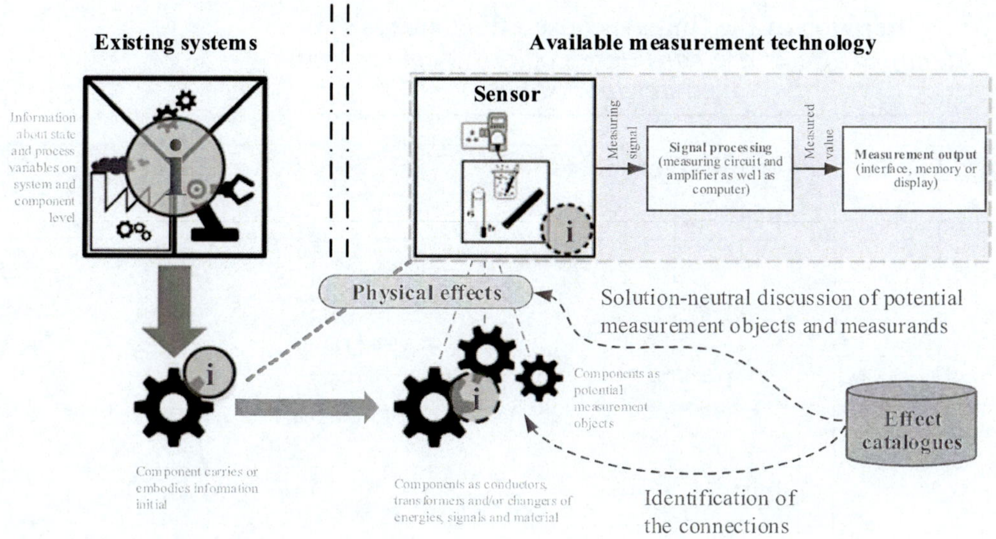

Fig. 8.5 Relationship between the quantity of interest and potential measurands–Identification of physical relationships and discussion of potential measurands as well as measurement objects (based on Vorwerk-Handing 2021)

catalogues is taken up and used in reverse to identify potential measurands (effects), starting from a quantity of interest (cause) (Vorwerk-Handing 2021).

Established effect catalogues, e.g. from Koller (1998) or Roth (2000), were developed to identify suitable physical principles for functions to be realised. This indicates the original aim of these catalogues: to establish connections between a desired effect and causes that might be usable for the intended function. However, two major limitations arise with regard to the intended identification of cause-effect relationships:

- Existing catalogues predominantly assume an effect to be realised. Due to the irreversibility of some physical effects, a strictly inverse approach of the existing catalogues is not permissible.
- According to the original aim of the effect catalogues, a consideration of design information is not intended. With regard to the addressed consideration of already existing technical systems, it is stated that the inclusion of the system's design is not sufficiently supported.

Vorwerk-Handing describes, based on existing catalogue systems and these two major limitations, the development of a physical catalogue system for the systematic identification of connections between a quantity of interest and potential measurands in an (existing) technical system (Vorwerk-Handing 2021). The basic idea of established effect catalogues is taken up and combined with the basics of multipole-based modelling. In accordance with

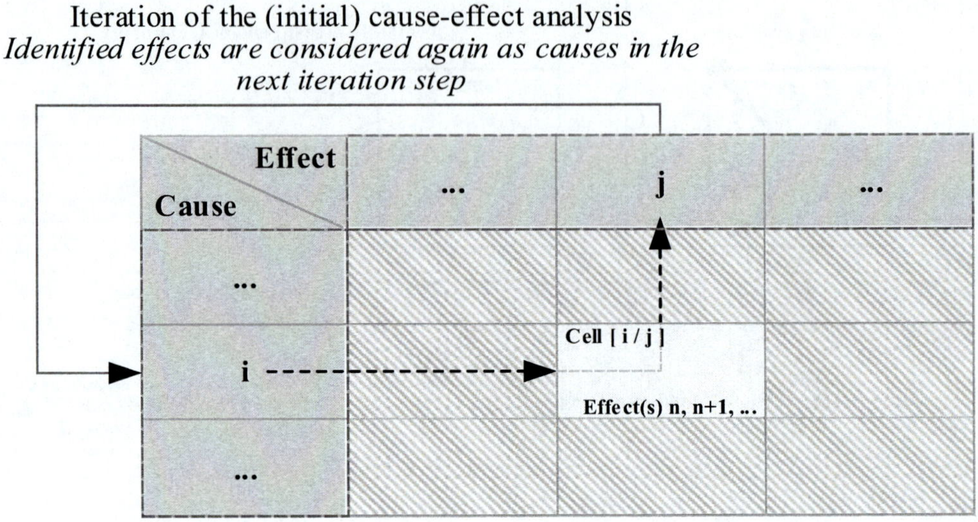

Fig. 8.6 Physical cause-effect consideration in the two-dimensional effect catalogue of Vorwerk-Handing (2021)

the intended purpose, a cause-effect view is applied throughout in order to achieve a physically and logically justified structure of the catalogue system (cf. Fig. 8.6).

Furthermore, a differentiation between functional variables and functionally relevant design parameters is introduced and implemented. The inclusion and structuring of design parameters is based on an extended consideration of multipole-based modelling and enables a systematic consideration of the (existing) technical system. Further information as well as a detailed description of how to systematically identify potential measurands starting from a quantity of interest can be found in Vorwerk-Handing (2021).

8.3 Identification and Consideration of Measuring Uncertainty

In the previous part of this chapter, the systematic identification of potential measurands and their associated measuring concepts were described using physical effect catalogues. Thereby, each measurand is part of a specific measuring concept in form of a chain of physical effects, the so-called effect chain. Typically, a large number of potential measurands and measuring concepts are identified that are potentially suitable to realise a desired measuring function. However, the individual measuring concepts differ in terms of their measuring location, the physical effects used in their effect chain as well as the required system and material properties. Consequently, each measuring concept is contrasted by a specific amount of uncertainty regarding, e.g. its dependency from environmental and boundary conditions that may affect its basic functionality negatively. Hence, the impact of the uncertainty connected to each measuring concept needs to be identified and considered already on the conceptual level to be able to discard the ones lacking functionality and enable an effective and efficient product development process. This allows an early and reasoned decision which measuring concepts should be pursued further (Vorwerk-Handing et al. 2020b).

"Uncertainty" is a term that is used in many scientific disciplines with different understandings and thus cannot be defined in a way that fits all the different understandings (Campos and de Henriques 2017; Hanselka and Platz 2010). For the field of product development, the ISO-Guide 73 (2009) defines uncertainty as "state, even partial, of deficiency of information related to, understanding or knowledge of, an event, its consequence, or likelihood". This state of deficiency can be qualitatively described and characterised by the quality of the therewith associated information (Pottebaum and Gräßler 2020). As defined in the DIN ISO 31000 (2018), uncertainty can have positive, negative or both impacts on a goal, the impact is generally referred to as risk. Hence, uncertainty is a neutral quantity that generally results in a risk and is only of direct interest on the conceptual level of the product development process, if it affects the system in such a negative way, that its functionality cannot be guaranteed. The limitation of the functionality of a system is generally referred to as failure. In terms of the integration of measuring functions into (existing) technical systems, failure not only refers to the actual breakdown of the measuring function but also impermissible measurement deviations. Thus, uncertainty can generally be analysed using established tools and models of risk and failure management.

Engelhardt et al. address the analysis of uncertainty and its consequences in the Uncertainty Mode and Effects Analysis (UMEA) (Engelhardt et al. 2011b). Due to its generality, it is also applicable to the analysis of uncertainty associated with measuring concepts. The holistic framework consists of multiple steps, each supported by standardised methods as well as models and is based on the risk management process in business economics (Engelhardt et al. 2009). The basic steps and the associated methods and models are depicted in Fig. 8.7 and explained in the following.

Environment and Goal Analysis The UMEA starts with a detailed and systematic investigation of the environment and the goals of the system under consideration. In order to carry out an analysis of the systems's environment, the system must be first delimited from its surrounding to be able to determine the influences acting on it from other systems or objects. Therefore, the system boundary already defined within the context of the systematic identification of potential measurands can be taken up. Moreover, relevant evaluation bodies (e.g. users, stakeholders or requirement groups) are identified and specified to analyse the goals of the considered system from different points of view, identify dependent variables (e.g. minimisation of costs or risk, maximisation of use or quality) and define the expected and tolerated border uncertainty. Therefore suitable methods are, e.g. SWOT-Analysis or Quality Function Deployment (Engelhardt et al. 2009, 2011b).

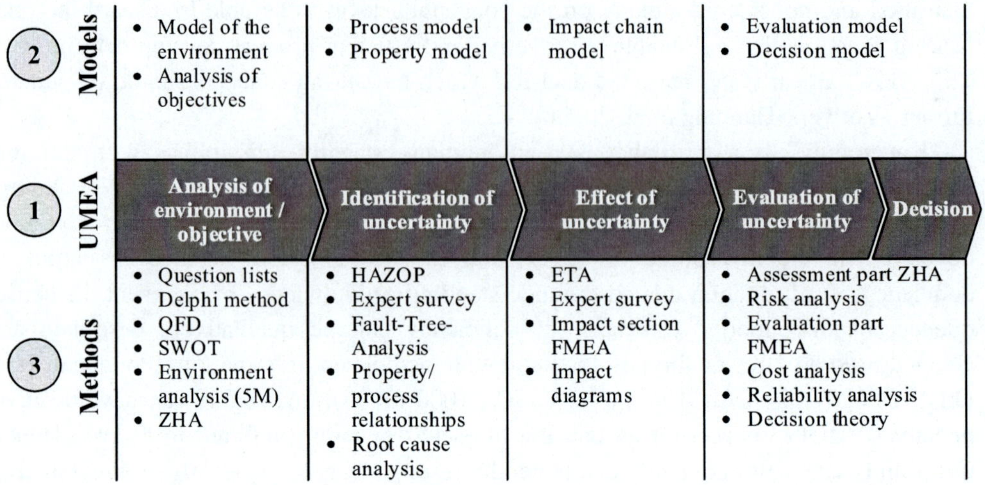

Fig. 8.7 Structure of the UMEA (translated from Engelhardt 2013)

Identification of Uncertainties and Their Causes In this step, relevant uncertainties and their causes are identified and described. In this context, relevant means that the uncertainty has an actual influence on the dependent variables, which were identified in the previous step. If possible, the uncertainties are quantified for the subsequent steps of the UMEA and the calculations therein. Especially in the early phases of the development process, a quantification may not always be possible. The uncertainties are then described quantitatively. Suitable methods for the identification of uncertainties are, e.g. the Fault Tree Analysis (FTA) or Hazard and Operability Study (HAZOP) (Engelhardt et al. 2009, 2011b).

In the context of the integration of measuring functions into (existing) technical systems, disturbance factors, as potential cause for uncertainty, are of great importance. Disturbance factors arise from the environment of the technical system or the technical system itself in form of secondary variables, and may cause the measuring concept to lack functionality. To identify these disturbance factors, e.g., checklists can be used to ensure that the majority of these factors are analysed in terms of their occurrence. An example for such a checklist is proposed by Vorwerk-Handing et al., which is based on the list of standardised disturbance factors from Mathias (Mathias 2016; Vorwerk-Handing et al. 2020b). Vorwerk-Handing et al. build upon this list by assigning the individual disturbance factors to the different physical disciplines, e.g., mechanics, thermodynamics or electricity and magnetism, and connecting them with their associated physical influencing variables. An extract of this checklist is shown in Fig. 8.8.

Physical influencing variables are variables that are caused by the individual disturbance factors and directly act on the considered technical system and consequently the measuring

Potential disturbance factor	Picto-gram	Influencing variable(s)	Occurrence	Quantification
Homo-geneous el. field		Force F El. field E_{el} El. charge q El. capacity C	☐	
Electromag. field (static)	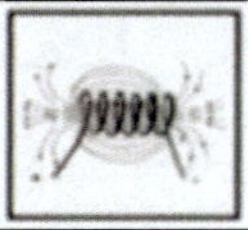	Force F Mag. flux density B	☐	
...	...	...	...	...

Abbreviations: el. – electric; mag. – magnetic

Fig. 8.8 Extract of the disturbance factor checklist by Vorwerk-Handing et al. (2020b)

function to be integrated. Based on the influencing variables of each occurring disturbance factor, caused unintended physical effects can be identified in a subsequent step using a suitable physical effect catalogue.

Detection of Effects of Uncertainties After the identification of occurring uncertainties, their interrelations as well as their caused effects are analysed. Thereby, special attention is paid to the impact of the individual uncertainties on the system defined in the first step of the UMEA. Suitable methods for the identification of potential effects are, e.g. the Event Tree Analysis (ETA) or the effect part of the Failure Mode and Effects Analysis (FMEA) (Engelhardt et al. 2011b). In addition, this process can be supported by suitable models, e.g. the Contact and Channel (C&C^2) Approach (cf. Grauberger et al. 2020).

Possible effects caused by uncertainties, especially the ones caused by the physical influencing variables of the previously identified disturbance factors, can be detected using physical effect catalogues. Therefore, the influencing variables are considered as input quantities of unintended physical effects, respectively interconnected physical effects, whose output quantities are the input and/or output quantities of physical effects already included, and thus intended, in the effect chain of the considered measuring concept. Moreover, unintended physical effects may occur between the different input, intermediate and/or output quantities of the effect chain. To identify this type of unintended physical effects, these quantities are considered systematically in pairs as input and output quantities of an unintended effect, respectively interconnected physical effects. Both types of unintended physical effects are visualised in Fig. 8.9 (Vorwerk-Handing et al. 2020b).

Evaluation of Uncertainty Effects In this step, the identified and analysed uncertainties are evaluated in order to establish a basis for the subsequent decision and identify the most

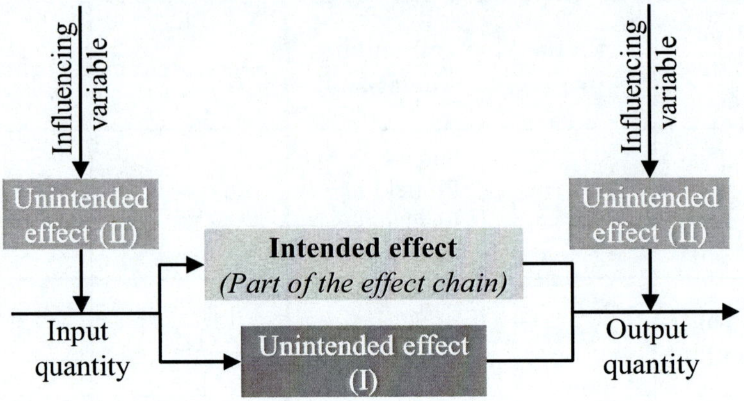

Fig. 8.9 Types of unintended physical effects

relevant uncertainties. Suitable methods for the evaluation of uncertainty are the evaluation part of the Zurich Hazard Analysis (ZHA) as well as the one of the FMEA or risk analysis (Engelhardt et al. 2009, 2011b).

For the evaluation of uncertainty, especially in the context of the integration of measuring functions, Vorwerk-Handing et al. propose a new evaluation method based on a FMEA. Therein, the identified uncertainties are evaluated using criteria, which are partly derived from those used in a classic FMEA (severity, probability, detection). This is due to the circumstance that the classic criteria for the evaluation of failures are not fully transferable to the evaluation of uncertainty. For example, evaluating uncertainty in terms of its probability of occurrence is not expedient because uncertainty always exists to a certain extend. Hence, Vorwerk-Handing et al. defined the following evaluation criteria for their method: severity, significance and controllability. The sevtrity of an uncertainty refers to the different levels of the uncertainty model of the Collaborative Research Centre (CRC) 805, which specify the amount of available and/or reliable information about an uncertainty. In this model, the spectrum of uncertainty ranges from “determinacy” (complete information and exact models) to “nescience” (model and parameters not sufficiently known). The uncertainty model of the CRC 805 is shown in Fig. 8.10 (Vorwerk-Handing et al. 2020b).

The significance of uncertainty describes the effect of an uncertainty in terms of the caused variance of the measuring chain’s output. The controllability refers to the ability to reduce the considered uncertainty, taking into account the effort required to achieve this. For each criterion, an evaluation scheme was defined, ranging from 1 (low criticality) to 5 (high criticality) (Vorwerk-Handing et al. 2020b).

Decision Based on the results of the evaluation of the previously identified and evaluated uncertainties, a decision has to be made whether the occurring uncertainties require further measures, e.g., to reduce them. It must be noted that the UMEA was developed just for the analysis of uncertainties, consequently, a subsequent last step for the development of

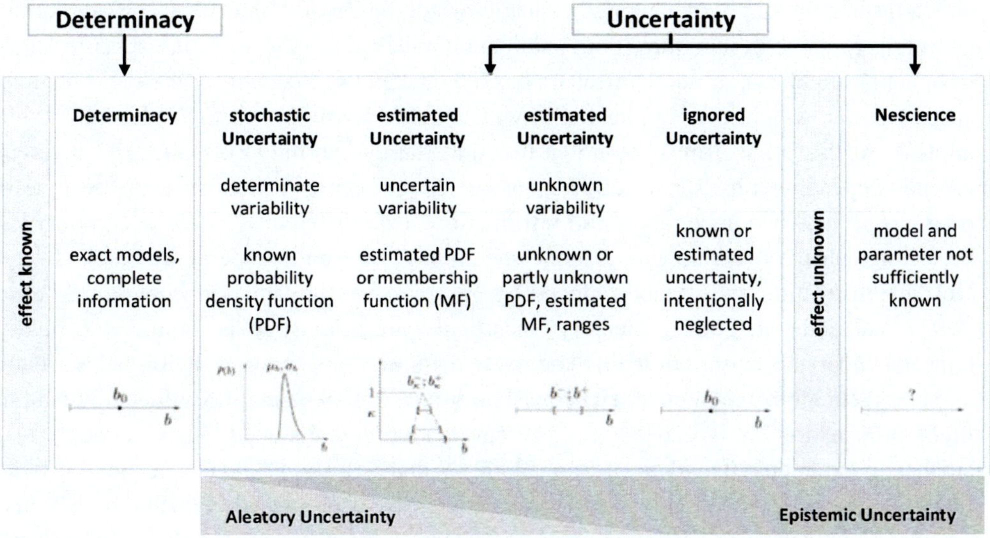

Fig. 8.10 Uncertainty model of the CRC 805 (translated from Lotz 2018, based on Engelhardt et al. 2010)

suitable measures to reduce uncertainty is not included within this framework and thus addressed separately in the subsequent section of this chapter (Engelhardt et al. 2011a).

In the context of the integration of measuring functions, the process of decision-making can be supported, e.g. by defining threshold values above which an uncertainty is considered critical for the functionality of the considered measuring concept. If an uncertainty is critical, the need for appropriate measures to reduce this uncertainty must be examined. However, other available measuring concepts, respectively measurands, must also be taken into account to prevent an early and unreasoned fixation on a measuring concept, respectively measurand (Vorwerk-Handing et al. 2020b).

Measures to Reduce Critical Uncertainty

For the development of suitable measures to reduce critical uncertainty, two fundamental approaches can be pursued, depending on the amount of available information regarding the critical uncertainty and its effect on the basic functionality of the considered measuring concept:

- Reduction of the uncertainty itself or
- Reduction of the effect of the uncertainty.

A reduction of uncertainty itself is required, if there is not enough information available to conduct the evaluation of an uncertainty. Hence, to reduce uncertainty itself, measures that result in an increase in information, understanding and/or knowledge are required. In engineering, this procedure of detailing or elaboration is generally referred to as concretisation (Vorwerk-Handing et al. 2020b). Examples for suitable measures for concretisation are experimental studies using prototypes, expert interviews or simulations as well as research in literature (Pottebaum and Gräßler 2020). Hereby, the specific effort-to-benefit ratio of each possible measure has to be taken into account for the individual use case to decide which, or if applicable in which order, possible measure(s) has/have to be applied. At this point it must be noted that uncertainty can only be reduced to a certain extend, depending on its nature. The nature of uncertainty is referring to the type of relationship between uncertainty and information. Two different types of uncertainty can be distinguished: epistemic and aleatory uncertainty. Epistemic uncertainty results from a lack of information and can be reduced by gaining more information (Oberkampf et al. 2002; Walker et al. 2003). Aleatory uncertainty, in contrast to epistemic uncertainty, appears under the assumption that the system under consideration is fully describable and complete information exist. Hence, aleatory uncertainty cannot be reduced by gaining more information. It is caused by inherent randomness induced by a variation of a stochastic parameter (Oberkampf et al. 2002; Walker et al. 2003). Consequently, if the identified uncertainty is mainly aleatory uncertainty, the usefulness of a further reduction of the uncertainty itself in terms of measures needs to be questioned critically (Vorwerk-Handing et al. 2020b).

Measures to reduce the effect of critical uncertainty on the basic functionality of a considered measuring concept can be developed based on the Robust Design strategies defined by Mathias et al. (Mathias 2016, based on Mathias et al. 2010). The different strategies are shown in Fig. 8.11.

The suitability of each strategy depends on the individual use case. Accordingly, it is not possible to define a general order in which the strategies should be considered. Similar to measures to reduce uncertainty itself, possible measures to reduce the effect of critical uncertainty have to be analysed and compared in regard of their individual effort-to-benefit ratio in the considered use case. Based on this comparison, the most effective measure (s) can be selected and applied to reduce the impact of critical uncertainty and thus ensure the basic functionality of the considered measuring concept.

8.4 Approach: Sensing Machine Elements

The previously presented methods can be used for the identification of potential measurands and measurement locations for an in situ measurement and to identify and consider uncertainties that occur therein. Sensing Machine Elements (SME) provide an opportunity for the integration of sensory functions into a technical system and allow in situ measurements. Vorwerk-Handing et al. (2020a) defined and classified SME as shown in Fig. 8.12, based on an initial description by Stücheli and Meboldt (2013).

SME are based on conventional machine elements, which are widely used in mechanical engineering due to their standardised design and dimensioning as well as their universal applicability in many different areas. Machine elements cannot be further disassembled in a non-destructive manner (Binz 2014). Examples for machine elements are rolling bearings, screws, seals or shaft couplings.

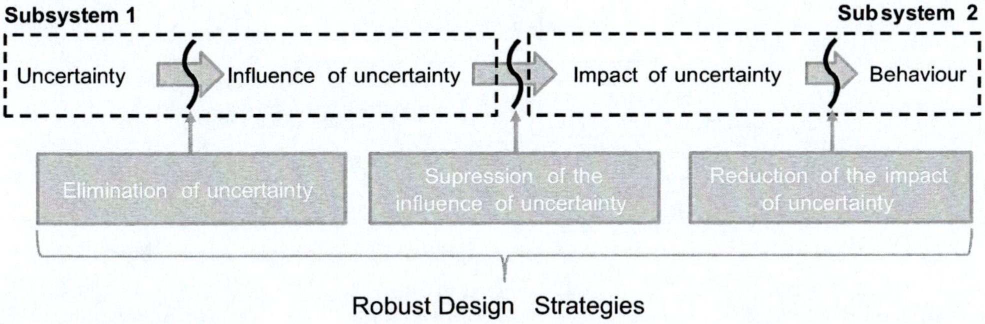

Fig. 8.11 Robust Design strategies (Mathias 2016, based on Mathias et al. 2010)

As an extension to conventional machine elements and their mechanical functions, SME provide additional sensory functions, illustrated in Fig. 8.12. Depending on the realisation of the implementation of this sensory function, SME can be subdivided into so-called Sensor carrying Machine Elements (ScME), Sensor integrating Machine Elements (SiME) and Sensory utilisable Machine Elements (SuME).

- ScME are machine elements with an integrated sensor whose measured quantity is independent from the primary mechanical function of the machine element. An example of a ScME is the measuring screw, shown in Fig. 8.12, that allows the measurement of the surrounding temperature at the screw end. This additional measuring function does not correlate with the primary function of the screw. Another example of a ScME is described by Ebner et al. (2020) who apply a thin film sensor on a gear tooth to measure the temperature in the active gear mesh.
- SiME are machine elements with an integrated sensor element, whose measured quantity correlates with the primary mechanical function of the machine element. For example, in the sensor-integrating timing belt (cf. Fig. 8.12), the sensor system allows the determination of the belt's pretension by measuring its eigenfrequencies using acceleration sensors.
- SuME are machine elements without an additional integrated sensor element. Instead, only the electric properties of the conventional machine element are used to fulfil the sensory function. An example is the load measurement of a rolling bearing based on its electrical impedance. The correlation between the measured impedance and the bearing load can be used for sensing as recently shown by Schirra et al. (2021). Originally, the capacitive properties of the lubricating film were used to model the behaviour of rolling bearings under the effects of damaging currents, cf. Bader et al. (2017) and Radnai et al. (2015).

The use of SME in technical systems is often accompanied by a reduction of the distance between the point of origin of a quantity of interest and the measurement location of the measured quantity (Vorwerk-Handing et al. 2020a). By integrating a sensor directly into

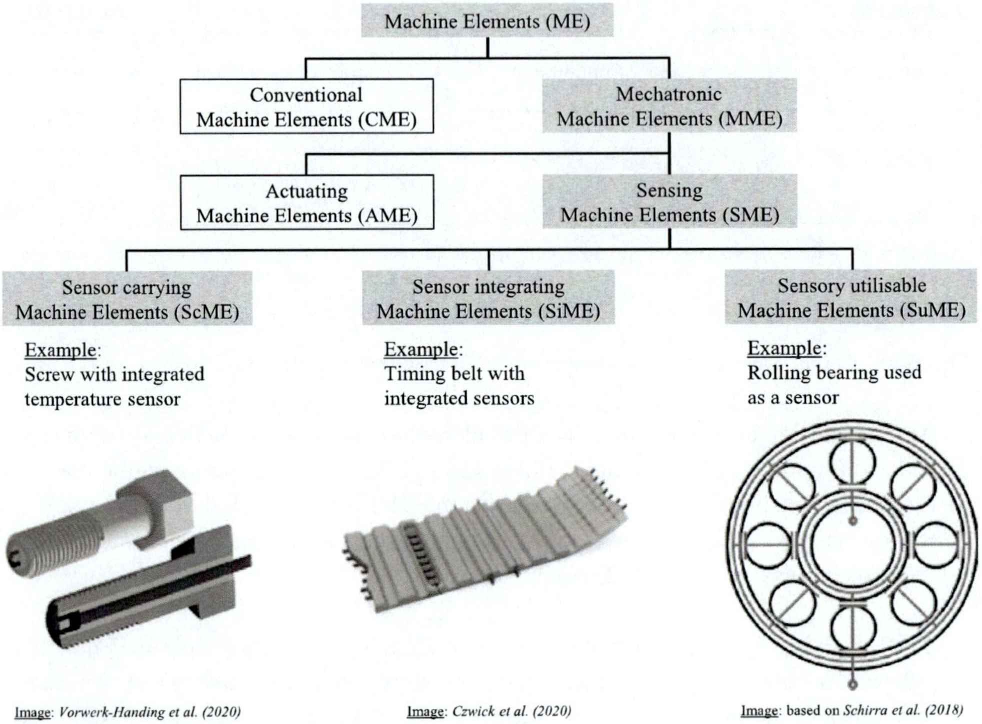

Fig. 8.12 Classification of Sensing Machine Elements (Hausmann et al. 2021; based on Vorwerk-Handing et al. 2020a)

the process, the uncertainty associated to the measurement is shifted from the system and the transfer path of the quantity of interest towards the SME. For example, the quantity of interest can be subject to fewer conversions and disturbances on its transfer path from its point of origin to the measuring location. At the same time, however, the uncertainty associated to the actual measurement in the SME as sensory element remains due to the scope of understanding of the underlying sensor principle and the disturbances acting on the SME. At the same time, new uncertainties may arise, e.g. due to new and innovative sensor principles, that need to be addressed using the approach described in Sect. 8.3 (Hausmann et al. 2021; Kirchner et al. 2018).

The decision to use SME can be made based on a number of potentials. First and foremost, SME offer the possibility of using standardised machine elements as sensor solutions with uniform dimensions and interfaces in new or existing technical systems without extensive changes of the design (Vorwerk-Handing et al. 2020a). This results in an installation space neutral application and thus an economic integration of measuring functions into (existing) technical systems. SME also have the potential to enable innovative and previously unknown measurement concepts based on new sensor principles. This includes, e.g. the bearing load determination through bearing impedance measurement (Schirra et al. 2019).

In addition to the potentials of an application of SME, current research is already investigating open challenges. For example, new sensor principles, the description of associated models and the therein occurring uncertainty are constantly analysed in order to improve their applicability. Furthermore, the electric description of the necessary energy and signal transfer paths to and from the SME are also in the scope of current research. In many cases, these paths are structure integrated, i.e. the signal runs directly through components of the technical system. The underlying models, transfer behaviours as well as the acting disturbances must therefore be described in detail to ensure a high-quality interpretation of the sensor data. For an application of SME in practice, there is still a lack of design rules and supporting design methods, e.g. with regard to the insulation of the energy and signal transfer path in the system by means of insulation sleeves or non-conductive coatings (Vorwerk-Handing et al. 2020a).

References

Akerkar R (2019) Artificial intelligence for business. Springer, Cham

Anderl R, Fleischer J (2016) Guideline Industrie 4.0: guiding principles for the implementation of Industrie 4.0 in small and medium sized businesses. VDMA Verlag, Frankfurt a. M.

Bader N, Furtmann A, Tischmacher H, Poll G (2017) Capacitances and lubricant film thickness of Rease and oil lubricated bearings. In: Mohseni H, Moghadam AD, Xiao H (eds) Proceedings of STLE 2017. Curran Associates, Red Hook, pp 311–314

Binz H (2014) Universitäre Lehre in der Produktentwicklung: Leitfaden der Wissenschaftlichen Gesellschaft für Produktentwicklung (WiGeP). Konstruktion 66(6):74–79

Campos AA de Henriques E (2017) Identification, Classification and Modeling Uncertainty in Early Stage of Manufacturing Systems: A Survey. In: Silva Gomes FF, Meguid SA (eds) Proceedings of M2D 2017

DIN ISO 31000 (2018) Risikomanagement. Deutsches Institut für Normung. Beuth, Berlin

Dörr M, Ries M, Gwosch T, Matthiesen S (2019) Recognizing product application based on integrated consumer grade sensors: a case study with handheld power tools. In: Putnik GD (ed) Procedia CIRP 2019. Elsevier, Amsterdam, pp 798–803

Ebner M, Ziegltrum A, Lohner T, Michaelis K, Stahl K (2020) Measurement of EHL temperature by thin film sensors: thermal insulation effects. Tribol Int 149:105515. https://doi.org/10.1016/j.triboint.2018.12.015

Engelhardt R (2013) Uncertainty Mode and Effects Analysis: Heuristische Methodik zur Analyse und Beurteilung von Unsicherheiten in technischen Systemen des Maschinenbaus. VDI-Verlag, Düsseldorf

Engelhardt R, Birkhofer H, Kloberdanz H, Mathias J (2009) Uncertainty-mode- and effects-analysis: an approach to analyze and estimate uncertainty in the product life cycle. In: Norell Bergendahl M, Grimhelden M, Leifer L, Skogstad P, Lindemann U (eds) Proceedings of ICED 09. Design Society, Glasgow, pp 191–202

Engelhardt R, Koenen JF, Enss GC, Sichau A, Platz R, Kloberdanz H, Birkhofer H, Hanselka H (2010) A model to categorise uncertainty in load-carrying systems. In: Heisig P, Clarkson PJ, Vajna S (eds) Modelling and management of engineering processes: proceedings of MMEP 2010, pp 53–64

Engelhardt R, Eifler T, Mathias J, Kloberdanz H, Birkhofer H, Bohn A (2011a) Linkage of methods within the UMEA methodology: an approach to analyse uncertainties in the product development process. In: Culley SJ, Hicks BJ, McAloone TC, Howard TJ, Chen W, Dong A (eds) Proceedings of ICED 11. Design Society, Glasgow, pp 433–442

Engelhardt R, Wiebel M, Eifler T, Kloberdanz H, Birkhofer H, Bohn A (2011b) UMEA: a follow-up to analyse uncertainties in technical systems. In: Culley SJ, Hicks BJ, McAloone TC, Howard TJ, Chen W, Dong A (eds) Proceedings of ICED 11. Design Society, Glasgow, pp 257–266

Fleischer J, Klee B, Spohrer A, Merz S, Metten B (2018) Leitfaden Sensorik für Industrie 4.0: Wege zu kostengünstigen Sensorsystemen

Grauberger P, Wessels H, Gladysz B, Bursac N, Matthiesen S, Albers A (2020) The contact and channel approach: 20 years of application experience in product engineering. J Eng Des 31:241–265. https://doi.org/10.1080/09544828.2019.1699035

Hanselka H, Platz R (2010) Ansätze und Maßnahmen zur Beherrschung von Unsicherheit in lasttragenden Systemen des Maschinenbaus. Konstruktion 2010(11/12):55–62

Hausmann M, Koch Y, Kirchner E (2021) Managing the uncertainty in data-acquisition by in situ measurements: a review and evaluation of sensing machine element approaches in the context of digital twins. Int J Prod Lifecycle Manag 13(1):48–65. https://doi.org/10.1504/IJPLM.2021.115700

Hirsch-Kreinsen H, Kubach U, Stark R, Wichert G von Hornung S, Hubrecht L, Sedlmeir J, Steglich S (2019) Key themes of industry 4.0: research and development needs for successful implementation of industry 4.0

ISO GUIDE 73 (2009) Risk management. International Organization for Standardization, Beuth, Berlin

Janschek K (2009) Systementwurf mechatronischer Systeme: Methoden - Modelle - Konzepte. Springer, Berlin

Kirchner E, Martin G, Vogel S (2018) Sensor integrating machine elements: key to in-situ measurements in mechanical engineering. In: Schützer K (ed) 23° Seminário Internacional de Alta Tecnologia: Desenvolvimento de Produtos Inteligentes: Desafios e vovos requsitos. Piracicaba, Brazil

Koller R (1998) Konstruktionslehre für den Maschinenbau : Grundlagen zur Neu- und Weiterentwicklung technischer Produkte mit Beispielen, 4th edn. Springer, Berlin

Löpelt M, Wilsky P, Ruffert J, Göhlert N, Prielipp R, Riedel R (2019) Sensorauswahl für Bestandsanlagen. Zeitschrift für wirtschaftlichen Fabrikbetrieb (ZWF) 114:273–276. https://doi.org/10.3139/104.112087

Lotz J (2018) Beherrschung von Unsicherheit in der Baureihenentwicklung. Dissertation, Technical University of Darmstadt, Germany

MacFarlane AGJ (1967) Analyse technischer Systeme. Bibliographisches Institut, Mannheim

Martin G, Schork S, Vogel S, Kirchner E (2018) MME – Potentiale durch mechatronische Maschinenelemente. Konstruktion 70:71–82

Mathias J (2016) Auf dem Weg zu robusten Lösungen. Dissertation, Technical University of Darmstadt, Germany

Mathias J, Kloberdanz H, Engelhardt R, Birkhofer H (2010) Strategies and principles to design robust products. In: Marjanović D, Storga M, Pavković N, Bojčetić N (eds) Proceedings of DESIGN 2010. Zagreb, Croatia, pp 341–350

Naeini EZ, Prindle K (2018) Machine learning and learning from machines. Lead Edge 37:886–893. https://doi.org/10.1190/tle37120886.1

Oberkampf WL, DeLand SM, Rutherford BM, Diegert KV, Alvin KF (2002) Error and uncertainty in modeling and simulation. Reliab Eng Syst Saf 75:333–357. https://doi.org/10.1016/S0951-8320(01)00120-X

Pottebaum J, Gräßler I (2020) Informationsqualität in der Produktentwicklung: Modellbasiertes Systems Engineering mit expliziter Berücksichtigung von Unsicherheit. Konstruktion 72:76–83

Radnai B, Gemeinder Y, Kiekbusch T, Sauer B, Binder A (2015) Schädlicher Stromdurchgang: Untersuchung des Schädigungsmechanismus und der zulässigen Lagerstrombelastung von Wälzlagern in E-Motoren und Generatoren verursacht durch parasitäre hochfrequente Lagerströme. FVA-Heft 1127

Roth K (2000) Konstruieren mit Konstruktionskatalogen: Band 1: Konstruktionslehre, 3rd edn. Springer, Berlin

Schirra T, Martin G, Neu M, Kirchner E (2019) Feasibility study of impedance analysis for measuring rolling bearing loads. In: Proceedings of STLE 2019. Curran associates, Red Hook, USA

Schirra T, Martin G, Puchtler S, Kirchner E (2021) Electric impedance of rolling bearings: consideration of unloaded rolling elements. Tribol Int 158:106927. https://doi.org/10.1016/j.triboint.2021.106927

Stücheli M, Meboldt M (2013) Mechatronic machine elements: on their relevance in cyber-physical systems. In: Abramovici M, Stark R (eds) Proceedings of CIRP design 2013. Springer, Berlin, pp 263–272

Tränkler HR, Reindl LM (2015) Sensortechnik: Handbuch für Praxis und Wissenschaft, 2nd edn. Springer, Berlin

Vogel S, Martin G, Schirra T, Kirchner E (2018) Robust design for mechatronic machine elements: how robust design enables the application of mechatronic shaft-hub connection. In: Marjanović D, Štorga M, Škec S, Bojčetić N, Pavković N (eds) Proceedings of design 2018. Zagreb, Croatia, pp 3033–3040

Vorwerk-Handing G (2021) Erfassung systemspezifischer Zustandsgrößen: Physikalische Effektkataloge zur systematischen Identifikation potentieller Meßgrößen. Dissertation, Technical University of Darmstadt, Germany

Vorwerk-Handing G, Martin G, Kirchner E (2018) Integration of measurement functions in existing systems: retrofitting as basis for digitalization. In: Ekströmer P, Schütte S, Ölvander J (eds) Proceedings of NordDesign 2018, Linköping, Sweden

Vorwerk-Handing G, Gwosch T, Schork S, Kirchner E, Matthiesen S (2020a) Classification and examples of next generation machine elements. Forsch Ingenieurwes 84:21–32

Vorwerk-Handing G, Welzbacher P, Kirchner E (2020b) Consideration of uncertainty within the conceptual integration of measurement functions into existing systems. Procedia Manufacturing 52:301–306. https://doi.org/10.1016/j.promfg.2020.11.050

Walker WE, Harremoës P, Rotmans J, van der Sluijs JP, van Asselt M, Janssen P, Krayer von Krauss MP (2003) Defining uncertainty: a conceptual basis for uncertainty management in model-based decision support. Integr Assess 4:5–17. https://doi.org/10.1076/iaij.4.1.5.16466

Weckenmann A, Sommer K-D, Siebert BRL (2006) Qualität für Messergebnisse: Messunsicherheit nach GUM. Tech Mess 73:189–199. https://doi.org/10.1524/teme.2006.73.4.189

Wellstead PE (1979) Introduction to physical system modelling. Academic, London

Zeller F-J (1995) Sensorplanung und schnelle Sensorregelung für Industrieroboter. Dissertation. Carl Hanser, Munich, Germany

Part III

Facing the Challenges in Product Development

Context-Adapted Methods of Modern Product Development: Recommendations and Best Practice Examples

Daniel Roth and Hansgeorg Binz

Abstract

Product developers continue to face many challenges when it comes to ensuring the most efficient and effective product development possible. In order to meet these challenges, appropriate support is required. But what do expedient supports look like? This contribution addresses the challenge of developing methods that are as flexible as possible and adapted to the context as a form of expedient support. To this end, general aspects that such a method should take into account are presented at the beginning, as well as the overarching question of what is basically understood by methods that are flexible and adaptable to the context. The core of this contribution is then formed by recommendations for the development of context-adaptable methods and supports. Examples from the institute's everyday life are used as best practices and briefly presented where possible. The aim is to provide future product developers with suggestions for, in particular, the development of flexible and adaptable methods and also to further minimize the reservations that still exist about this.

9.1 Introduction

Today's product development has become increasingly decentralized, global and distributed all over the world, enabled by enormously improved information technology support (Lindemann 2016; Ehrlenspiel and Meerkamm 2017; Bender and Gericke 2021).

D. Roth (✉) · H. Binz
University of Stuttgart, Institute for Engineering Design and Industrial Design, Stuttgart, Germany
e-mail: daniel.roth@iktd.uni-stuttgart.de; hansgeorg.binz@iktd.uni-stuttgart.de

D. Krause, E. Heyden (eds.), *Design Methodology for Future Products*,
https://doi.org/10.1007/978-3-030-78368-6_9

Companies are faced with the challenge of providing their customers with the right products of the right quality in an acceptable time. The underlying effectiveness and efficiency of product development have a great influence on the success of companies. However, reliable, fast and company-specific product development processes are required in order to be successful, which holistically include both the product and the later target market, and thus the customer. The performance of the design methods and tools used in this process is a decisive factor. Therefore, it is no wonder that intensive research work has been carried out for many years on the development of suitable engineering design methods, as well as their transfer and application in industrial practice. Nevertheless, there are still reservations about "academic" methods and approaches, which are often still perceived as too complex and time-consuming (Binz et al. 2011).

This is aggravated by the fact that it is no longer sufficient to know the current requirements for methods; it is equally important to have knowledge about future requirements and framework conditions. How different application scenarios of future methods can look like is shown with four conceivable scenarios in Albers et al. (2017).

An understanding of the necessary method components as well as the inclusion of the addressed application environment will have an influence on a successful application. Hence, the core of this contribution is the presentation of methods adapted to the context, as they have been developed at the Institute for Engineering Design and Industrial Design (IKTD) and which can also be used for product development in general. The aim is to offer support for product developers in the future—in addition, general recommendations are made.

With the underlying assumption that a significant obstacle for the introduction of (new) methods is their lack of adaptability to the respective application context and thus non-flexible use in different situations, views or opinions on the most important terms of this contribution are presented in the following in a brief form.

9.2 Clarification of Terms and Situation Analysis

Within the contribution under the term method all supports are subsumed, which hold a processual describing basic idea, as well as an improving one. This includes classic procedures that describe the organizational framework and processes and contain individual process steps that frequently overlap or have to be run through in parallel—known in the domain of mechanical engineering as design methodology—but also terms such as tools, guidelines, frameworks and the like. A clear change can be seen, in the further development of all these supports, especially away from rigid constructs to more flexible models with more application relevance—as can be seen, for example, in Gräßler et al. (2018) as well as Graessler and Hentze (2020). A very catchy example is the further development of the VDI guideline 2221 (VDI Guideline 1993) to the VDI guideline 2221 (VDI Guideline 2019 Part 1), in which a significantly stronger process orientation is present as an essential innovation. Particularly worth mentioning are the exemplary product development processes in different contexts contained in part 2 (VDI Guideline 2019 Part 2) and the explanation of activities depending on defined process phases.

This rethinking in the design of support emphasizes the necessity of a context-adapted method development, which was named as an assumption before. From the author's point of view, however, the term "context-adaptation" covers much more than the mere reference to the operational environment. Without claiming to be exhaustive, some central aspects are named below. First, the context itself must be considered. This ranges from the operational environment of the future product, the company developing the product (SME, large companies), the existing method and product development knowledge itself to the consideration of available resources. How many personnel can be deployed, what budget is available, what software and what machines can be used?

On the other hand, adaptability of the method to the context itself becomes necessary: Methods must be able to act flexibly and variably according to the situation. In addition to the dissolution of rigid, sequential models, methods are expected to provide situationally appropriate steps. Steps that are not appropriate must not be applied and the awareness for changing circumstances must be sharpened. The next problem under consideration may require a completely different solution path. Thus, a context-adapted method is understood as a support that acts flexibly depending on the context—for example, by omitting steps, adapting sets of criteria, selecting methods that fit to the situation. Often, the time factor is a decisive criterion for selection in everyday industrial life.

9.3 Superordinate Aspects of a Method Development

Aligned with the development of a method to support a situation, the development of the "right" method must be ensured, as discussed in the introductory chapters. The right method is understood as whether an expedient support can be offered. "If and how an engineering design methodology can provide this support in reality has to be assessed using appropriate criteria" (Binz et al. 2011). As early as 2009, criteria in the form of requirements for methods were defined in a contribution, whereby their consideration supports the targeted development of appropriate development methodologies (see Fig. 9.1). Five aspects are distinguished, which can be divided into 19 further requirements (from 8 groups): Normativity, Didactics, Uncertainty, Competitiveness as well as Match & Limit. "The objective (of this work) was to define a set of requirements on engineering design methodologies that provides a mean to assess the outcome of the development of methodologies (...). Interdependencies between the requirements, if existent, have been reasoned and analyzed." (Keller and Binz 2009).

This set of criteria is always used at IKTD when developing new methods. In addition to the requirements addressed there, a focus of the IKTD's method development is on small and medium-sized enterprises. As a consequence, emerging methods should be able to be used without (expensive) software applications.

Aspect	Group Description	Grouped requirements
Normativity	*Revisability* by appropriate and accepted means	Validation Verification
	Scientific soundness by backing up the hypotheses of a methodology	Objectivity Reliability Validity
Didactics	*Comprehensibility*	Comprehensibility Repeatability Learnability Applicability
Uncertainty	Providing a *structure* for complex tasks and problems and *compatibility* with different environments	Handling complexity Problem solving cycle Structuring Compatibility
	Providing *flexibility* for the designer using degrees of freedom when applying a methodology	Flexibility
Competitiveness	*Practical relevance and competitiveness* by satisfying a need for a methodology	Innovativeness Competitiveness
	Usefulness	Effectiveness Efficiency
Match & Limit	Problem *specifity* allowing links between an assignment and a matching methodology, and defining the application limits of a methodology	Problem specifity

Fig. 9.1 Aspects of a methodology, description of related groups, and grouped requirements on engineering design methodologies (Keller and Binz 2009)

9.4 Best Practice Examples of Methods for Developing Context-Appropriate Support

In the subchapter on situation analysis and clarification of terms, a wide variety of aspects were named whose consideration has an influence on the design of context-appropriate methods. But how can they actually be taken into account?

One recommendation could be to base the development of such methods on the general steps of the Design Research Methodology (DRM) by Blessing and Chakrabarti (2009). According to this, four steps are to be followed in the development of a purposeful support. Starting with a Research Clarification (RC) to determine the research objective, a build-up of understanding—for example, through intensive literature reviews or even supplementary analyses of empirical data—takes place in Descriptive Study I (DS I). In the

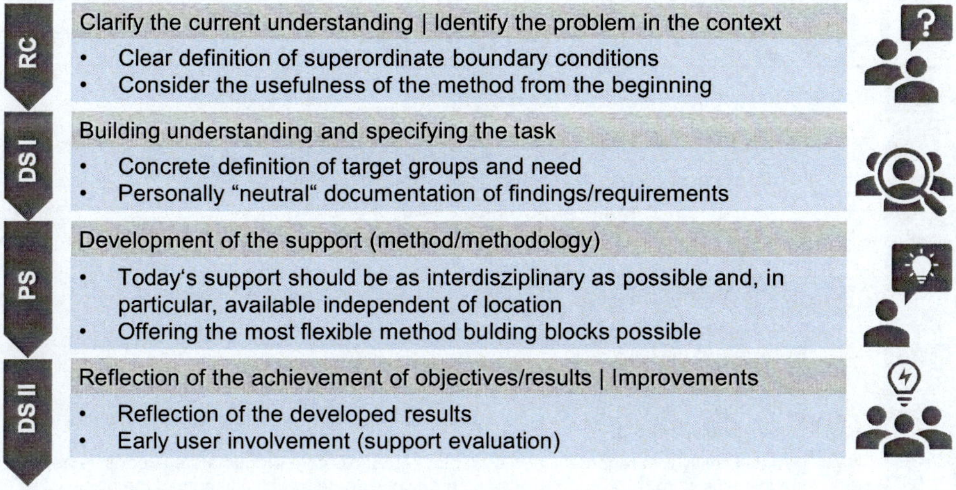

Fig. 9.2 Recommendations for the development of flexible context-adapted methods

subsequent Prescriptive Study (PS), the synthesis of expectations and experiences ultimately leads to the development of a support for a previously identified problem, which is finally reflected upon and evaluated in the Descriptive Study II (DS II) stage. These four stages can be run through in variable depth and repetition depending on the research question to be answered.

This procedure is transferred to the development of context-adapted methods and shown in Fig. 9.2. Here recommendations are named, whose consideration can be necessary for reaching an appropriate context adaptability.

Based on the previous findings, possible support is presented in the following in the form of best practice examples from the IKTD. These are listed in a chronological order in which they are applied in a product development process.

9.4.1 Best Practice for Generating and Documenting Appropriate Problem Ideas

Frequently, the development of appropriate support already fails when it comes to formulating the actual need for support. In doing so, identifying and defining problems worth pursuing can be a challenge. "We fail more often because we solve the wrong problem than because we get the wrong solution to the right problem" (Ackoff 1974). However, the IKTD's emoji method (see Fig. 9.3) provides support that can be used to generate application-oriented and thus user-focused, new problem ideas in a structured manner (Binz et al. 2019).

"Problem ideas" represent, in our view, defined tasks for the early stages of product development. The goal of encouraging the product developer to identify and formulate new

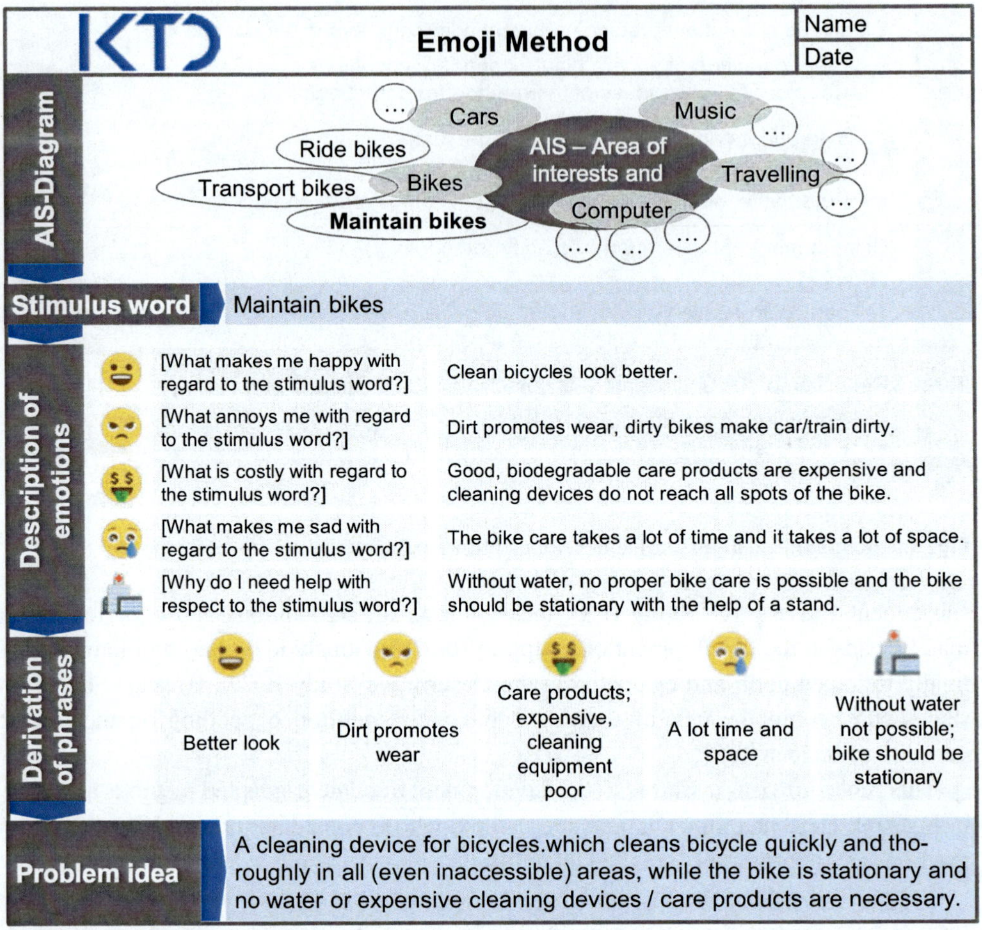

Fig. 9.3 Emoji Method adapted to Herrmann et al. (2019) and Binz et al. (2019)

problem ideas consists of several steps. First, the establishment of an Areas of Interest and Skills diagram (AIS) takes place, supported by a verb catalog similar to the main feature list of a requirements list. Then, a stimulus word is identified. This can relate to an individual person as well as to an issue in a company.

The emoji method thus specifies problems, needs and open solution fields for collected interests and skills. The next step is the description of emotions regarding the stimulus word. The five emojis used should represent what makes the user happy, sad and angry. In addition, costs and necessary equipment are analyzed. In a further analysis, the most important points of the described emotions are then summarized in short textual core modules. A final synthesis of the derived phrases leads to a clearly defined problem, the problem idea (Herrmann et al. 2019).

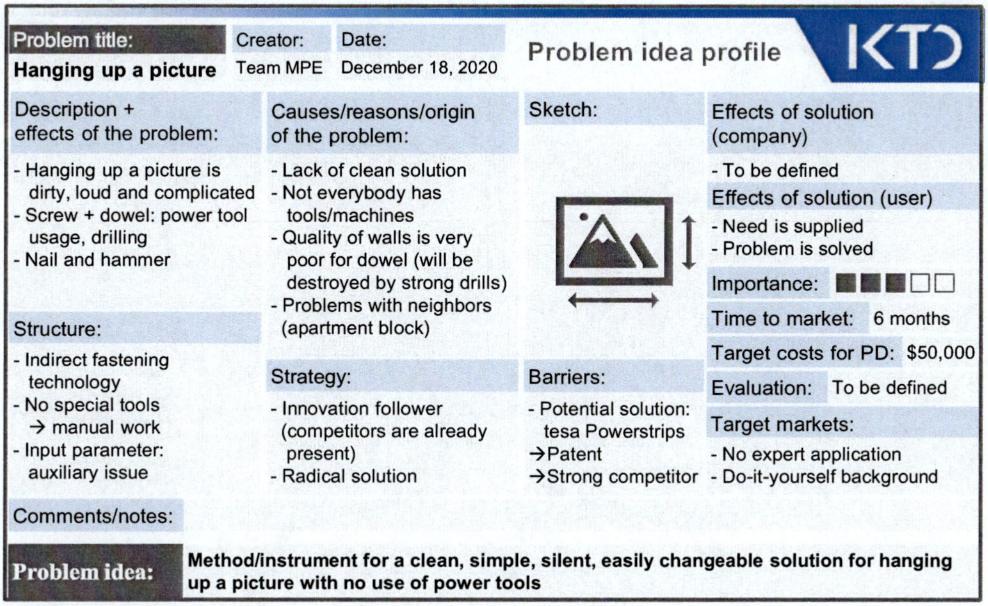

Fig. 9.4 Problem idea profile (adapted to Herrmann et al. 2017)

Based on initial applications and a discussion of the information to be stored, provided and distributed, a "problem idea profile" (see Fig. 9.4) was subsequently developed that can be used for documented problem analysis (Herrmann et al. 2017).

Summarizing, the following can be stated. There is often unsatisfactory communication regarding the expectations of the product to be developed. Information that addresses a customer's need or problem or the benefit of a new product or process is not clearly stated. As a result, new ideas and new products fail because they do not address the real problem. The challenge is therefore to ensure precise documentation and comprehensible provision and distribution of information. With the presented supports, this can be made possible and a higher "context adaptability" can be achieved.

9.4.2 Best Practice for the Selection of Methods Appropriate to the Situation

Product development and design methods have been an integral part of development processes for decades. Numerous method collections exist for the representation of a large variety of methods existing there, in which methods are classified by means of the most diverse categories and criteria. In the DFG (German Research Foundation) Collaborative Research Center SFB 1244 "Adaptive Envelopes and Structures for the Built Environment of Tomorrow" another method map has been developed. This is particularly suitable for the centrally combined representation of the largest possible number of

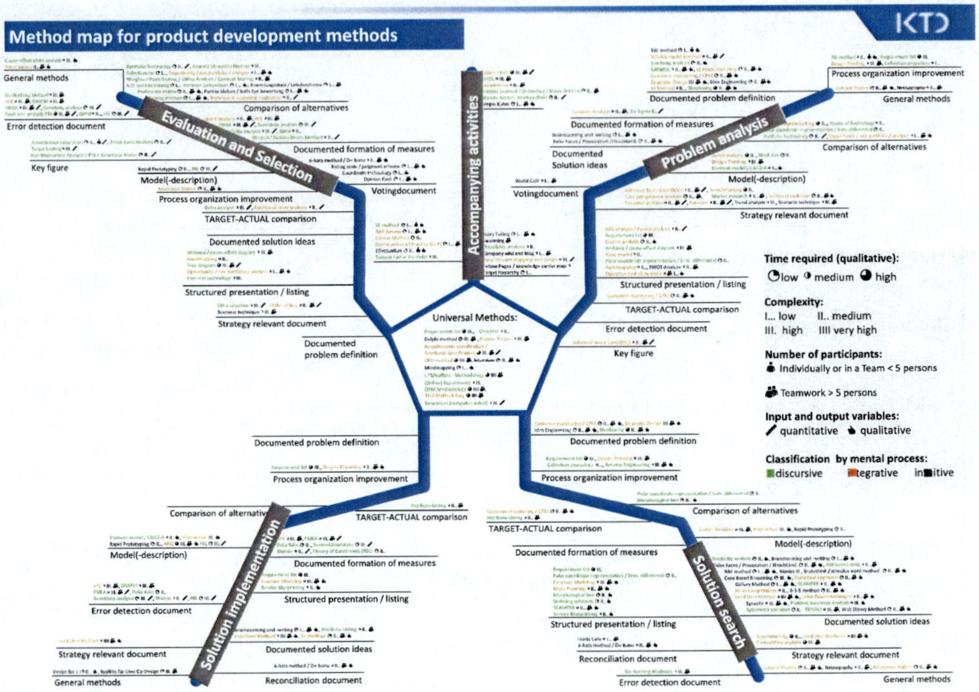

Fig. 9.5 Method map (Honold et al. 2019)

methods in method classes and at the same time supports the selection of the correct method in an interdisciplinary context. The result is shown in Fig. 9.5 (excerpts). The complete method map is included in Honold et al. (2019).

The method map contains 133 methods and is divided into five branches, each representing one of the activities problem analysis, solution search, solution implementation, evaluation and selection, and accompanying activities. The branches in turn bundle methods with comparable results. The structure chosen in this way makes it possible for the user to be guided purposefully to a suitable method with the help of the branches via the activity. The additional information provided in the form of symbols ensures that the methods can be used in the appropriate context. The right method can be selected depending on the available time, a permissible method complexity depending on the expertise of the users involved, available personnel capacities, or also desired input and output variables (quantitative or qualitative) as well as different forms of thought processes.

In addition to the provision of such selection representations, further, more context-specific method provision is also conceivable. One such best practice example is a catalog of knowledge management solutions for the product development process, which was created within a higher-level development of a product development-specific knowledge management procedure for SMEs (Laukemann et al. 2017). A particular feature is the examination of the company context—here the product development process—in which

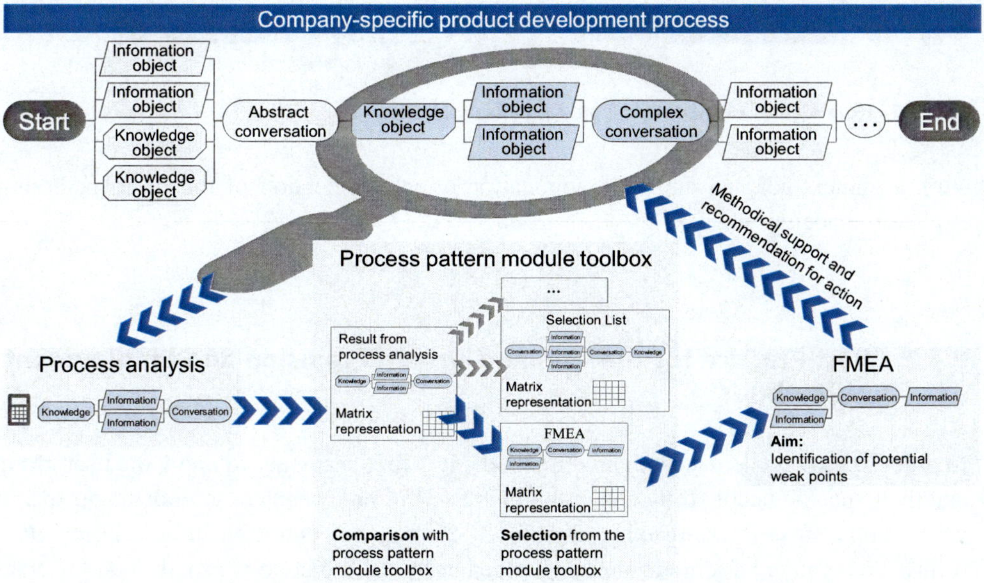

Fig. 9.6 Analysis and procedure of a process-adapted method provision in the context of knowledge management at SMEs (adapted to Laukemann et al. 2015)

the support is to take place, as well as the methods that are to be used for support. Methods can be used particularly well adapted to their context if this is very well known. This has resulted in a procedure in which the company-specific product development process is analyzed with regard to the need for support with the aid of the "Knowledge Modeling and Description Language" according to Gronau (2009). In this analysis, the methods modeled in the same way with KMDL are used, which are assigned situationally to the correct points in the development process by means of process pattern analysis. The basic conceptual sequence is shown in Fig. 9.6 and described in Laukemann et al. (2015, 2017).

A similar approach has proven to be successful in the selection of knowledge elicitation methods appropriate to the situation (cf. Roth 2020) as well as in the situational support of the feedback of test knowledge into product development (cf. Karthaus et al. 2015). In an abstracted form, the underlying idea can be represented as sketched in Fig. 9.7.

The core idea is to match certain abilities of a method with a situation (e.g., a process), i.e., to achieve the best possible fit. In this context, methods have inherent abilities, which should be assigned as best as possible to characteristics that are yet to be operationalized from a subject matter. This is an analogy to the procedure shown in Laukemann et al. (2015).

In conclusion to these examples, a clear need for the situation-oriented provision of methods can be determined. Different best practices were presented for this purpose. In addition, the method map provides a practical method for visualizing and selecting suitable support. With suitable selection criteria (required time, necessary number of participants,

Fig. 9.7 Core idea of a situation-adapted method selection

etc.), a context-adapted and thus application-oriented selection of the right methods is explicitly made possible.

9.4.3 Best Practice for the Demand-Driven Provision and Employment of Methods

In order to provide demand-oriented methods, it is first necessary to know the user groups and their specific needs. If the general aim is to determine needs or to understand specific user groups, surveys are usually suitable. These can be conducted in a wide variety of forms. To define a basic understanding, a generally valid picture is usually first generated by means of intensive literature research before this is compared with industrial practice (cf. as well for example Blessing and Chakrabarti 2009).

To carry out such a systematic literature search, the following four-stage procedure has proven successful at the IKTD:

1. Analysis and determination of all relevant synonyms of the subject under investigation.
2. Execution of a search in previously defined indexed electronic databases—use of the synonyms from (1)
3. Initial selection of the found contributions
4. Detailed analysis of the relevant contributions

The result of such a search is shown in Fig. 9.8 as an example of how radical product ideas are defined in literature. The procedure offered support, starting from a very large amount of data, to systematically arrive at a manageable information situation. This is particularly important if a fundamental issue is to be well understood—for example, if a suitable support is to be offered for a problem to be defined in more detail.

If the core is then about the provision of such support, it is necessary to know the need of the later user. How this need can be determined and how it can also look is demonstrated in Hommel et al. (2020). The survey conducted there examines user needs as well as obstacles in the application of aluminum foam sandwich. The possibilities named for providing information in a method-supported manner are conspicuous. The answers range from paper-based documents to design catalogues or design guidelines as well as an online platform. From this, a clear necessity can be derived that the provision of methods should also be tailored to the needs of the subsequent users. In particular, digital forms of information provision seem to be preferred (Hommel et al. 2020). This supports the

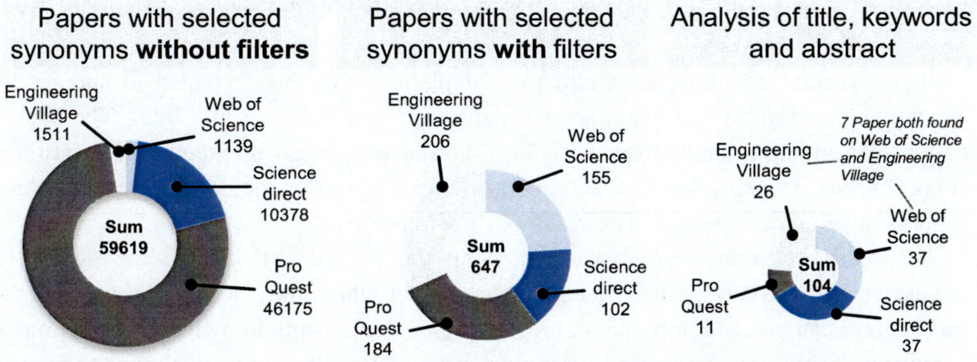

Fig. 9.8 Result of a structured determination of relevant contributions (cf. Herrmann et al. 2018)

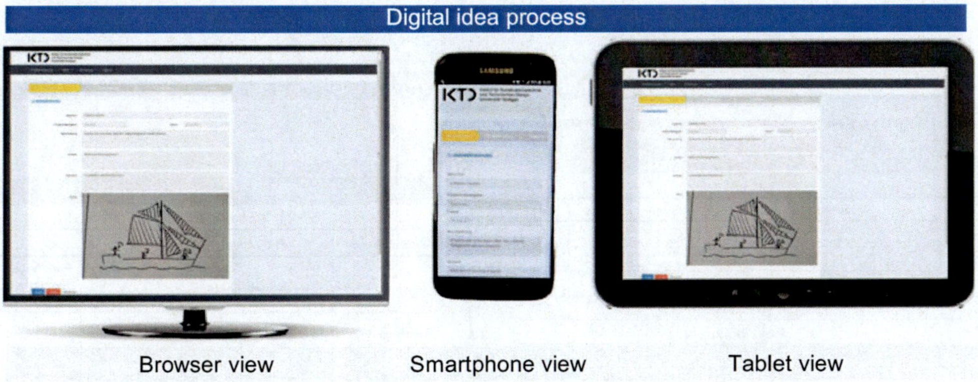

Fig. 9.9 System-independent web app for targeted support of the idea process (Binz et al. 2017)

premise formulated at the beginning of the situation analysis that support should be provided as flexibly as possible and independent of location.

If the focus is not only on the form of provision, but also on support at the appropriate point in time in the product development process, a solution can be seen in digitized methods. Two small best practice examples of successful digitization will be briefly presented.

The first example shows the digital idea process implemented at the IKTD via web app (see Fig. 9.9).

This is intended to provide targeted support for the management of ideas. For example, initial idea sketches can be digitally assessed in the evaluation process and stored and used in a targeted manner as part of systematic knowledge management. Furthermore, a digital profile for the purposeful recording of solution ideas, an evaluation logic for digital evaluation, as well as a digital report are available. The implementation of this support as a web app provides a device- and system-independent application that does not require a local installation on the user's device. Communication with the program is purely server-based via a web browser.

In the second example, an internet-based platform has been created to support the development of additively manufactured parts (see also Weiss et al. 2018). The content of the platform is grounded on the collection and evaluation of previously determined support needs. The structure of the support offered is based on the time sequence during a product development project. The result is shown in Fig. 9.10.

It should be emphasized that when using the platform, the users can decide for themselves in which steps they need support and in which they do not. This means that this support can be used in particular depending on the existing knowledge of the product developer. Therefore, the support offered adapts situationally to the level of experience.

Thus, a special focus in this chapter was on the user-oriented development of a support. In order to make this possible, it is imperative that the needs of the user as well as the

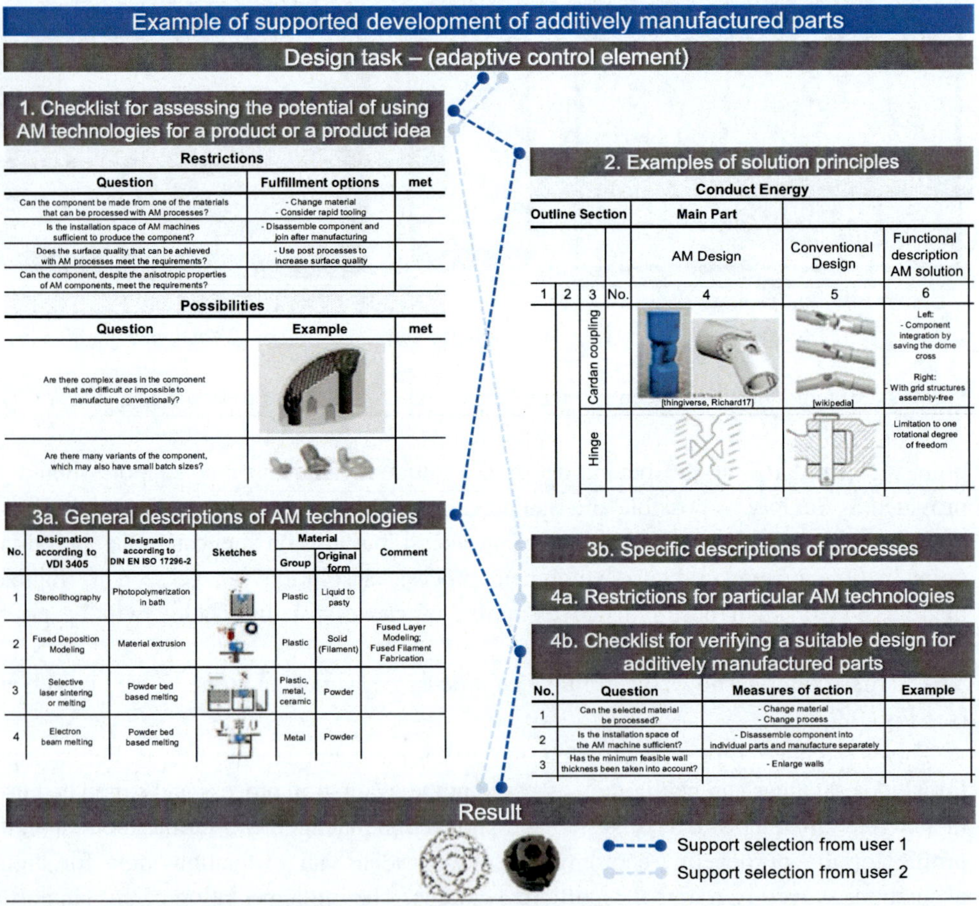

Fig. 9.10 Knowledge and demand-oriented method provision (adapted to Weiss et al. 2018)

environment itself are considered in advance. Methods that are developed can often be offered in digital form and should be able to be used flexibly, depending on the situation.

9.4.4 Best Practice for the Company-Adapted Implementation of Processes and Methods in Companies

The introduction of methods is accompanied by an ongoing debate regarding their applicability in practice (Gericke et al. 2013). In Messerle et al. (2014), the question of which problems exist in particular and how they can be solved was therefore investigated in the field of idea processes. The result is a process for the introduction of idea processes in companies, which was subsequently tested and validated in practice. The aim of this process, illustrated in Fig. 9.11, is to support the adaptation of methods and processes so that they fit to the specific context and circumstances of the respective company.

In the process, four basic steps with their respective sub-steps can be identified. First, the introduction is planned (Preparation Phase), based on the necessary recognition that existing processes or conditions require a change. This is followed, among other things, by steps to check the fit with the corporate strategy, to determine the implementation team and to define the project more precisely.

In the second phase (Diagnosis Phase) of the implementation process, it is then necessary to assess the existing situation and processes (framework conditions, problems, etc.). The result of this diagnosis phase is the development of a first, rough concept. Within the framework of this rough concept, questions are asked such as "Which methods or which tools best support the user in his individual work in everyday life?" and "Which

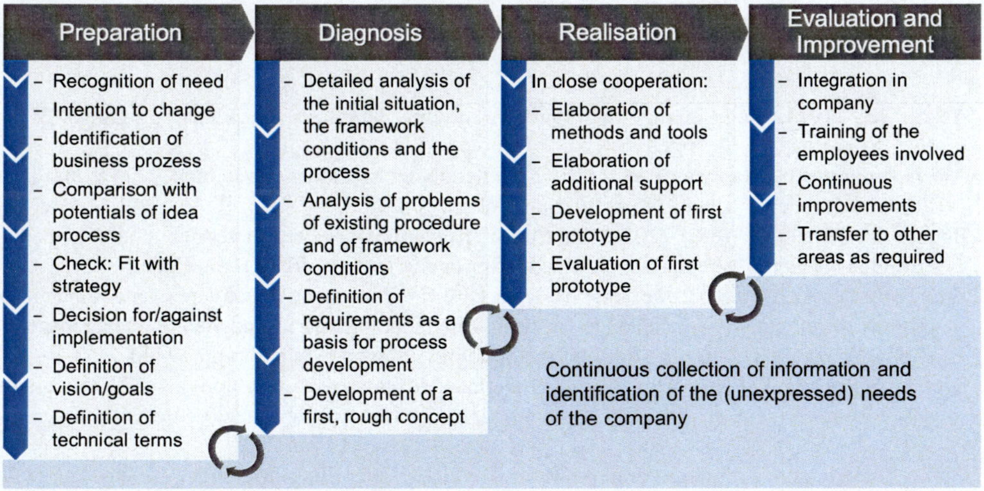

Fig. 9.11 Company-adapted implementation of processes and methods in accordance with Messerle et al. (2014)

process steps, or process sub-steps, are necessary and thus useful, or which can be omitted?"

The subsequent phase of realization concretizes the concept. In addition, a first prototype is created and initial evaluations, in this case of the idea process, take place.

In the final phase of evaluation and improvement, the new process is embedded in the corporate landscape. This is followed by employee training and regular review loops in which the new process is checked and further optimized.

Transferred to the overall context of this contribution, the procedure presented provides elementary steps for the introduction of new processes as well as new methods in companies. A very central aspect is the constant reflection of the corporate context and thus a focus on the application field as well as the users themselves.

9.5 Conclusion

The central element of this contribution are the best practice examples presented. These examples show how the necessary aspects of context-adapted methods mentioned in the chapter on clarification of terms can be achieved in an appropriate way. The goal of all these considerations is to provide methods and thus support that are accepted in the industrial environment and contribute to the success of the company in a way that adds value. Among other things, the available resources of a company must be kept in mind, company-specific processes must be known, and the future user must be kept in mind. Today's methods should be as intuitive, flexible and location-independent as possible.

References

Ackoff RL (1974) Redesigning the future. A systems approach to societal programs. Wiley, New York

Albers A, Bursac N, Marthaler F et al (2017) Szenarien der Methodenanwendung – Ein White-Paper für die Methodenforschung. Konstruktion Vol 69:72–77

Bender B, Gericke K (eds) (2021) Pahl/Beitz Konstruktionslehre. Methoden und Anwendung erfolgreicher Produktentwicklung, 9th edn. Springer Vieweg Verlag, Deutschland

Binz H, Keller A, Kratzer M, Messerle M, Roth D (2011) Increasing effectiveness and efficiency of product development – a challenge for design methodologies and knowledge management. In: Birkhofer H (ed) The future of design methodology. Springer, London, pp 79–90

Binz H, Herrmann T, Laukemann A (2017) Digitaler Ideenprozess: web-app des IKTD unterstützt das Ideenmanagement. In: WiGep-news 2/2017. http://www.wigep.de/fileadmin/download/wigep/WiGeP_News_Ausgabe2_2017_web.pdf. Accessed 14 Jan 2021

Binz H, Herrmann T, Laukemann A, Roth D (2019) Die Emoji-Methode: Neuer Ansatz zur Identifizierung und Formulierung von Problemideen. In: WiGep-News 1/2019. http://www.wigep.de/fileadmin/download/wigep/WiGeP-News_1_2019.pdf. Accessed 14 Jan 2021

Blessing LTM, Chakrabarti A (2009) DRM, a design research methodology. Springer, London

Ehrlenspiel K, Meerkamm H (2017) Integrierte Produktentwicklung. Denkabläufe, Methodeneinsatz, Zusammenarbeit. 6. vollständig überarbeitete und erweiterte Auflage. Carl Hanser Verlag, München

Gericke K, Meißner M, Paetzold K (2013) Understanding the context of product development. In: Lindemann U, Venkataraman S, Kim YS, Lee SW, Reich Y, Chakrabarti A (eds) DS75: proceedings of the 19th international conference on engineering design 2013 (ICED 2013), design for Harmonies, Seoul, August 2013. The Design Society, Glasgow

Graessler I, Hentze J (2020) The new V-model of VDI 2206 and its validation. Automatisierungstechnik 68:312–324. https://doi.org/10.1515/auto-2020-0015

Gräßler I, Hentze J, Bruckmann T (2018) V-Models for interdisciplinary systems engineering. In: Marjanović D, Storga M, Skec S, Bojcetić N, Pavković N (eds) DS92: proceedings of the DESIGN 2018. 15th international design conference, Dubrovnik, May 2018. Faculty of Mechanical Engineering and Naval Architecture, University of Zagreb, Croatia and The Design Society, Glasgow, pp 747–756

Gronau N (2009) Wissen prozessorientiert managen: Methode und Werkzeug für die Nutzung des Wettbewerbsfaktors Wissen in Unternehmen. Oldenbourg, München

Herrmann T, Binz H, Roth D (2017) Tool for creating a defined task as preparation for a target-oriented generation process. In: Maier A, Skec S, Kim H, Kokkolaras M, Oehmen J, Fadel G, Salustri F, Van der Loos M (eds) DS87: proceedings of the 21th international conference on engineering design (ICED17), Vancouver, august 2017. The Design Society, Glasgow

Herrmann T, Roth D, Binz H (2018) Derivation of criteria for radical product ideas. In: Marjanović D, Storga M, Skec S, Bojcetić N, Pavković N (eds) DS92: proceedings of the design 2018. 15th international design conference, Dubrovnik, May 2018. Faculty of Mechanical Engineering and Naval Architecture, University of Zagreb, Croatia and The Design Society, Glasgow

Herrmann T, Laukemann A, Binz H, Roth D (2019) The emoji method: a new approach for identifying and formulating problem ideas. World Acad Sci Eng Technol: Int Res Conf Proc 12 (12):1351–1366

Hommel P, Roth D, Binz H (2020) Deficits in the application of aluminum foam sandwich: an industrial perspective. In: Marjanović D, Storga M, Skec S, Bojcetić N, Pavković N (eds) DS102: proceedings of the DESIGN 2020. 16th international design conference, Dubrovnik, May 2020. Cambridge University Press, Cambridge

Honold C, Fischer S, Roth D, Binz H (2019) Method map for the selection of product development methods in an interdisciplinary context. In: Binz H, Bertsche B, Bauer W, Riedel O, Spath D, Roth D (eds) Stuttgarter Symposium für Produktentwicklung 2019. Fraunhofer Verlag, Stuttgart

Karthaus C, Binz H, Roth D (2015) Methodenbaukasten zur situativen Unterstützung der Rückführung von Erprobungswissen in die Produktentwicklung. In: Binz H, Bertsche B, Bauer B, Roth D (eds) Stuttgarter Symposium für Produktentwicklung 2015. Fraunhofer Verlag, Stuttgart

Keller A, Binz H (2009) Requirements on engineering design methodologies. In: Bergendahl N, Grimheden M, Leifer L, Skogstad P, Lindemann U (eds) DS58: proceedings of ICED 09. 17th international conference on engineering design, Palo Alto, August 2009, The Design Society, Glasgow

Laukemann A, Binz H, Roth D (2015) Konzept eines produktentwicklungsspezifischen Wissensmanagementverfahrens für kleine und mittlere Unternehmen. In: Binz H, Bertsche B, Bauer W, Roth D (eds) Stuttgarter Symposium für Produktentwicklung 2015. Fraunhofer Verlag, Stuttgart

Laukemann A, Binz H, Roth D (2017) Katalog von Wissensmanagementlösungen für den Produktentwicklungsprozess. In: Binz H, Bertsche B, Bauer W, Spath D, Roth D (eds) Stuttgarter Symposium für Produktentwicklung 2017, Stuttgart, Fraunhofer Verlag

Lindemann U (ed) (2016) Handbuch Produktentwicklung. Carl Hanser Verlag, München

Messerle M, Binz H, Roth D (2014) Implementation of idea processes in the specific context of business practice. In: Marjanović D, Storga M, Pavković N, Bojcetić N (eds) DS77: proceedings of the DESIGN 2014. 13th international design conference, Dubrovnik, May 2014. Faculty of Mechanical Engineering and Naval Architecture, University of Zagreb, Croatia and The Design Society, Glasgow

Roth D (2020) Analyse und Bewertung von Wissen in der Produktentwicklung. Dissertation, University of Stuttgart. https://doi.org/10.18419/opus-10846

VDI Guideline 2221 (1993) Systematic approach to the development and design of technical systems and products. Beuth, Berlin

VDI Guideline 2221 Part 1 (2019) Design of technical products and systems – model of product design. Beuth, Berlin

VDI Guideline 2221 Part 2 (2019) Design of technical products and systems – configuration of individual product design processes. Beuth, Berlin

Weiss F, Roth D, Binz H (2018) Content and functions of an internet-based platform for supporting development of additively manufactured parts. In: Marjanović D, Storga M, Skec S, Bojcetić N, Pavković N (eds) DS92: proceedings of the DESIGN 2018. 15th international design conference, Dubrovnik, May 2018. Faculty of Mechanical Engineering and Naval Architecture, University of Zagreb, Croatia and The Design Society, Glasgow

Check for updates

10

An Approach to Develop Designer-Centred Methods: Illustrated by an Example on How to Overcome Cognitive Bias in Product Development

Sven Matthiesen, Thomas Nelius, and Matthias Eisenmann

Abstract

Although the designer has been recognized to have a particularly important role in product development, design method development still shows a lack of putting the designer in the focus. This contribution presents an approach to develop designer-centred methods in three steps: Assessment of designer thinking, design method synthesis and design method validation. To illustrate the approach, the development of a design method to overcome cognitive bias in product development is presented using the approach. Cognitive biases are systematic deviations from rational decisions that lead to illogical interpretations of information or data. By taking cognitive bias and designer thinking into consideration, data driven design can be better supported. The method development here addresses the confirmation bias, a particularly difficult bias. With the Design-ACH, a method is presented that supports design engineers in reducing the confirmation bias, which is shown by results of multiple studies. Implications for future development of designer-centred methods include data driven method development, which can be made possible by automated quantitative measurement of designer thinking. Also, there is a need for studies in both laboratory and field to ensure designer-centred method development.

S. Matthiesen (✉) · T. Nelius · M. Eisenmann
Karlsruhe Institute of Technology (KIT), IPEK – Institute of Product Engineering, Karlsruhe, Germany
e-mail: sven.matthiesen@kit.edu

D. Krause, E. Heyden (eds.), *Design Methodology for Future Products*,
https://doi.org/10.1007/978-3-030-78368-6_10

10.1 How to Develop Designer-Centred Methods?

Despite of many years of continuous development, design method usage remains limited in design practice. One reason for this limited uptake are shortcomings in putting the designer as a human being in the focus of design methods. This view aligns with the position of Badke-Schaub et al. (2011) who argued for a designer-centred methodology more than a decade ago. Since then, there has been only little change in research approaches to foster the development of designer-centred methods in product development. However, the demand for consideration of rational and unconscious thinking in product development is coming to the front (Ehrlenspiel 2020). To develop designer-centred methods, empirical investigations are necessary, which focus on the designer during product development.

In a first step designer thinking and its influence on the process of designing need to be understood and quantified. However, **making designer thinking accessible for assessment is particularly difficult**. It lies in the nature of ways of thinking that they often remain unconscious or even subconscious. But to be designer-centred, methods need to address those ways of thinking which therefore need to be made explicit. In the view of the authors, this is hindered by several challenges in the field of empirical design research. Those challenges are laid out in the following.

The main issue as stated by Üreten et al. (2020) is the **development of a valid operationalisation**. Operationalisation determines how the aspect under investigation is made observable or measurable. A proper operationalisation is a necessity to enable assessment of designer-related aspects, which are relevant for product development.

Operationalisation also strongly relates to another challenge: the **high effort and resources required for data collection, analysis and interpretation** needed in empirical design research (Üreten et al. 2020). The more the operationalisation focusses solely on the relevant aspects and the more it enables automatisation of assessment and analysis, the lesser the effort and resources required. For example, to assess when a designer encounters a problem during development, research methods originating from the social sciences such as retrospective interviews or concurrent think aloud can be used to assess conscious designer thinking. This is very resource intensive as all utterances have to be transcribed and interpreted afterwards. Also, measures to mitigate bias in interpretation have to be taken. Therefore the designer's problems should be operationalised by objectively measurable variables. By using bio signals such as heart rate or eye-tracking metrics as variables and identifying threshold values, cognitive processes can be studied objectively (Lohmeyer and Meboldt 2016). By using algorithms, the evaluation can be accelerated through automation (Wolf et al. 2018).

To develop operationalisations, which enable quantitative measurement of aspects of designer thinking, a detailed understanding of relevant aspects to be measured is necessary. To achieve this, explorative and qualitative investigations of design in practice are needed as prerequisite. However, in current investigations concerning designer thinking, data acquisition and analysis remain on a qualitative and interpretative level. This results in a lack of comparability of results and extensive measures to ensure objectivity have to be undertaken. Objectivity could be raised by advancing to develop standardised instruments for quantitative data acquisition comparable to IQ tests. This standardisation can also be achieved via evaluation algorithms, which inherently enable an objective and reliable evaluation.

Summing up, future investigations of designer thinking should enable quantitative data acquisition and analysis, which is necessary for data driven development of design methods. This raises comparability of results and at the same time reduces bias of interpretation as well as the resources needed for research.

In order to develop design methods, Blessing and Chakrabarti (2009) suggest three steps after the research clarification within their research framework in DRM—a design research methodology: Understanding design, Developing support and Evaluation (see Chap. 9). To enable development of designer-centred methods, this approach needs to be focussed on methods supporting designer thinking. Additionally, design method development should aim at producing quantitative results concerning the method impact operationalised through proper variables, because this is needed for a comprehensive validation.

We therefore propose three steps of method development connected by designer thinking (see Fig. 10.1) aiming at a quantification of effects by fitting operationalisation: (1) Assessing ways of designer thinking, (2) Design method synthesis, (3) Design method validation. Those three steps are further described in the following. An example of how to conduct designer-centred method development is elaborated in Sect. 10.2.

10.1.1 Assessing Ways of Designer Thinking

In order to develop designer-centred methods the current situation needs to be understood by empirical investigation of designer thinking. The goal of these investigations should be the identification of best practices or problems and their causes (see Fig. 10.1). An important part in this step is to identify situations in which design engineers really have a need for support by a design method. A lack of subjective need of support is one of the main reasons why design methods are seldom applied in design practice (Eisenmann and Matthiesen 2020). Additionally, by analysing the circumstances in which difficulties occur, a deeper understanding can be gained about the underlying causes. Those causes are needed for goal oriented method development in the second step (Sect. 10.1.2) as well as for validation in the third step (Sect. 10.1.3). This first step can be divided in a qualitative and a quantitative phase.

In the qualitative phase a detailed understanding of problems and their causes in designer thinking in practice should be gained. It is advisable to use research methods from social sciences in this phase. Those range from already established methods like protocol analysis and think aloud over focus group interviews up to seldom used human subject experiments to investigate alterations in thinking and behaviour. These research methods aim at putting the human being in the focus of the investigation, thus enabling the

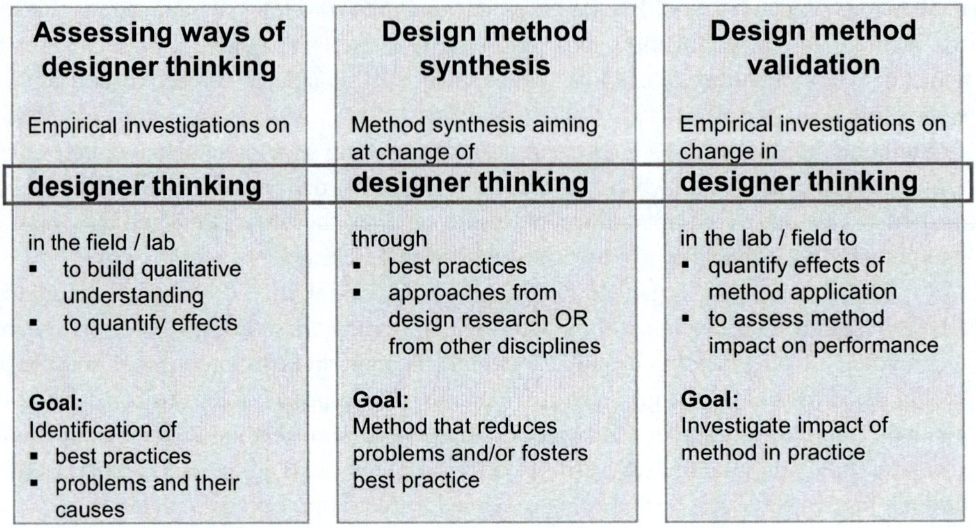

Fig. 10.1 "designer thinking" as central element connecting the three steps in development of designer-centred methods

researcher's view to become designer-centred. Those investigations should start in a real design context to increase the external validity of identified problems and best practices.

In the following quantitative phase, the identified results should be verified by quantitative assessment. This should initially be conducted in a laboratory context, because such a more focused and less influenced environment enables higher numbers of participants and therefore statistical data analysis of occurring effects. Also, a laboratory context makes it possible to evaluate objectivity and reliability of the chosen operationalisation. While the focus on a limited number of aspects is necessary to verify the occurrence of effects, study designs should aim at being as realistic as possible to include aspects relevant for practice.

Summing up, by using research methods from social sciences, ways of designer thinking can be assessed in order to identify best practices and problems as well as their underlying causes in the first step. Quantification should then be used to verify the chosen operationalisation. This enables focussing the following design method development on relevant aspects in designer thinking.

10.1.2 Designer-Centred Method Synthesis

Designer-centred *method synthesis* focuses on the question of how to overcome causes for problems in the design process by focusing on the designer to achieve a better result than without the method. As mentioned in Sect. 10.1.1, this can either be achieved by using best-practices as a starting point or by searching for ways to influence designer thinking to overcome the occurring problems.

If the first step identified best practices, those should be made as explicit as possible to make them accessible for other designers. If the first step did not yield any best practices, it should be carefully considered if there are established approaches in design research to overcome the occurring problems. For example, when dealing with problems in design decision making, existing methods and approaches of this area should be reviewed for their potential value in the current situation.

In many cases, causes of problems in designer thinking relate to more fundamental aspects of human thinking, like logical reasoning or interpretation of information. Especially in those cases, existing approaches from other professional disciplines should be considered for design method development. Like the use of research methods from social sciences, approaches from other disciplines often require careful analysis and modification for the use in a design context (Bender et al. 2002).

To ensure not only the method's influence on designer thinking but also cause a certain impact of the design process' result, possible positive as well as negative effects should be thought ahead. For example, an ideation method to support short-term designer thinking takes too long to apply for short-term memory and therefore loses its effect.

Summing up, designer-centred methods should be synthesised combining expertise in the process of product development with methods from social sciences or other disciplines and the usage of existing approaches. An example of design method synthesis using approaches from psychology and intelligence analysis is elaborated in Sect. 10.2.2.

10.1.3 Design Method Validation

The third step *design method validation* aims at investigating whether the developed design method has the desired impact. In the case of designer-centred methods, this impact has to be investigated on two levels: (1) Does the design method influence designer thinking as anticipated? (2) Does the change in designer thinking lead to a better performance?

Marxen and Albers (2012) suggest to investigate design methods through experimental research before implementing them in practice. In design method validation, it is advisable to use human subject experiments, because they enable the generation of causal relationships between design method application and its effects on designer thinking. Experiments are set up by comparing two groups of participants solving a design task which represents originally occurring problems and enables a reproducible data acquisition. The test group is using the newly developed method while the other group–the control group–works intuitively or with a benchmark method. Üreten et al. (2019) summarise different aspects to consider when investigating design methods in experiments in the form of a concept map.

To investigate the influence of the developed design method on designer thinking, operationalisations created in the first step (see Sect. 10.1.1) for quantitative measurement can be used. Like this, the change of designer thinking caused by method application can be quantified. The assessment of performance is a general challenge in design research, because it is concerned with the actual impact in practice. The assessment in practice is hampered by a multitude of disturbances. Design processes in companies are unique which strongly reduces comparability: time required and costs vary significantly between similar design projects, caused by a multitude of influences by different stakeholders. The investigation in a controlled laboratory environment is therefore to be seen as a necessary step to enable a later transfer on the context in practice. A challenge is to acquire design engineers for such human subject studies.

One aspect to raise performance is the reduction of occurring problems, which were detected in the first step. It needs to be verified if the change in designer thinking indeed reduces those problems. For a comprehensive design method validation, the impact of this reduction of problems on performance needs to be investigated as well. Performance can relate to multiple different aspects ranging from time required for a task over quality of generated solutions up to the reduction of costs.

Summing up, to develop designer-centred methods three steps are needed which are connected by designer thinking as a central element (see Fig. 10.1). In the first step, ways of designer thinking are assessed in order to understand and quantify occurring problems and their causes. The second step then aims at developing a design method to influence designer thinking to overcome problems by using best practices or approaches from design research or other disciplines. In the final step, the design method is validated by quantifying its impact on designer thinking and consecutively on design performance. For those steps, researchers need competences in engineering design as well as in research methods of the social sciences to put the designer in the centre of design method development.

10.2 Method Development to Overcome Cognitive Bias in Product Development

In the following, the three steps of **designer-centred** method development are demonstrated by means of an example.

Designing can be understood as an iterative problem-solving process (Albers and Braun 2011). To solve the design problem, design engineers have to acquire and interpret different data and information (see Chap. 11). Especially under high uncertainty, challenges arise in the interpretation of results in product development (Pottebaum and Gräßler 2020). Misinterpretations of information can lead to lengthy and expensive iterations in the design process. In psychology, systematic misinterpretations in human thinking are called cognitive bias. Although misinterpretations in design have a severe impact, cognitive bias in design have hardly been considered so far. Studies here mostly investigate the influence of cognitive bias on the synthesis of new ideas (cf. design fixation (Neroni et al. 2017)) or cognitive heuristics (cf. (Bursac et al. 2017; Bursac et al. 2018; Tanaiutchawoot et al. 2019)). In the following, development of a method is described that aims to support designer thinking **to overcome cognitive bias** during the failure analysis in engineering design. The design method is developed, following the three steps of the approach presented in 10.1.

10.2.1 Assessing Ways of Designer Thinking: Identifying the Influence of Confirmation Bias on Designers' Understanding of Problems

The first step of design method development started with a qualitative field study (see Nelius et al. 2021a) to capture challenges and their causes in a realistic setting. A problem-solving workshop addressing an actual problem occurring in a company was used as a research environment. The aim of the workshop was to identify the cause of the failure of a construction machine and to develop technical solutions to resolve the failure. Through observation of and reflection with the participants, several challenges were identified in the workshop. For the participants, the main challenge was to assess whether an identified cause of the failure was the actual one. It could be observed during the workshop that mostly, information was explored that explained or supported the suspected cause of the problem. Information that contradicted the suspected cause of the problem was rarely searched for actively. Therefore, false failure causes were pursued several times over long periods of time until disconfirming information was found unintentionally. Because of this challenge, technical solutions were developed several times that did not solve the real problem (Nelius et al. 2021a).

The pursuit of false causes of problems could be traced back to *search for confirmatory information* as the root cause. This mind set is known as confirmation bias–one of the before mentioned cognitive bias. Confirmation bias describes the tendency to seek and interpret information in a way that confirms one's own views (Nickerson 1998) and is to be seen as a particularly serious cognitive bias.

Whereas the influence of confirmation bias has already been studied in many disciplines (e.g. psychology, law, medicine, informatics), its influence in engineering design has not yet been investigated. It was therefore necessary to investigate the influence of confirmation bias on the search for and interpretation of information in the failure analysis of designers in a laboratory study. In this way, occurring problems, their causes and best practices of design engineers could be made accessible to investigation to support method synthesis.

Laboratory Study on the Confirmation Bias: Data Collection and Analysis[1]

In order to replicate the challenges of the field study, a laboratory study was set up that was as close to reality as possible. The goal of the laboratory study was to quantify the influence of confirmation bias on the perception and interpretation of information by design engineers. The task depicts a real failure from the design department of a power tool

---

[1]The results of the laboratory study were published in Nelius et al. (2020).

manufacturer. During the development of a power tool, a premature component failure occurred in the prototype phase, which led to the failure of the entire device. The study participants had access to the usual information sources of a responsible developer: the entire power tool, the worn parts, a 3D model of the affected assembly and the technical drawing of the assembly. The participants had the task of analysing the cause of the failure and sketching a suitable technical solution. For the evaluation of the study, 12 students and 8 designers (with more than 8 years of professional experience) were considered.

Confirmation bias was expected to have an impact on both the interpretation of information and the search for information. These aspects of confirmation bias were operationalised as follows:

- To capture participant misinterpretation, concurrent think aloud was used, in which the participants expressed their thoughts aloud during the task. The participants' assumptions on the cause of the failure were recorded over the course of the task and what information they used for their analysis. The information was assessed as to whether it confirmed or disconfirmed the failure cause from the participant's subjective perspective. In addition, it was assessed whether the information also confirmed, disconfirmed, or was unrelated to the failure cause from an objective perspective. The coding was reviewed by a second person in order to obtain objective results. It is considered a sign of confirmation bias if misinterpretations occur more frequently in the confirming direction (neutral and disconfirming information is interpreted as confirming) than in the disconfirming direction (neutral and confirming information is interpreted as disconfirming).
- The perception of information during task processing was recorded via eye tracking. Here, we examined how long participants looked at confirming, neutral, and disconfirming information concerning the failure cause being tracked. By combining this with participant statements, it was also possible to capture whether misinterpretations were related to low visual attention.

Laboratory Study: Results and Discussion[2]

The results of the study (see Fig. 10.2) show that the participants use confirmatory evidence much more frequently to check their assumptions than disconfirming evidence. The occurring misinterpretations take place almost exclusively in the confirmatory direction. Almost one third of the evidence the participants used as confirming evidence for argumentation is misinterpreted compared to the objective view of the evaluators. Participants wrongly interpreted neutral information (i.e. projection error) as well as disconfirming evidence (i.e. interpretation error). The evidence participants used as disconfirming is almost completely interpreted correctly. Due to the dominance of the subjectively

[2]The results of the laboratory study were published in Nelius et al. (2020).

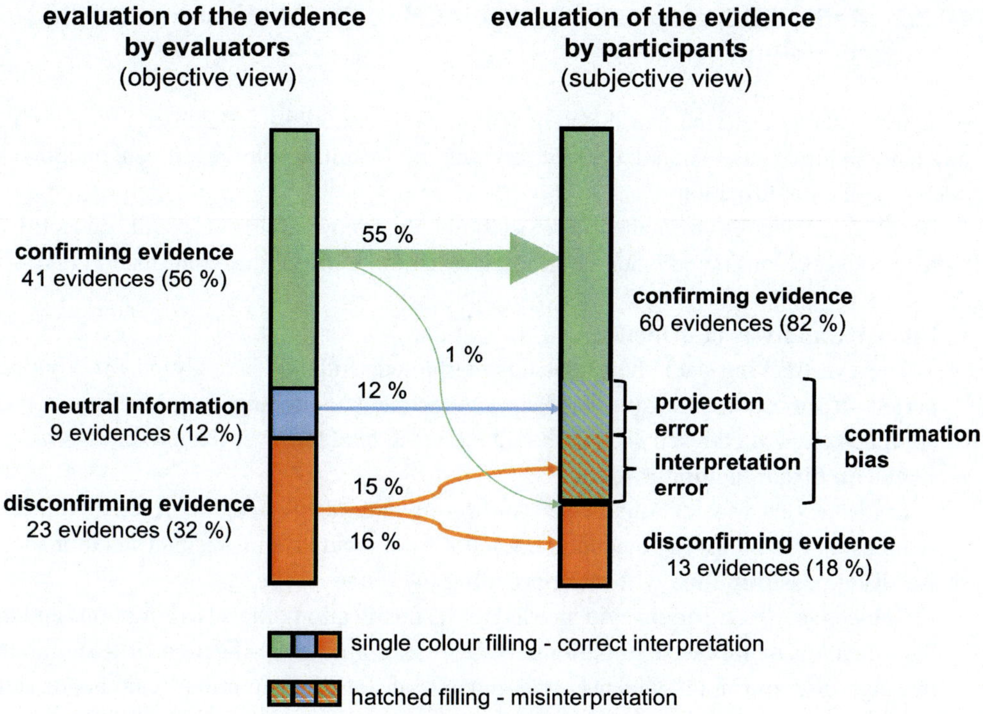

Fig. 10.2 Quantification of the confirmation bias within the participants' reasoning: misinterpretations like projection and interpretation error (hatched areas) occur systematically more often in the confirmatory direction (Nelius et al. 2020)

confirming evidence, most participants keep their assumed cause of the problem, even if it is wrong.

The eye tracking data show that the visual attention of the participants is significantly longer on objectively confirming evidence than on objectively disconfirming evidence. Misinterpretation of information is therefore also associated with low visual attention.

In many preceding studies in the state of the art, the higher frequency of statements of confirming evidence are seen as an indication of confirmation bias. However, since the difficulty of discovery and amount of evidence is unknown, misinterpretation is a more appropriate operationalisation of confirmation bias. Through this operationalisation and the use of eye tracking, it was possible to quantify the influence of confirmation bias on both reasoning and visual attention during failure solving. It could be shown, that the confirmation bias often leads to a wrong understanding of the problem. This incorrect understanding of the problem would have led to the development of unsuitable solutions and lengthy and expensive iterations in industry. The intensive use of disconfirming evidence can be understood as best practice, which can be used for the synthesis of methods. Information identified as disconfirming was used correctly more often. In addition, disconfirming evidence more often led to the rejection of false assumptions.

10.2.2 Method Development: Design-ACH to Avoid Misunderstanding of Design Problems[3]

Existing methods describe that different failure causes should be identified and the most probable failure cause should be selected, but no specifications are given on how to overcome the confirmation bias.

Based on the findings from the laboratory study, the following aspects could be identified, which should be considered when developing a method to overcome confirmation bias:

- **Intensive analysis of evidence**
  The eye tracking data show that misinterpreted evidence is analysed for a shorter period of time. An intensive analysis and detailed modelling (e.g. with the C&C^2 approach (e.g. Matthiesen 2021)) should therefore lead to fewer misinterpretations.
- **Focus on disconfirming evidence**
  Evidence that was identified as disconfirming had mostly been interpreted correctly. A focus on disconfirming evidence can reduce the incidence of misinterpretation.
- **Falsifying assumptions with disconfirming evidence**
  Subjective disconfirming evidence led to the falsification of false assumptions and the identification of further assumptions. By focusing on the falsification of assumptions through disconfirming evidence, the pursuit of false assumptions can be directly counteracted.

Typical engineering design methods for failure analysis do not include the implications arising from confirmation bias. An approach from another professional discipline could be identified to address the confirmation bias. The Analysis of Competing Hypotheses (ACH) is a method developed by Heuer (1999) for Intelligence Analysis. Its goal is the objective evaluation of multiple hypotheses for observed data. The ACH method was developed taking into account insights from cognitive psychology, decision theory and philosophy of science. The aim is to overcome or at least minimise the analyst's weaknesses and thinking errors. (Heuer 1999) The goal of the ACH method covers the identified needs for methodical support for designers in failure analysis in large parts. Since, in contrast to engineering design, no experimental investigations concerning the evaluation of hypotheses are possible in intelligence analysis, the ACH does not provide any support in this regard. Therefore, the ACH was further developed to the Design-ACH for the application in engineering design. For this purpose, the former eight steps of the ACH method were simplified to three steps and a step for defining efficient hypothesis testing was added.

The approach of the Design-ACH includes four steps (see Fig. 10.3, left). In the first step, several hypotheses and circumstantial evidence are identified. In this process,

[3]The method development was published Nelius et al. (2021b).

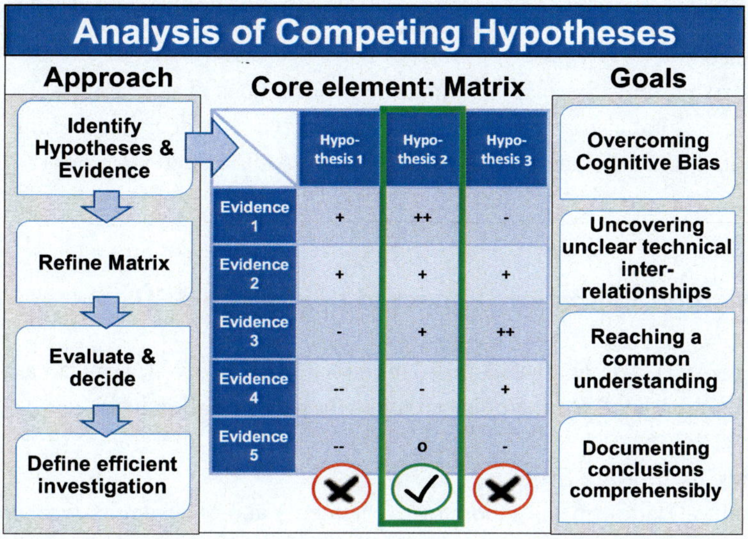

Fig. 10.3 Developed Design-ACH for root cause identification in engineering failure solving (Nelius et al. 2021b)

hypotheses on the cause of the failure are generated, if possible, in an interdisciplinary team. For all collected hypotheses evidence is collected. The hypotheses and evidence are compared in a matrix (see Fig. 10.3, center). In the cells, it is recorded whether the evidence confirms or disconfirms the hypotheses, or whether no statement can be made. In a second step, this matrix is refined. The evaluation of the hypotheses is done row by row, that means one piece of evidence after the other. This fosters the *intensive analysis of evidence* as each piece of evidence is analysed in the light of all hypotheses. By constantly switching between the hypotheses, the commitment to one hypothesis and thus the confirmation bias should be reduced.

In the second step, the matrix is refined. Here, findings from step 1 are used to combine similar hypotheses, establish new hypotheses and evidence. Evidence that does not allow prioritisation of the hypotheses (e.g. because this evidence confirms all hypotheses) is removed from the matrix.

In the third step evaluation and decision takes place. Column by column, the probability of each hypothesis is evaluated. Here, the focus is on *disconfirming evidence* in order to *falsify hypotheses* that do not apply.

If none of the hypotheses can be selected as most probable based on the available information, an efficient hypothesis test is to be defined in step 4. In this step, investigations are defined with which remaining hypotheses after step 3 can be falsified. For this purpose, the possible investigation results are preconceived and included in the matrix as circumstantial evidence. By evaluation of the possible investigation results in relation to the hypotheses, the significance of the investigations can be estimated. After promising investigations have been carried out, the matrix is updated and the most probable hypothesis is selected.

The matrix presents the analysis results and conclusions in a clear and standardised way. It is therefore suitable for reviewing the conclusions, as a decision-making template for follow-up investigations and documentation of the failure analysis.

10.2.3 Method Validation: Impact of the Design-ACH

The Design-ACH was evaluated in a laboratory study as well as in a case study in an industrial environment. Both studies are described in the following.

Laboratory Evaluation Study[4]

For method validation in the laboratory, 7 students and 5 designers were trained with a simplified version of the Design-ACH. After the theoretical part of the training, the participants applied the method in a practical exercise under the guidance of a moderator. The moderator answered questions concerning the method and ensured the correct application of the method. The previously presented task (Sect. 10.2.1) was used for data collection, which the participants worked on individually.

The impact of the Design-ACH was operationalised through the reduction of the confirmation bias. By applying the Design-ACH, participants generated more hypotheses and used more evidence. The students benefited particularly by using twice as much confirming evidence and three times as much disconfirming evidence with the Design-ACH. The proportion of misinterpreted evidence was reduced by 24% across all participants. Without the method, 27% of all evidence was associated with confirmation bias. With the method, this value was reduced to 17%. In order to record the acceptance of the method, the participants were questioned by survey to rate the benefit of the method on a scale from 1 (low) to 7 (high). Although the students benefited significantly more from the application of the method, they did not rate the benefit of the method as high (4.9/7) as the designers (5.8/7).

The Design-ACH resulted in confirmation bias occurring less frequently. False assumptions were rejected more often and the proportion of misinterpreted circumstantial evidence decreased. However, difficulties were still observed in the application of the method. The proportion of disconfirming circumstantial evidence increased only slightly. In addition, it was observed that despite subjectively disconfirming circumstantial evi-

[4]The results of the laboratory evaluation study were published in Nelius and Matthiesen (2019).

dence, some assumptions were not discarded. Both difficulties should be reduced by moderation.

Case Study[5]

In the case study, the Design-ACH was used in a problem-solving workshop. The aim of the case study was to qualitatively evaluate the applicability and usefulness of the method in an industrial context. After training the workshop participants, the Design-ACH was applied to a real failure occuring in the participants' company: In the workshop, the cause of a failure of a production machine was to be identified. The 13 participants applied the Design-ACH under moderation.

Initially, about half of the participants were convinced of one cause of the failure. Alternative causes of the failure were hardly considered. At the beginning of the Design-ACH application, existing information was collected. It became clear that not all information was known beforehand, even by the employees involved in the failure solving. The existing assumptions were transformed into four testable hypotheses. Collected information was transformed into meaningful evidence that allowed a statement on the probability of the hypotheses. The collection and discussion of existing information was described by the participants as an important step, as it made it possible to achieve a uniform understanding of the problem at hand. In addition, the amount of information was compressed to a manageable level by narrowing down the information through an evaluation in relation to the hypotheses.

The intensive discussion of the hypotheses while applying the Design-ACH led to a significant impact by reducing the original fixation on individual causes of the failure. The application of the Design-ACH also influenced the performance: Through the structured evaluation within the framework of the Design-ACH, two of four hypotheses on the cause of the failure could be excluded through the identification of clear disconfirming evidence. To further narrow down the cause of the failure, precise follow-up investigations could be defined through the use of the Design-ACH. The use of the Design-ACH was evaluated very positively in a survey and a reflection of the participants. The greatest benefit was found in the moderated application of the method.

To sum up, by developing the Design-ACH using approaches from psychology and intelligence analysis, an impact on both designer thinking as well as performance could be achieved. Designer thinking could be guided to a thorough evidence analysis and to focus on disconfirming evidence in order to falsify hypotheses. This impact on the confirmation bias could be quantified in a laboratory study. A change in performance could then be assessed in a case study, where positive effects of the Design-ACH on development could be identified.

Summing up, the three steps *assessing ways of designer thinking*, *design method synthesis* and *design method validation* to develop designer-centred methods presented in

[5]The results of the case study were published in Nelius et al. (2021b).

Sect. 10.1 could be successfully applied to develop a design method in order to overcome confirmation bias in product development. Ways of designer thinking were assessed in a field study and the occurring problems attributed to be caused by the confirmation bias. This was then verified by operationalisation of the bias and quantification of its effects in a laboratory study. By using best-practice approaches identified in the initial field study and adapting the ACH method originating from another discipline, the Design-ACH could be developed. The design method was then validated in a laboratory study, which resulted in a quantified reduction of the confirmation bias and an increased success in failure solving. This success could be qualitatively reproduced in a case study concerned with solving a company's real problems in a workshop.

The presented research in Sect. 10.2 illustrates that the confirmation bias as an example of cognitive biases has a considerable effect on how data is interpreted and used for decisions in engineering design. Because product development is becoming more and more data driven, cognitive biases are of high relevance for design engineers as they negatively influence data interpretation. Future design methods should therefore enable design engineers to objectively interpret the growing amount of data in product development.

10.3 Implications for Future Method Development

In Sect. 10.1 we have described an approach on how designer-centered methods should be developed. Currently, the following points are often given too little attention:

- The designer as a human being with his/her abilities and limitations is not considered enough in the development of design methods. **Designer thinking** as means of putting the designer in the focus is seldom used.
- Many investigations in design research are **limited to qualitative statements**, where the requirements of objectivity, reliability and validity are not fulfilled.
- Design methods are often evaluated either only in case studies, without the possibility to replicate results, or only in laboratory studies, without providing statements about applicability in practice. To combine the advantages of both, relevant aspects of design practice need to be included in the laboratory. One particularly relevant aspect relates to the development process. In real product development, design engineers are able to test the system under development for its functionality. This leads to **iterations**, which **are currently seldom represented in laboratory studies**.

In Sect. 10.2 we have shown the development of a designer-centred method, which takes into account the previously mentioned points. By means of a suitable operationalisation, it was possible to quantify the occurrence of the confirmation bias and its effects. Through the use of eye-tracking, a further quantification through measurement was also possible. Investigations in both practice and the laboratory enabled both

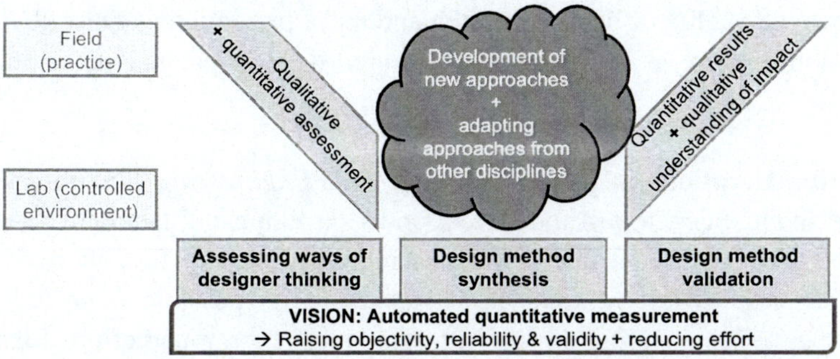

Fig. 10.4 V-shaped process of designer-centred method development between field and lab accompanied by the vision of automated quantitative measurement to support development

requirements for replicability and significance for practice to be taken into account. The following describes derived implications for future research on design methods (additionally summarised in Fig. 10.4).

Understanding Designer Thinking Up to now, the capabilities and limitations of human thinking have rarely been taken into account in the development of design methods. However, this is a necessity to develop suitable methods. For this purpose, insights from the social sciences and especially psychology must be increasingly taken into account and closer cooperation with these disciplines must be established. Consequently, research methods should therefore also be used which make understanding and quantifying aspects of designer thinking possible. Since it is often unknown which influences affect human thinking and action, it is mandatory to investigate design methods both in an environment that is as uninfluenced as possible in the field and in a replicable way under unchanging framework conditions in the laboratory.

Enabling Automated Measurement Current studies in design research often use qualitative methods, which rely on the interpretation by a coder. For the most objective results possible, the coding must be done by additional coders. Current quantitative methods provide additional insights. However, as shown in Sect. 10.2, the data must be combined with qualitative data to allow an interpretation. This also requires a high effort, which often limits the number of participants considered. The development of research methods that enable automated measurement could increase the objectivity of study results. Due to the reduced evaluation effort, more participants could be examined with the same resources.

To enable data driven design method development, automated measurement methods are necessary. For this, indicators must be identified that can be captured via existing data from design (project plans, CAD data) or bio signals (heartbeat, eye movement, brain waves, muscular tension). Indicators must be developed through a combination of established qualitative methods and automated measurement methods since meaningful relationships can only be identified through a prior qualitative understanding. This process will initially involve a high effort. In the long term, however, such automated approaches will improve the quality of design research and make previously unfeasible investigations possible. With the use of automatically measured data, investigations in practice should also become possible.

Representing Iterations in the Laboratory with *Engineering Simulators* In order to enable design engineers to test the system under development, laboratory studies need to provide the possibility of actual manufacturing and testing. Through rapid prototyping technologies such as laser cutters or 3D-printers, it is possible to include this highly relevant aspect of engineering design without requiring too much effort. Like this, development processes from ideation over detailed design up to commissioning of the product can be simulated in a short timespan (Matthiesen et al. 2016). Study designs that include iterations by functional testability of products in a laboratory context are called *Engineering Simulators*. By including iterations in laboratory studies, it is possible to simulate the most relevant aspects of practice.

The use of *Engineering Simulators* bears additional advantages resulting from the integration of iterations: By being able to actually test the designed system, study participants can reflect on the causes of functionality or lack of functionality of their design. On the one hand, this motivates participants to design a functioning product. On the other hand, this represents designer behaviour and thinking during design processes in practice more realistically. Additionally, researchers can use the manufactured systems for performance evaluation. The functionality of the technical system has no longer to be evaluated by experts but can be measured by indicators of functionality predefined in the design task. *Engineering Simulators* can range from short tasks on small systems enabling multiple iterations in several hours as described in Matthiesen et al. (2016) up to development processes spanning several days to develop more intricate systems (Omidvarkarjan et al. 2020).

Future study designs should aim at integration of relevant aspects such as iterations in order to bring practice to the reproducible laboratory context. *Engineering Simulators* are a fitting way to include those aspects.

The presented approach bears multiple potentials for future research. It focusses on the identification of problem causes and best practices in designer thinking. Like this, design method development becomes more targeted on the designer. When properly operationalised, designer thinking also fosters assessment of method impact. By using the presented approach, operationalisation aiming at quantifying effects is supported. This fosters comparability of results in design method validation. Consequently, methods which are validated in this way are more likely to be taken up in industry on a wider scale.

References

Albers A, Braun A (2011) A generalised framework to compass and to support complex product engineering processes. Int J Prod Dev (IJPD) 15(1/2/3):6–25. https://doi.org/10.1504/IJPD.2011.043659

Badke-Schaub P, Daalhuizen J, Roozenburg N (2011) Towards a designer-centred methodology: descriptive considerations and prescriptive reflections. In: Birkhofer H (ed) The future of design methodology. Springer, London

Bender B, Reinicke T, Wünsche T, Blessing LTM (2002) Application of methods from social sciences in design research. In: DS 30: proceedings of DESIGN 2002, the 7th international design conference, Dubrovnik, pp 7–16

Blessing LTM, Chakrabarti A (2009) DRM, a design research methodology. Springer, London

Bursac N, Rapp S, Albers A, Breitschuh J, Tanaiutchawoot N (2017) Entscheidungsheuristiken in der PGE – Produktgenerationsentwicklung. In: Krause D, Paetzold K, Wartzack S (eds) DFX 2017: proceedings of the 28th Symposium design for X, 4-5 October 2017, Bamberg, Germany: Beiträge zum 28. DfX-Symposium, Oktober 2017. Tutech Verlag, Hamburg

Bursac N, Tanaiutchawoot N, Rapp S, Albers A, Breitschuh J, Eckert C (2018) Decision heuristics in PGE – product generation engineering. In: Horváth I, Suárez Rivero JP, Hernández Castellano PM (eds) Proceedings of TMCE 2018, 7–11 may, 2018, Las Palmas de gran Canaria, gran Canaria, Spain. TU Delft, Delft, pp 385–394

Ehrlenspiel K (2020) Das Wechselspiel zwischen Unbewusstem und Rationalität in der Produktentwicklung. In: Corves B, Gericke K, Grote K-H, Lohrengel A, Löwer M, Nagarajah A, Rieg F, Scharr G, Stelzer R (eds) 18. Gemeinsames Kolloquium Konstruktionstechnik 2020: Nachhaltige Produktentwicklung: KT 2020. DuEPublico, University of Duisburg-Essen, Germany, Essen, pp 1–26

Eisenmann M, Matthiesen S (2020) Identifying reasons for a lack of method application in engineering design practice – an interview study. In: DESIGN 2020, vol 1. Design society, pp 2495–2504

Heuer RJ (1999) Psychology of intelligence analysis. Center for the Study of Intelligence, Washington, DC

Lohmeyer Q, Meboldt M (2016) The integration of quantitative biometric measures and experimental design research. In: Cash P, Stanković T, Štorga M (eds) Experimental design research. Springer International Publishing, Cham, pp 97–112

Marxen L, Albers A (2012) Supporting validation in the development of design methods. In: DESIGN 2012. Design society

Matthiesen S (2021) Gestaltung – Prozess und Methoden. In: Bender B, Gericke K (eds) Pahl/Beitz Konstruktionslehre, vol 22, 9th edn. Springer, Berlin, pp 397–465

Matthiesen S, Nelius T, Schmidt S, Cersowsky S, Schneider T (2016) Entwicklungssimulator – Erleben von Problemsituationen zur Steigerung der Methodenakzeptanz. In: Kammasch G (ed) Anwendungsorientierung und Wissenschaftsorientierung in der Ingenieurbildung Wege zu technischer Bildung: Referate der 10. Ingenieurpädagogischen Regionaltagung 2015 an der Fontys University of Applied Sciences 5.-7. November 2015, Berlin

Nelius T, Matthiesen S (2019) Experimental evaluation of a Debiasing method for analysis in engineering design. In: Wartzack S, Schleich B (eds) DS 94: proceedings of the design society: 22nd international conference on engineering design (ICED19). Cambridge University Press, Cambridge, pp 489–498

Nelius T, Doellken M, Zimmerer C, Matthiesen S (2020) The impact of confirmation bias on reasoning and visual attention during analysis in engineering design: an eye tracking study. Des Stud 71:100963. https://doi.org/10.1016/j.destud.2020.100963

Nelius T, Eisenmann M, Grauberger P, Matthiesen S (2021a) Tatort Technik (Teil 1) – Herausforderungen bei der Problemanalyse in der Konstruktion. Konstruktion 73(3):70–74. https://doi.org/10.37544/0720-5953-2021-03-70

Nelius T, Eisenmann M, Grauberger P, Matthiesen S (2021b) Tatort Technik (Teil 2) – Unterstützung bei der Problemanalyse in der Konstruktion mit der Design-ACH Methode. Konstruktion 73 (4):70–74. https://doi.org/10.37544/0720-5953-2021-04-70

Neroni MA, Vasconcelos LA, Crilly N (2017) Computer-based "mental set" tasks: an alternative approach to studying design fixation. J Mech Des 139:071102-1-071102-10. https://doi.org/10.1115/1.4036562

Nickerson, RS (1998) Confirmation bias: a ubiquitous phenomenon in many guises. Rev Gen Psychol 2(2):175–220. https://doi.org/10.1037/1089-2680.2.2.175

Omidvarkarjan D, Conrad J, Herbst C, Klahn C, Meboldt M (2020) Bender – an educational game for teaching agile hardware development. Procedia Manufacturing 45:313–318. https://doi.org/10.1016/j.promfg.2020.04.023

Pottebaum J, Gräßler I (2020) Informationsqualität in der Produktentwicklung: Modellbasiertes systems engineering mit expliziter Berücksichtigung von Unsicherheit/information quality in product engineering: model-based systems engineering in explicit consideration of uncertainty. Konstruktion 72(11–12):76–83. https://doi.org/10.37544/0720-5953-2020-11-12-76

Tanaiutchawoot N, Bursac N, Rapp S, Albers A (2019) Investigating the influence of the decoy effect in pairwise comparison in terms of idea selection in the product development process. Proc Int Conf Eng Des 1(1):1195–1204. https://doi.org/10.1017/dsi.2019.125

Üreten S, Eisenmann M, Nelius T, Cao S, Matthiesen S, Krause D (2019) A concept map for design method experiments in product development – a guideline for method developers. In: DFX2019. Design society

Üreten S, Eisenmann M, Nelius T, Garrelts E, Krause D, Matthiesen S (2020) Current challenges and solution approaches in empirical engineering design research – a workshop for empirical research. In: DESIGN 2020, vol 1. Cambridge University Press, Cambridge, pp 61–70

Wolf J, Hess S, Meboldt M, Lohmeyer Q, Bachmann D (2018) Automating areas of interest analysis in mobile eye tracking experiments based on machine learning. J Eye Mov Res 11:1–11. https://doi.org/10.16910/jemr.11.6.6

1 Data and Information Flow Design in Product Development

Kristin Paetzold

Abstract

The value of product development lies in the generation of data and information to develop technical systems. The development knowledge of companies is stored in processes and methods. In order to be able to use this knowledge, a target-oriented availability of data and information is required. These data and information flows are anchored in the development organization and in the IT structures. This article presents methods for identifying the necessary requirements, analyzing and designing the distribution and use of data and information. In addition to the IT tools, however, the developer in particular is also an important carrier and transmitter of data and information, and must be included in such considerations. Agile development methods address this aspect of supporting communication, co-op-ration and collaboration between developers in order to use the developers' tacit knowledge on the one hand and to force learning in the development organization on the on the other, thus contributing to knowledge retention. An efficient use of data and information in the development requires that developers use also the available methods. Method acceptance is not given per se. This problem is addressed in the article, and approaches are shown to explain and support data and information flows from both a tool-oriented and a human-oriented perspective.

K. Paetzold (✉)
UniBw Munich, ITPE – Institute for Technical Product Development, Neubiberg, Germany
e-mail: kristin.paetzold@unibw.de

D. Krause, E. Heyden (eds.), *Design Methodology for Future Products*,
https://doi.org/10.1007/978-3-030-78368-6_11

11.1 Introduction

The added value in product development lies in generating data and information in order to be able to transform a product idea into a functional and high-quality product. Solution approaches from various disciplines are available for the implementation of functionalities in technical systems. These solution approaches must be concretised and synergetically integrated in the sense of implementing a defined system behaviour. The use of expertise from different domains is not only accompanied by a high degree of division of labour in order to comply with development times. Consequently, there are also interfaces for the transfer and use of data and information, which must also be taken into account in the process design.

A number of different stakeholders have an interest in the development activity. In addition to the customer or the user, aspects such as legal requirements or social interests, which manifest themselves in values, standards and guidelines, must also be taken into account. Internal stakeholders such as sales, production or procurement interpret corporate strategies according to their area of responsibility, which on the one hand leads to additional requirements, but on the other hand also determines the scope for action and decision-making (Paetzold 2017). Customers and stakeholders do not only generate input for the development. They may also be linked to expectations regarding the flow of information during development and the development results.

Product development follows a logic that is stored in process models for development (e.g. Pahl and Beitz 2020) and, in addition to the temporal-logical sequence, also includes a problem-solving cycle (Ehrlenspiel and Meerkamm 2013). The process models are supplemented by methods for concrete problem solving, whereby methods as instructions for action (Lindemann 2007) support the execution of individual development steps through the targeted provision of knowledge and information. Generic processes and activities can therefore be specified for product development.

Development processes are also characterised by a certain uniqueness. Not only because of variable customer requirements, but also because of ever-changing development boundary conditions, development processes are ultimately never the same. This leads to a paradox (Kline 1995): even with a development process that has been carried out several times, the same result cannot be expected because of differences in the boundary conditions. For reasons of effectiveness and efficiency, however, a clear and pre-structured procedure is prescribed for development, which is also associated with certain standardisations (Paetzold 2017). For the data and information flows, this means that although they are predefined via process descriptions, additional information arises in the application from the boundary conditions that shape the actual scope for action and decision-making (Gericke et al. 2013). In addition, development is a highly creative process that must allow room for innovative ideas. This underpins the paradox because it has repercussions on the design of data and information flows and the associated communication channels.

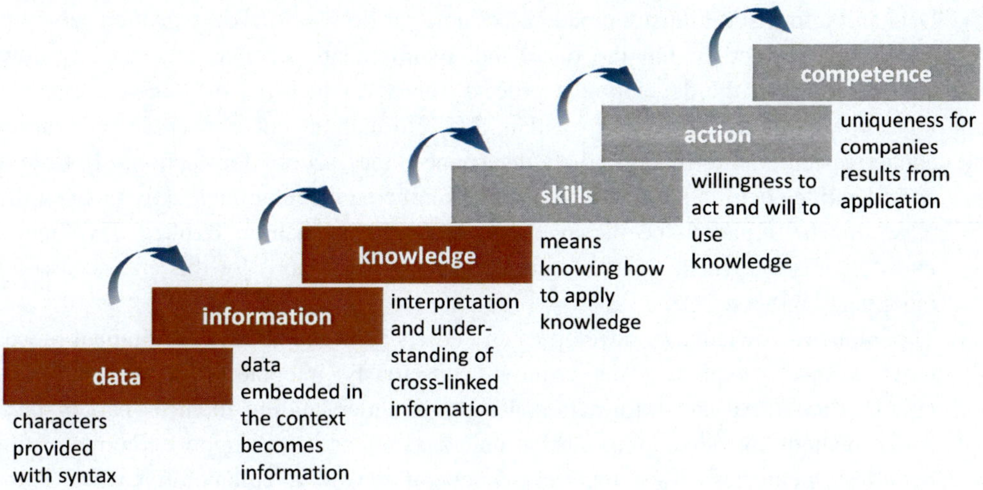

Fig. 11.1 North's knowledge staircase in the context of development (North 2016)

Data and information flows ultimately form the basis for knowledge in the development organisation and are manifested, among other things, by the processes and methods used (Schönwald et al. 2017). This knowledge is essential for the competitiveness of the company. At the operational level, this means making knowledge available for concrete development projects (North and Maier 2018). The associated value creation is a step-by-step process that is represented by a knowledge staircase according to North (North 2016) (Fig. 11.1).

In the course of the various development activities (synthesis/analysis), the developers first generate data that have a defined meaning, from which information emerges. This information is the basis for decisions, on the basis of which actions are carried out. Goal-oriented action of a developer describes his ability to select information in a targeted manner. Knowledge is the result of conscious or unconscious information processing, which in turn manifests itself in actions. Knowledge can be perpetuated in the use of routines or also explain a defined problem-solving behaviour (Rutz 1985). These mechanisms in turn are embedded in organisational structures, processes and practices for development. Both organisational and individual competencies for development are reflected in the ability to act appropriately in specific situations. This requires that knowledge can be mobilised in specific contexts (Probst et al. 2012). The processes in development and the data and information flows attached to them are therefore gaining in importance not only because they are carriers of development knowledge, but also because the information stored there becomes available to developers, interpretable and contextualised in order to make its potential accessible (Paetzold 2018).

Three challenges therefore arise for the preparation and use of data and information in development:

- Data and information must be made accessible for decision-making, as they serve as a framework for action. On the one hand, requirements are derived from top-down considerations of the development process, through which not only the development logic flows into the structuring of data and information, but also organisational and context-specific information that determines the scope for action. Bottom-up considerations help to specify data needs for concrete development activities and thus to secure the situation-specific availability of data for action. Explicit development knowledge is thus made available via IT structures. Methods for this are discussed in more detail in Sect. 11.3.
- The implicit knowledge of developers and others involved in the development process must be made explicit. This requires consciously integrating people into these organisational data and information flows and understanding them as part of them. Agile methods for developing mechatronic systems not only support communication, but also form the basis for moving from cooperative work to collaborative work. This is discussed in more detail in Sect. 11.4.
- The generation and use of knowledge in development does not only require making data and information available. It is also necessary to integrate methods and processes into these data and information flows, to develop them further and to apply them in order to bring the action potential of the development organisation to bear in the best possible way. This makes method acceptance an important factor for knowledge retention and thus competence maintenance in development. Section 11.5 shows the correlations and challenges.

11.2 Basic Considerations and Framework Conditions for Data and Information Flows in Development

In the progress of development, a large number of product artefacts are created, in digital or physical representation, which contain data and information about the emerging product in a more or less structured form. These product artefacts are not only the result of individual development steps, they also trigger development activities and serve as input for downstream development steps. Further information from the development environment, residual friction, boundary conditions and disturbance variables determine the scope for action and decision-making. In terms of effective and efficient development, it is necessary not only to know the data and information flows, but also to design them in the sense of achieving the goals. The basis for this is knowledge of the development logic or the development processes derived from it. Strategic and operational process models form the basis for the design of development processes in companies, but implications for the data and information flows can also be derived from this, which are described in more detail below.

11.2.1 Implications from Macrologic

Process models in product development are based on systemic approaches (Haberfellner et al. 2015). The starting point is formed by system descriptions, which are initially only notions or ideas about technical systems or the expected system behaviour, but which are then concretised within the framework of the development process. Process design in the macrological sense follows a phase structure that can be subdivided into the four phases of planning, designing, developing and integrating, following (Pahl and Beitz 2020; Paetzold 2017).

The starting point for the development is the target system description, which summarises the wishes and needs of the stakeholders. This determines not only the expected system behaviour, but also the requirements description and thus serves as initial information for the development. In terms of **technical-physical development**, a functional analysis is carried out on the basis of this information. Sub-functions and their dependencies are identified, for which solution principles are determined, which in turn can lead to a detailed description of the function. Through this alternation of function and principle synthesis (Andreasen 1980), the target system is successively concretised into a product or its description.

Characteristic for the development is that one has to deal with uncertain and incomplete data and information (Freisleben and Schabacker 2002). Product-describing data and information are successively generated and concretised. Uncertainties result for the data and information not only from the fact that requirements change and become more concrete, but also because situations can arise in the course of development that change the scope for action. Especially in the early phases, assumptions have to be made due to missing data and information, which have to be checked via iterations. Thus, the quality of the available data and information is strongly dependent on the development progress (Reitmeier and Paetzold 2012).

Functions or solution principles are combined into modules in product architectures, which are further detailed by different development teams and finally integrated into a complete system. The product architectures and the way they are handled are strongly dependent on the development organisation in the company and the typical development tasks. Consequently, the definition of the system architecture is also accompanied by a division of labour, which provokes process-related and organisational interfaces in the data and information flows.

Another challenge for the design of the data and information flow arises from the fact that development tasks are being parallelised for reasons of efficiency. This requires the coordination of data and information exchange, which in turn requires precise knowledge of process-related and organisational interfaces (Kößler et al. 2014a, b). As a result, more and more effort is being put into **technical management** in development, not only to bring together individual development teams and ensure overall system integration. Rather, associated tasks such as project, risk, requirements and assurance management are becoming increasingly necessary in order to meet development deadlines, reduce development risks and ensure the quality of the products. The data and information generated in technical-physical development form the basis for this, but are supplemented by corresponding task-dependent information.

11.2.2 Implications from Micrologic

In addition to the phase structure, it is also important to consider the processes at the level of concrete task processing. This is based on generic procedures from the psychology of thinking (Miller et al. 1973), which Ehrlenspiel and Meerkamm take up with the problem-solving cycle (Ehrlenspiel and Meerkamm 2013). After a task has been clarified, the search for solutions takes place in the synthesis, which are then analysed and evaluated with regard to their suitability, leading to a permanent interplay between synthesis and analysis.

For considerations of the data and information flows, some implications arise from these considerations. Of course, synthesis and analysis are connected via the data and information flows, but according to Weber (2005) there are two categories with regard to the characterisation of the data:

- **Characteristics** as a result of synthesis are defined by the developer, they define the product and at the same time are adjusting screws to manipulate the system behaviour. Therefore, they form the input for the analysis.
- **Properties** as a result of the analysis are established on the basis of selected characteristic values. They cannot be influenced directly, but they must fulfil target specifications according to the requirements.

Input for the synthesis are (target) properties, for the fulfilment of which characteristics are identified as output. These in turn provide the input for the analysis, which examines whether the desired properties actually occur (Weber 2005). An example of the application of the approach to support simulation planning can be found in (Reitmeier et al. 2015). For system description, models are used that represent both structural and functional aspects of the system to be designed. Models structure the data and information and depict relationships between the product-describing parameters. Formal interfaces in the data and information flow result on the one hand because individual parameters can be used in different models. On the other hand, the data models are created with corresponding tools and are linked to tool use. (e.g. Kößler et al. 2014a, b). Integration in the data and information flow requires not only formal interfaces. In order to avoid breaks in development, it is always necessary to consider which information can be extracted from existing data and what a task-oriented preparation must look like in order to support the development situation (Forsteneichner et al. 2015). This requires knowledge of process-related information needs.

Consequently, two aspects need to be considered when looking at data and information flows:

- As a rule, more data is generated than is directly needed, which is only used further if this is known and requested.
- The developer is involved in the process of generating information from the data, and the content and depth of the information is thus also dependent on the developer's competences, skills and experience, which gives the developer an important role in the flow of data and information.

11.2.3 Implications from the Organisation of Development Processes

The data and information created during development are generated using various tools, both via models and as data sets. PLM systems, in which the data and information are stored directly or as semantic information, are used for centralised collection (Eigner and Stelzer 2009). The data and information are made available via the IT structures in the company. These are usually company-specific and are not only oriented to the workflow organisation and the processes for solving typical development tasks. They are therefore highly context-specific. In these IT structures, there are usually a large number of data storage systems, some of which are available centrally, but some of which are only used on a department-specific basis (Schönwald et al. 2019). In addition, development activities are supported by specific workbenches that link tools and data storage systems to specific tasks via sub-processes. Examples include simulation workbenches or those for tolerance management (e.g. Forsteneichner et al. 2015).

The developer needs corresponding information for his work, which is made available to him by these IT structures. It is therefore necessary to coordinate the data and information flows in such a way that they are not only made available, but also that the results of development activities are specifically fed back into the IT structures. In addition, temporal-logical dependencies must be taken into account in this coordination, which can be derived from maturity considerations.

The definition of data and information flows is usually done via the process descriptions in the company. Nevertheless, there are always breaks in development. Tasks are carried out twice in ignorance of the availability of data and information. There is often a lack of clarity about where these are stored and who is responsible for them. Last but not least, such breaks in the data and information flows also result from the fact that generated data and information are not passed on or stored correctly, so that they are not accessible (Schönwald et al. 2018a, b). In order to better deal with the challenges in the organisation and coordination of data and information flows, a methodology was developed at our Institute ITPE to capture data and information needs in a bottom-up approach and to evaluate them so that they can be aligned more strongly with needs. This is described in the following chapter.

11.3 Methodology for the Analysis and Assessment of Data Needs

The focus of the analysis methods for development processes to be described here is on recording data and information needs and understanding mechanisms for transforming data into information. Based on this, it is necessary to map data and information flows (Paetzold 2015). The aim of bottom-up process mapping is to specify the data and information requirements for individual development activities and at the same time to include the developers as actors as well as the tools used in order to make their significance for the data and information flows visible. Therefore, the recording of development processes is carried out from two perspectives (illustrated in Fig. 11.2):

- Business Process Modelling and the notation used there (BPMN) serve as a modelling method for recording the temporal-logical relationships in the processes (Freund and Rücker 2016). On the one hand, this allows secondary process knowledge about actors, required tools and product artefacts to be recorded as carriers of data and information; on the other hand, typical process analyses can be carried out and weak points in the temporal-logical sequence can be detected.
- The use of coordination theory (Malone and Crowston 1990) serves to identify characteristics and to concretise the data and information requirements (Song et al. 2016). Coordination theory generally describes the coordination between activities, goals and actors, assuming that there are interdependencies between all of these. The approach is accompanied by the assumption that a defined input is required and a defined output is expected to execute an action step, which manifests itself in product artefacts, i.e. any kind of documentation on the product in the development process. As a result, communication and cooperation mechanisms can be specified.

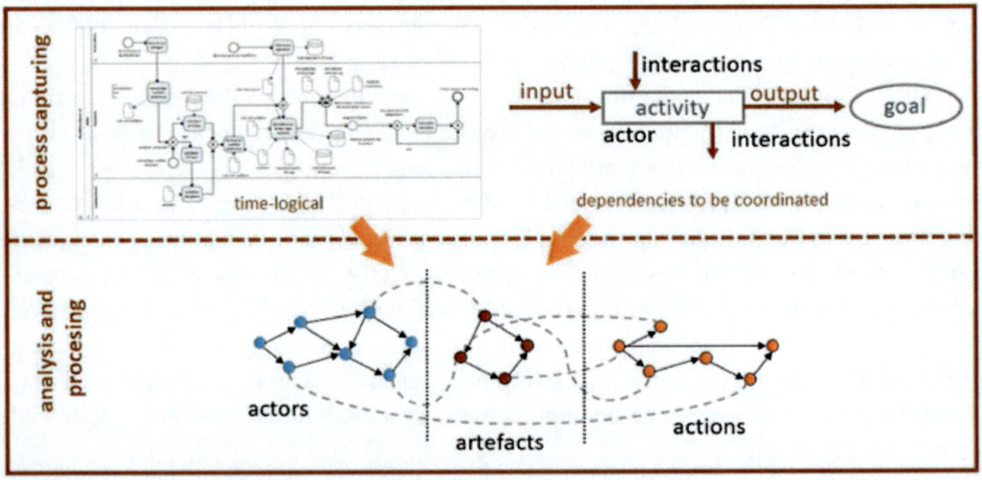

Fig. 11.2 Framework for the analysis of hedging measures

The information collected in this way is processed using graph theory. Initially, sub-networks are created that map dependencies between users, between product artefacts, between activities and between engineers. These can then be aggregated into a multilayer network that can be evaluated using mathematical methods (Chahin et al. 2016). The key figures that can be determined via graph-theoretical investigations must be interpreted and placed in the context of the modelled processes. An example of the application in hedge management can be found in (Chahin et al. 2016).

In this way, individual activities in the development can first be evaluated in detail and with regard to determining characteristics, e.g. organisational integration, data availability for activities or use of results. The individual activities are also linked to the development process by identifying to which subtasks and milestones results are contributed. For this purpose, the analysed development activities have to be classified in the top-down processes generally existing in the companies and evaluated with regard to the availability of data and information. Reference processes are not only those that map the development process itself, but also those that address specific process aspects such as risk or quality management. The associated change of perspective helps to evaluate the data and information needs holistically (Schönwald et al. 2018a, b).

One important finding from this research is that a large number of process descriptions are generally available for development in companies today, both for overarching contexts and for specific tasks (e.g. simulation data management). With these detailed process descriptions on different levels, the developers can thus draw on a broad knowledge of optimal processes and best practices. The difficulty in using these process descriptions was often that they were not known to the developers. Often, availability is not guaranteed because these process descriptions were thematically assigned to other tasks. Conversely, this means that the developers do not sufficiently assess their relevance for their own work.

Consequently, it becomes clear that the developer plays a major role in the use and design of data and information flows. The data and information flows tied to developers are difficult to make explicit and can, however, be significantly supported via communication and cooperation strategies (Fleskes et al. 2018).

11.4 Integration of Designer in Data- and Information Flow by Using Agile Methods

Research in product development today aims to make product development processes more efficient and effective, with an increasing awareness that context-specific structuring of development processes (Gericke et al. 2013) and the use of methods are proven means of making knowledge explicit in development organisations (Roth and Binz 2011).

The approach described above for recording, analysing and structuring data and information flows focuses very strongly on IT structures. Last but not least, developers are also information and knowledge carriers who need to be taken into account in such considerations. In order to transfer information into knowledge and action competence and to build it up and make it available for development in a targeted way, the role of people must be more strongly included in the considerations. Agile development methods can make a valuable contribution here (Schmidt and Paetzold 2016).

11.4.1 Basic Considerations on Agile Working

Product development processes as discussed above are characterised by a planned approach, which ensures their effectiveness and efficiency. Nevertheless, the development process is always characterised by changes in the framework conditions (Schmidt et al. 2016). Uncertainties arise not only from the fact that target system descriptions and requirements can be ambiguous or are changed and specified in the course of development. The plannability of development tasks is determined by the availability of knowledge or the ability of the development organisation to build this up for new tasks, requirements or technologies. However, the predictability of development processes is increasingly characterised by dynamic development environments (market, industry). Decisions are made on the basis of uncertain and incomplete information, making iterations an integral part of development processes. (Schmidt et al. 2018).

Agility is understood as the "*...ability of a development team to react continuously and quickly in a dynamic environment to expected and unexpected changes, to accept them and to use them to its advantage.*" (Böhmer et al. 2015, p. 4).

The associated paradigm shift, namely instead of anticipating risks at an early stage and considering countermeasures, to simply allow and accept them, requires above all adjustments in the management of development activities. However, this also means that the developers themselves are given greater responsibility in the collection and interpretation of data and information from the environment and are empowered to incorporate this information and their specific knowledge into the development process.

11.4.2 Meaning and Working Methods

Agile development methods, which initially emerged from software development, are based on newly formulated values for the implementation of development projects. Accordingly, agile development methods primarily support technical management; the procedures and methods that support technical-physical development naturally retain their importance and continue to be applied.

The values underlying agile methods (Beck et al. 2001) primarily address project implementation:

- *Individuals and interactions are more important than processes*
- *Functioning products are more important than detailed documentation*
- *Collaboration with customers is more important than contract negotiations*

Scrum	Framework within which various methods and techniques are used; provides iterative incremental approach to optimising product forecasting and risk ***Cycle-driven approach***
Kanban	Based on visualised workflows, tasks are broken down and presented in such a way that the progress of fulfilment is recognisable ***Flow optimisation***
Design Thinking	Procedure for finding solutions for complex systems; rather rudimentarily defined framework for early phases ***Creativity support***
Extreme Programming	Framework for using engineering principles to produce high-quality products; based on cooperation to verify results directly ***Quality assurance***

Fig. 11.3 Methods for implementing agile working methods [overview of methods in (Schmidt 2019)]

- *Openness to change is more important than following strict plans*

Certainly, these statements polarise in their reductiveness and it proves necessary to specify them in the development context (Atzberger et al. 2020). Among other things, 12 principles serve this purpose, which help to translate these values into concrete instructions for action (Boehm and Turner 2006), which can also be transferred to the development of mechatronic systems (Atzberger et al. 2020). Examples of agile approaches are shown in Fig. 11.3.

However, the explanations make it clear that methods of technical-physical development are not replaced, but are supplemented by further methods for data and information exchange in the sense of technical management (Schrof et al. 2018).

11.4.3 Impact of the Use of Agile Methods on Data and Information Flows

The special features and specific benefits of agile methods are explained below with a focus on data and information flows. The findings compiled here are based on regular annual surveys on the benefits and expectations of agile methods for the development of mechatronic systems, which are documented in (Study-Series 2020).

Agile methods focus on the direct exchange of information between stakeholders in the development. As a consequence, this leads to intensive communication with the customers and all other stakeholders in the development, which helps to avoid misunderstandings and to build a common understanding.

The way development teams are integrated into the process promotes and supports their self-organisation, which is an essential characteristic of agile development. For the team members, this means actively integrating themselves into the data and information flows necessary for development, which consequently contributes to a high level of transparency and social commitment. Team work is no longer characterised by cooperation (i.e. individual team members work on different subtasks quasi in parallel, the results of which must then be integrated), but by collaboration (all members of the team work together on the task to be fulfilled). As a consequence, all team members have access to the same data and information, which also significantly reduces interfaces in the data and information flows.

Teamwork is based on direct communication. This not only helps to create trust and thus reduce uncertainties in the data and information flows. At the same time, the importance of information exchange between developers is emphasised and seen as an important complement to the tool-driven data and information flows. Agile working thus clearly goes beyond classical approaches.

Working in defined cycles allows the permanent evaluation of development results. Combined with elements of reflection, not only does the gain in knowledge increase, but also its frequency. This is accompanied by a consolidation of development knowledge among staff, which is consequently also reflected in the goal- and task-oriented adaptation of methods and processes.

An essential element of agile working, regardless of which methods are actually used, is the clocking in which tasks are to be processed. The original intention of this was to generate self-contained tasks that could be processed by self-sufficient teams and which, as a consequence, has a strong effect on the division of labour (Schrof and Paetzold 2020). In the context of the development of complex mechatronic systems, however, challenges arise here: the complete accomplishment of tasks in the context of the development of complex technical systems by a team in a reasonable time is not realistic. This requires linking autonomous teams, which in turn entails organisational interfaces to link the individual tasks. This is the subject of scaled agile approaches (Dingsøyr et al. 2014; Dikert et al. 2016). With regard to scaling agile methods, there is still a clear need for research. From the point of view of mastering data and information flows, it is important to investigate the extent to which the design of product architectures support collaboration.

Agile approaches are also associated with a change of perspective: one no longer tries to foresee risks, but to constantly reevaluate the development situation through frequent iterations. This means that uncertainties and their effects do not have to be anticipated in advance, which is a challenge in itself because this can never be done completely. Uncertainties are clarified through product concretisation, which reduces the corresponding effort in risk assessment. Conversely, this changes the procedure, especially in process detailing. Development activities are not completely defined in advance because they are linked to expectations, but arise in detail during the course because the situation requires it.

In sum, agile methods provide the prerequisite for reacting flexibly to unforeseen situations or changes. They thus provide the basis for taking better account of uncertainties and unforeseen events in the processes, which also has an effect on the provision of data and information. Above all, when using agile approaches, the data and information flows are no longer viewed purely mechanistically at the tool level. Rather, the role of the developer with his or her skills and competences is strengthened and direct communication is appreciated as an important element in the data and information flows. The knowledge and especially the learning capacity of the development organisation is made explicit.

11.5 Importance of Methods and Process Acceptance

A major challenge for companies is to make knowledge permanent and available within the company. Supporting knowledge management in the company means networking information on an operational level so that knowledge, action and competences can be built up from this (North 2016). As discussed in Chap. 2, procedural models and methods are used in development to support the developer in his activities in the sense of technical assistance. Due to their prescriptive character, these are to be understood as instructions for carrying out certain activities (Birkhofer et al. 2005). Their application goes far beyond the simple mapping and use of data and information. With the rules of action, which are stored in the methods and process models, a goal-oriented use and transformation of data and information takes place. Rather, the focus is on networking information in such a way that new and complementary data and information are created in the sense of product concretisation.

From a scientific point of view, methods result from the processing and generalisation of knowledge for the solution of specific problems. The challenge for industrial application lies in adapting these methods to the specific development boundary conditions. However, method development is also part of the development work in the company: in the confrontation with problems, workarounds are established, which manifest themselves or are also specifically established and thus contribute to the consolidation of knowledge. On the one hand, methods serve to increase efficiency: best practices are transformed into action routines and made accessible for broad application. However, methods can also expand the skills of developers and thus their competences by providing specific information and knowledge.

The development of methods can be well mirrored in the knowledge creation process (adapted from North 2016):

- The exchange of tacit knowledge between developers (socialisation), because proven procedures or routines are initially adopted by individuals.
- If tacit knowledge is documented, i.e. prepared in the sense of best practices, methods and procedures, externalisation takes place. The knowledge is then available to the development organisation and serves as a basis for action.
- These procedures and methods are in turn taken up by the developers as individuals, applied, but also adapted to specific problems and situations. Explicit knowledge becomes implicit again (internalisation) because it contributes to the creation of new routines of action.

However, the use of methods is essential in the sense of knowledge building for development organisations in the sense of internalisation. Method acceptance can thus be seen as an active willingness to participate in knowledge consolidation. It is known from methodological research that many methods are not accepted in practice or are not used (Birkhofer et al. 2005). The reasons are manifold and have not yet been sufficiently researched:

- Reasons can lie with the developers, if these are understood as an intrusion into their own competences (Ropohl 1983) or do not fit into action routines.
- Reasons can lie in the method itself, if the tasks to be mastered are not sufficiently addressed by the method in the sense of a method-task fit (Goodhue and Thompson 1995).
- Reasons can lie in the development organisation. If one looks at this as a social system, communication and coordination processes also play a role in acceptance (Rogers 2003).

Figure 11.4 illustrates the connections between the aspects. The developer is the acceptance subject, the method the acceptance object. Figure 11.4 makes it clear that the developer must identify with the method and adopt it in his repertoire of actions in order to improve his performance in task fulfillment on this basis. Whether a fit is achieved here also depends on how well the methods are not only tailored to the task fulfilment, but also to what extent individual abilities and skills of the developers are taken into account (method-task fit). The development organisation forms the context for acceptance, whereby this is to be understood as a system of decisions, which thus forms an ordering framework for communication, within which utilisation decisions and their organisational reaction to them are structured (Rogers 2003).

Acceptance of methods/procedures therefore turns out to be a multi-layered concept. Consequently, there are two starting points for evaluating and influencing method development in product development:

- At the individual level, it is important to consider the behaviour and attitude of the developer. His decision to use the method is based on his assessment of the benefit of the method.
- The organisational level becomes significant because individual behaviour is not only shaped by the organisational context, but is also evaluated at this level. Methods are subject to a diffusion process that is characterised by communication, to which individual decisions can be attributed.

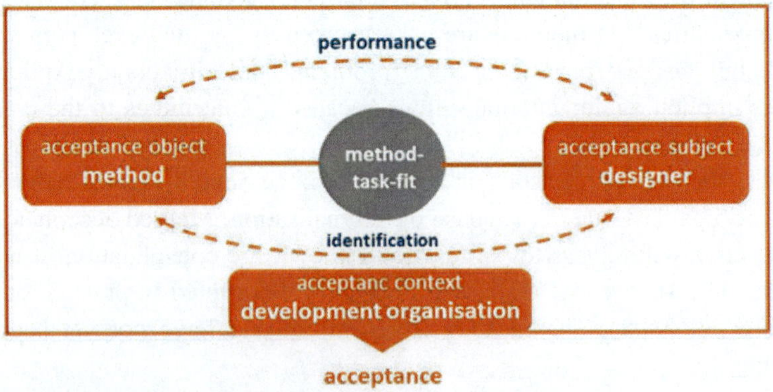

Fig. 11.4 Framework for method acceptance based on (Lucke 1995)

In order to identify conditions for acceptance formation that can be used to support the successful design of methods and their implementation in product development, the acceptance models identified as relevant to development were examined for their influencing variables in a literature review on acceptance research, these were extracted and transferred to the object of product development (Nicklas et al. 2021). The central fields of action method, developer, method-task-fit and context can be identified as starting points for acceptance-building measures (Wallisch et al. 2021). The question of how a (latent) positive attitude can be transformed into concrete behaviour remains challenging. The conceptually elaborated results of the article suggest that persuasive communication can play a central role here. However, in order to be able to make a reliable statement on this, it is necessary to empirically test the assumptions developed here with regard to possible cause-effect relationships. The first question must be whether the factors discussed here from acceptance research also prove to be empirically relevant for method acceptance in the development context. In addition, hypotheses on the cause-effect relationships for phenomena of acceptance formation must be derived and tested empirically. It seems worthwhile to draw on some theoretical concepts from behavioural research as a basis for formulating effect hypotheses, in order to make concrete strength and direction effects for all four fields of action simultaneously accessible and comparable, in order to be able to understand the structural conditions immanent to them in their overall effect context.

References

Andreasen MM (1980) Machine design methods based on a systematic approach. Dissertation, Lund University Sweden

Atzberger A, Wallisch A, Nicklas S, Paetzold K (2020) Antagonizing ambiguity – towards a taxonomy for agile development. Procedia CIRP, South Africa, pp. 464–471

Beck K, Beedle M, van Bennekum A, Cockburn A, Cunningham W, Fowler M, Grenning J, Highsmith J, Hunt A, Jeffries R, Kern J, Marick B, Martin R, Mellor S, Schwaber K, Sutherland J, Thomas D (2001) Manifesto for agile software development. Online. Available: agilemanifesto.org. Accessed Jan 2021

Birkhofer H, Jänsch J, Kloberdanz H (2005) An extensive and detailed view of the application of design methods and methodology in industry. In: Samuel AE, Lewis W (eds) Proceedings of the international conference on engineering design, pp 518–523

Boehm B, Turner R (2006) Balancing agility and Disziplin: a guide for the perplexed. Addison-Wesley, Boston

Böhmer A, Beckmann A, Lindemann U (2015) Open innovation ecosystem – makerspaces within an agile innovation process. In ISPIM Innovation Summit

Chahin A, Saad Noori N, Vilasis X, Paetzold K (2016) A practical approach to structure the product development process using network theory. In: Marjanović D, Štorga M, Pavković N, Bojčetić N: Proceedings of the DESIGN 2016, 14th International Design Conference

Dikert D, Paasivaara M, Lassenius C (2016) Challenges and success factors for large-scale agile transformations: a systematic literature review. J Syst Softw 119:87–108

Dingsøyr T, Fægri TE, Itkonen J (2014) What is large in large-scale? A taxonomy of scale for agile software development. In: Jedlitschka A, Kuvaja P, Kuhrmann M, Männistö T, Münch J,

Raatikainen M (eds) Product-focused software process improvement, vol 8892. Springer International Publishing, Berlin, pp 273–276
Ehrlenspiel K, Meerkamm H (2013) Integrierte Produktentwicklung – Denkabläufe, Methodeneinsatz, Zusammenarbeit, 5th edn. Carl Hanser, München
Eigner M, Stelzer R (2009) Product lifecycle management. Ein Leitfaden für Product Development und Lifecycle Management. Springer Verlag, Berlin
Fleskes J, Schönwald J, Paetzold K (2018) Vollgas –von Anfang an Koordinationstheorie als Tool zur Verbesserung von Entwicklungsprozessen In: QZ –Qualität und Zuverlässigkeit Bd 10
Forsteneichner C, Paetzold K, Metschkoll M (2015) Methodische Vorgehensweise zur Verifikation und Validierung komplexer Systeme. In: Krause D, Paetzold K, Wartzack S (eds) Design for X. Beiträge zum 26. DfX-Symposium. TuTech Verlag, Herrsching
Freisleben D, Schabacker M (2002) Wissensbasierte Projektnavigation in der Produktentwicklung. In: Proceedings of the 20th CAD-FEM users' meeting 2002, international conference on FEM-technology, vol 2, pp 1–10
Freund J, Rücker B (2016) BPMN ? die Notation im Detail. In: Praxishandbuch BPMN. Carl Hanser Verlag GmbH & Co. KG, Munich, pp 27–114
Gericke K, Meißner M, Paetzold K (2013) Understanding the context of product development. In: Proceedings of the 19th international conference on engineering design 2013, 19–22. August 2013, Seoul, Korea
Goodhue DL, Thompson RL (1995) Task-technology fit and individual performance. MIS Q 19 (2):213
Haberfellner R, de Weck O L, Fricke E, Vössner S (2015) Systems Engineering: Grundlagen und Anwendung. Orell Füssli Verlage; 13. Auflage, 2015
Kline SJ (1995) Innovation is not a linear process. Res Manage 26(2):36–45
Kößler J, Reitmeier J, Kristin P (2014a) Umsetzung einer eigenschaftsbasierten Simulationsplanung in einem PDM-System In: Entwerfen, Entwickeln, Erleben 2014. – Dresden: TUD Press; Kongress EEE 2014, 26–27 Juni 2014, Dresden
Kößler J, Reitmeier J, Paetzold K (2014b) Implementation and automation of the characteristic-property-concept within a PDM-system. In: Proceedings of NordDesign 2014, Aalto design factory Aalto University
Lindemann U (2007) Methodische Entwicklung technischer Produkte – Methoden flexibel und situationsgerecht anwenden, 2nd edn. Springer, Berlin
Lucke D (1995) Akzeptanz: Legitimität in der Abstimmungsgesellschaft. Springer Verlag
Malone T W, Crowston K (1990) What is coordination theory and how can it help design cooperative work systems? In: Proceedings of the 1990 ACM conference on Computer-supported cooperative work – CSCW '90, no. April, pp. 357–370
Miller GA, Galanter E, Pribram KH (1973) Strategien des Handelns – Pläne und Strukturen des Verhaltens. Ernst Klett, Stuttgart
Nicklas S, Wallisch A, Kristin P (2021) Why methods dont't work: Herausforderung Methodeneinsatz – Erklärungsansätze aus der Akzeptanzforschung. In: Stuttgarter Symposium für Produktentwicklung. (submitted)
North K (2016) Wissensorientierte Unternehmensführung. Springer Fachmedien Wiesbaden, Berlin
North K, Maier R (2018) Wissen 4.0 – Wissensmanagement im digitalen Wandel. HMD 55:665–681
Paetzold K (2015) Network-based product- and process-management for product development – model-based solutions. In: 2015 International conference on engineering, technology and innovation (ICE), 22–24 June 2015, Belfast, Northern Ireland
Paetzold K (2017) Product and systems engineering/CA* tool chains. In: Biffl S, Lüder A, Gerhard D (eds) Multi-disciplinary engineering for cyber-physical production systems. Springer, Berlin

Paetzold K (2018) Workflowunterstüzung in der Produktentwicklung. In: Rieg F, Steinhilper R (eds) Handbuch Konstruktion, 2nd edn. Hanser Verlag, München

Pahl G, Beitz W (2020) In: Bender B, Gericke K (eds) Konstruktionslehre – Grundlagen erfolgreicher Produktentwicklung. Methoden und Anwendung, 9th edn. Springer, Berlin

Probst G, Raub S, Romhard K (2012) Wissen managen – Wie Unternehmen ihre wertvollsten Ressourcen optimieren. Verlag Springer Gabler, Wiesbaden

Reitmeier J, Paetzold K (2012) Reduction of uncertainty by sensitivity analysis and context-specific data processing within virtual property validation. In: Kyvsgård HP, Rasmussen J, Jørgensen KA, Tolletrup C (eds) Proceedings of the 9th Norddesign

Reitmeier J, Chahin A, Paetzold K (2015) An approach to the property-based planning of simulations. In: Proceedings of the 20th international conference on engineering design (ICED15), Vol. 5: design methods and tools – part 1, 27–30.07.2015, Milan, Italy

Rogers EM (2003) Diffusion of innovations, 5th edn. Free Press, New York

Ropohl G (1983) Allgemeine Technologie – Eine Systemtheorie der Technik. Universitätsverlag Karlsruhe, Karlsruhe

Roth D, Binz H (2011) Entwicklung einer Methode zur Bewertung von Wissen in der Produktentwicklung. In: Spath D, Binz H, Bertsche B (eds) Stuttgarter Symposium für Produktentwicklung 2011. Konvergenz und Divergenz – Produktentwicklung im Wandel. Fraunhofer Verlag, Stuttgart, pp 122–123

Rutz A (1985) Konstruieren als gedanklicher Prozess. Dissertation, TU München

Schmidt T (2019) Towards a method for agile development in mechatronics. A lead-user-based Analysis on how to cope with the constraints of physicality. Dissertation, UniBw München

Schmidt T, Paetzold K (2016) Agilität als Alternative zu traditionellen Standards in der Entwicklung physischer Produkte: Chancen und Herausforderungen. In: Krause D, Paetzold K, Wartzack S (eds) Design for X. Beiträge zum 27. DfX-Symposium. TuTech Verlag, Herrsching

Schmidt T, Guaragni F, Paetzold K (2016) Demerging agility, leanness, flexibility, Reconfigurability and changeability: towards a clear distinction. In: 2016 international conference on engineering, technology and innovation (ICE), June 2016, Trondheim, Norway

Schmidt T, Weiss S, Paetzold K (2018) Expected vs. real effects of agile development of physical products. In: Proceedings of the DESIGN 2018 15th international design conference, Dubrovnik

Schönwald J, Fleskes J, Forsteneichner C, Paetzold K (2017) Analyse der Daten- und Informationsflüsse im Produktentstehungsprozess eines OEMs. In: Krause D, Paetzold K, Wartzack S (eds) Design for X. Beiträge zum 28. DfX-Symposium. S. 2–12

Schönwald J, Fleskes J, Forsteneichner C, Paetzold K (2018a) A method for a detailed analysis of verification and validation processes in product development. In: DS92: Proceedings of the DESIGN 2018, 15th international design conference, vol. 3: Design Methods, Dubrovnik, S. 1313–1324

Schönwald J, Vahrenhorst D, Chojnacki A, Zschako S, Sundarp L, Paetzold K (2018b) Integration von Versuchs- und Simulationsdatenmanagement in der Produktentwicklung. In: Krause D, Paetzold K, Wartzack S (eds) Design for X. Beiträge zum 29. DfX-Symposium

Schönwald J, Forsteneichner C, Vahrenhorst D, Paetzold K (2019) Improvement of collaboration between testing and simulation departments on the example of a motorcycle manufacturer. In: Proceedings of 22nd international conference on engineering design, ICED19 Delft

Schrof J, Paetzold K (2020) Agile product development in an increasingly dynamic automotive industry: potentials and limits. In: Wiener Motorensymposium, April 2020, Wien

Schrof J, Schmidt T, Paetzold K (2018) Eignungsanalyse agiler Prinzipien für die Entwicklung physischer Produkte, München. In: Krause D, Paetzold K, Wartzack S (eds) Design for X. Beiträge zum 29. DfX-Symposium

Song H, Rawat DB, Jeschke S, Brecher C (2016) Cyber-physical systems: foundations, principles and applications, 1st edn. Morgan Kaufmann, Burlington

Study-Series (2020) ITPE Uni Bw Munich: Agile Development of Physical Products – An Empirical Study about Potentials, Transition and Applicability, Neubiberg, Universität der Bundeswehr München. Since 2018, actual version 2020

Wallisch A, Nicklas S, Paetzold K (2021) Method accepted – fields of action for increasing methods application in product development. In: 23th international conference on engineering design, Gothenburg

Weber C (2005) CPM/PDD – an extended theoretical approach to modeling products and product development processes. In: Bley H, Jansen H, Krause F-L, Shpitalni M (eds) Proceedings of the German-Israeli symposium on advances in methods and systems for development of products and processes. Fraunhofer Institut Stuttgart, Stuttgart

Part IV

Model-Based Engineering in Product Development

Model-Based Systems Engineering: A New Way for Function-Driven Product Development

12

Stephan Husung, Christian Weber, and Atif Mahboob

Abstract

Since the 1950ies Design Theory and Methodology has described product and system development as a process that goes through the stages of requirements definition, considerations of functions, allocation of solution principles and, finally, detailing. Computer support of this process has more or less evolved backwards: Starting with manufacturing information (i.e. supporting processes *after* product development), going through geometric modelling to simulation—which is basically our present state. Functional modelling and requirements management has been very difficult to realise with conventional methods. Today, Model-Based Systems Engineering (MBSE) offers new ways to come to a holistic coverage of the product development process providing and—very importantly—linking model elements for all of its stages. Starting from the needs, as seen from Design Theory and Methodology, this article describes the current state of MBSE as a new, integrative approach for product and system development and identifies needs for further progress in this field.

S. Husung · C. Weber (✉) · A. Mahboob
Technische Universität Ilmenau, Institute for Design and Precision Engineering, Product and Systems Engineering Group, Ilmenau, Germany
e-mail: stephan.husung@tu-ilmenau.de; christian.weber@tu-ilmenau.de

D. Krause, E. Heyden (eds.), *Design Methodology for Future Products*,
https://doi.org/10.1007/978-3-030-78368-6_12

12.1 Introduction and Motivation for Function-Driven Product Development

Product development has to meet numerous challenges from the market, legislators, etc. An important challenge is the increased and still increasing customer focus on solutions for specific demands (need-orientation) (Albers et al. 2018a; Stark et al. 2010). This leads to an increased and still increasing complexity of products (e.g. transition from purely mechanical to mechatronic, maybe even cyber-physical systems) (Eigner et al. 2014) and increased and still increasing requirements for quality, reliability, safety and security in the last years (Aptiv et al. 2019). In addition, there is a demand to collect information from a product's later life stages using digital twins (WiGeP 2020) with the aim of using it for the extraction of information supporting ongoing or future development processes, for maintenance prediction or for identification of new business models (Moyne et al. 2020). Within the companies, these external drivers lead to changed corporate culture and cooperation during the development process.

Since a long time, one answer to meet current challenges in product development is the increased use of information and communication tools. However, the computer support of product creation processes[1] ("CAx", Computer-Aided x with x = design, engineering, optimisation, manufacturing, etc.) until today bears a significant gap: modelling of the functions and their relations. Even though science made several proposals to integrate functional information into CAx, almost nothing of it found the way into practical application.

In the last years, Model-Based Systems Engineering (MBSE), among others along with the standardised modelling language SysML (Systems Modeling Language) offers new opportunities in this respect.

The focus of this article is to show the theoretical base of functional modelling (coming from Design Theory and Methodology, DTM), to outline its role in product development processes and to investigate the contribution of MBSE/SysML to it.

12.2 Types and Applications of Functional Descriptions in Product Development

12.2.1 Functions in Design Theory and Methodology

Design Theory and Methodology (DTM) became an independent topic of research after World War II. Its task was and is to explore how much of product development is art (i.e. based on intuition which supposedly could not be taught and trained) and how much

[1] Product creation: Complete process starting with the design request and ending with the finished product, i.e. comprising of product development, production planning ("production development") and production execution.

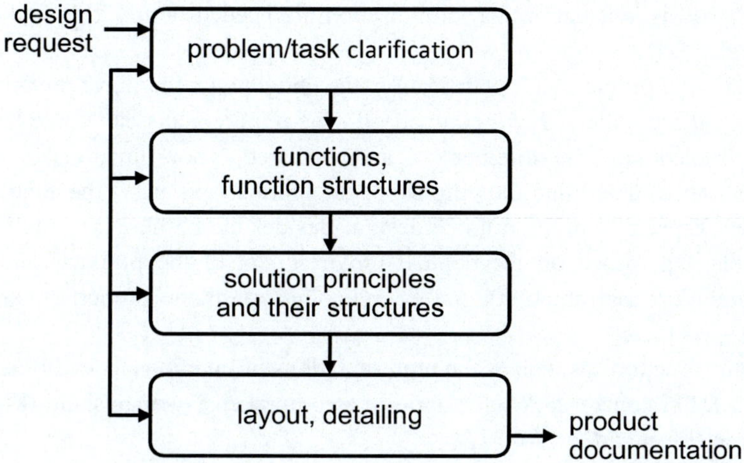

Fig. 12.1 Basic concept of product development (condensed from (VDI 2019))

can be systemised, supported by methods and tools (maybe even automated), based on scientific findings and concepts which make the activity of designing things teachable and trainable.

The most widely spread approach to DTM is—in much condensed form—shown in Fig. 12.1. It goes back to early works at the Technische Universität Ilmenau (at that time: "Hochschule für Elektrotechnik", "College of Electrical Engineering") by Bischoff and Hansen (Bischoff and Hansen 1953; Hansen 1955) and was later picked up, extended and specialised by many other authors in Germany, Scandinavia and Switzerland—too numerous to be listed here in detail. Because of translations into the English language, best known in the international context are the VDI guideline 2221 (VDI 1987 and 2019), the books of Pahl and Beitz (1983 and 2007; Bender and Gericke 2021) and of Hubka and Eder (1996).

The general procedure starts with the design request and leads to the description of the solution (product documentation) via the phases problem/task clarification, functional considerations (***what*** is needed), assignment of solution principles and their structures (***how*** are the functions realised on the principle level) to the description of the physical and logical structure of the solution (layout, detailing).

12.2.2 Some Notes

The base of the concept according to Fig. 12.1 is the insight that products can be described on different levels of abstraction—namely requirements, functions, solution principles, layout and detail information—and that product development processes pass through these phases, i.e. from abstract to concrete and, at the same time, from undetailed to detailed. This process concept is often misunderstood as linear, but is in reality highly iterative: Circles of synthesis, analysis, evaluation and optimisation are needed both within and between the phases (Weber 2014).

In the phase of problem/task clarification, the initial requirements of the product/system to be designed are collected and concretised. The requirements can come from different sources: Functional requirements and related non-functional requirements (e.g. performance) determine the purpose of the product and are at the centre of the next design steps, especially in the original design. Besides these, there is a multitude of other requirements, e.g. based on the context (environment of the product) and concerning manufacturability, assemblability, safety, reliability, aesthetic properties, environmental friendliness, cost, etc.

The term "function" as well as the representations of functions have different meanings in different DTM concepts. A selection (see a comprehensive analysis in (Eisenbart et al. 2013) and in (Eisenbart 2014)):

- In the traditional "European" literature, a function is an abstract object within a product that transforms one or more input values into one or more outputs. The usual representation method for functions is the use of blocks and the function structure is a block diagram (see (VDI 2019)). This concept can be directly transferred to MBSE and is therefore the core concept addressed in this paper.
- (Hubka and Eder 1996) have the same approach, but place a so-called transformation system above it: Any product/system is part of an execution system to accomplish a certain transformation process of material and/or energy and/or information (e.g. shaping a workpiece or communicating with other people). Decisions concerning this transformation process determine the functions of the product/system to be designed (e.g. a machine tool or a mobile phone). Also, this concept can easily be adapted into MBSE, is therefore mentioned here and included in the considerations (even if not addressed explicitly in the following).
- Some contributions in the US-American literature use the term "function" for all types of requirements, i.e. including, besides functions in the "European" sense, manufacturing, assembly, aesthetic, and other criteria. This more general concept of "function" can only partially be represented by block diagrams and is therefore excluded in this paper.

Basically, two ways of working can be distinguished in product development:

1. Way 1 comprises a pure top-down approach, whereby the system is designed from scratch in a top-down fashion: from functional requirements and other stakeholder demands right down to the specifications of sub-systems and components. The pure top-down approach stands for new product development. It is rare to non-existent in industry, but an important extreme case in science and teaching (where practically everything is "new" for students).

2. Way 2 is the normal case in practice and is addressed in several research activities (Hanna and Krause 2017; Schindel and Peterson 2013; Trujillo and Madni 2020). In this approach, existing sub-systems are used to meet the current development task (see also Chap. 2), i.e. the process starts from functions that have already been implemented, but extensions and additions can be made as required. For this approach, Albers et al. coined the term "Product Generation Development" (Albers et al. 2015) or "Product Generation Engineering" (Albers et al. 2018c). Based on the requirements, the necessary system functions are decomposed until they can be realised by functions of the sub-systems that have already been implemented—for example: the measuring functions of precision sensors (As-Is functions, see also CPM/PDD approach (Weber 2014) and solution pattern (Weber and Husung 2016)). Depending on the product and project, the results of the analysis and assessment between required and As-Is functions determines the next (synthesis) steps in the process (Weber and Husung 2011).

In the following, we will start with way 1. The As-Is description and the analysis and assessment from way 2 are then supplemented.

12.2.3 Role and Applications of Functional Descriptions

The role and applications of functional descriptions (following concepts 1 and 2, as explained above) in product development are:

- They decompose the functional requirements of a product or system into internal steps necessary to transform the inputs into the required outputs.
- They have a temporal and logical dependency, so functions can be executed sequentially, in parallel or under certain conditions.
- They are on the elementary function level the base information to search for and assign solution principles which, in turn, answer the how question.
- The solution principles carry the basic information for the next activities within product development, i.e. layout and detail design of all sub-systems and components of the final solution. For example, in the case of physical components of a product/system, they determine the geometric features that are necessary from the functional viewpoint (in DTM sometimes called "working surfaces") which, of course, have to be complemented by details that stem from other requirements (e.g. manufacturing, aesthetics, etc.).
- The preceding points show that functional modelling is indispensable if full traceability is required: Without it, the question of why something looks like it looks like cannot be traced back via the underlying solution principles to the functions, concretisations incl. decisions and, finally, to the functional (and other) requirements.
- Besides other things, traceability that includes complementary functional information and related parameters is an important prerequisite to support both re-use of design

knowledge and engineering change management: If a function and/or its solution principle is changed or added (which is often the case in practical development projects) it is important to find out the affected sub-systems or components, in an ideal case right down to the local geometry specification. Or vice versa: If a detail of a solution is changed for other than functional reasons (e.g. manufacturing) an integrated traceability concept will immediately show the affected solution principles, functions and requirements.

12.2.4 Computer Support and Early Attempts of Functional Modelling

Computer support of product creation processes (CAx) has its origins in the late 1940ies, i.e. it is roughly of the same age as Design Theory and Methodology. The first activities in this field aimed at manufacturing (Computer-Aided Manufacturing, CAM), namely NC programming. In the late 1950ies the term "Computer-Aided Design (CAD)" was coined (Ross 1956, 1960); at the same time, new concepts of computer-supported advanced simulation techniques (Computer-Aided Engineering, CAE) were developed (Argyris 1955, 1960), leading to the Finite Element Method as the first usable method. As an extension of CAE, Computer-Aided Optimisation (CAO) came in the 1980ies. The—still expanding—set of these tools were introduced into the industry on a broad scale in the late 1970ies and early 1980ies. About 10 years later Product Data Management (PDM), nowadays extended to Product Life-Cycle Management (PLM), was the next step of computer support in product creation.

The tools named in the last paragraph still dominate the CAx landscape today. It may be noted that the computer support of product creation and product development processes has more or less evolved backwards compared to the prevailing concepts from Design Theory and Methodology (see Fig. 12.1).

For a long time, especially the integration of computer-supported functional modelling has been a big problem. Starting some 20–30 years ago, several attempts were made to create software support for functional modelling, however, so far without transfer into engineering practice. A selection of early proposals:

- A very early contribution that was part of a joint effort of several universities in the United Kingdom was the Schemebuilder project (Bracewell et al. 1992). It may be noted that this software was already linked to the layout phase of product development in form of a CAD-system ("Layout" part of the overall project).
- (Schulte et al. 1993) tried to use the—then quite new—Feature Technology to link functional and layout, even detail design information. In this project, the top-down and the bottom-up approaches (ways 1 and 2 of product development, as described above) were combined, however with limited success (the top-down approach was given up later).

- (Grabowski et al. 1998) propose to develop a Universal Design Theory (UDT) as a base for integrated computer support, including functional modelling.
- (Koch and Meerkamm 2002) show a new attempt to link functional modelling with CAD as an extension of an already existing Design System ("Konstruktionssystem mfk") towards early phases of product development.
- In a way, Modelica as a quite popular language for modelling physical systems (Schamai et al. 2009) can be seen as a related approach. However, it addresses functions only indirectly but instead transfers equations of elements that, seen from a DTM perspective, stem from solution principles realising functions.

Most of these (and some more) approaches date back to times before MBSE with SysML or were developed in parallel (like Modelica). Having MBSE/SysML today, new developments may be envisaged, and, in consequence, functional modelling has come back into the focus of DTM as well as computer scientists, resulting in an increasing number of publications—too many to be mentioned here in detail.

12.3 Overview over MBSE and SysML as Modelling Language

12.3.1 Systems Engineering (SE)

Systems Engineering (SE) has its origins in the 1940ies and 1950ies (in a way in parallel to the origins of Design Theory and Methodology) as an approach to model and handle complex and increasingly multi-disciplinary engineering systems and their behaviour (in the beginning many of them in military, air and space applications). In 1990, in the USA the National Council on Systems Engineering (NCOSE) was formed by a number of companies and research bodies in order to spread and improve SE practices and education. As a result of growing interest from outside the USA, the movement was re-organised as the International Council on Systems Engineering (INCOSE) in 1995 which is today a large, active and influential society publishing guidelines and hosting conferences.

Today, SE is a well-known approach that is described in the ISO/IEC/IEEE 15288:2015 and the INCOSE Systems Engineering Handbook (Walden et al. 2015). SE includes processes (technical processes, management processes, agreement processes and organisational process support) and related methods, with the aim "to enable the realisation of successful systems" (Walden et al. 2015). The described technical methods largely cover the steps during the development process. A wide spectrum of methods is described in (Haberfellner et al. 2019). In order to meet the current challenges in product development, in addition to the SE processes and methods, models are useful to consistently and continuously represent the information that has been generated during the development process (based on synthesis decisions and analysis results) (Huth and Vietor 2020; Albers et al. 2018b; Kleiner et al. 2017).

12.3.2 Model-Based Systems Engineering (MBSE) and SysML

The enhancement of SE with models is called Model-Based Systems Engineering (MBSE). The most widely used modelling language in MBSE is the Systems Modeling Language (SysML) (Friedenthal et al. 2015; OMG 2018). SysML is a semi-formal graphical modelling language to model the product on the mechatronic system levels as well as the system context (where the stakeholder needs and demands as well as the surrounding systems are described). The SysML model can additionally facilitate analysis, verification and validation activities on the design (Husung et al. 2018). In the following, the term "system" is used due to the model reference.

For the definition of the **model elements** as well as for the use of model elements in further development steps, **diagrams (views** on the system model) are used which enable to show a context- and task-specific selection of model elements. Figure 12.2 shows a few example relations between model elements (incomplete) and selected SysML diagrams in which the specific model elements are visible. The diagrams in SysML that are used to

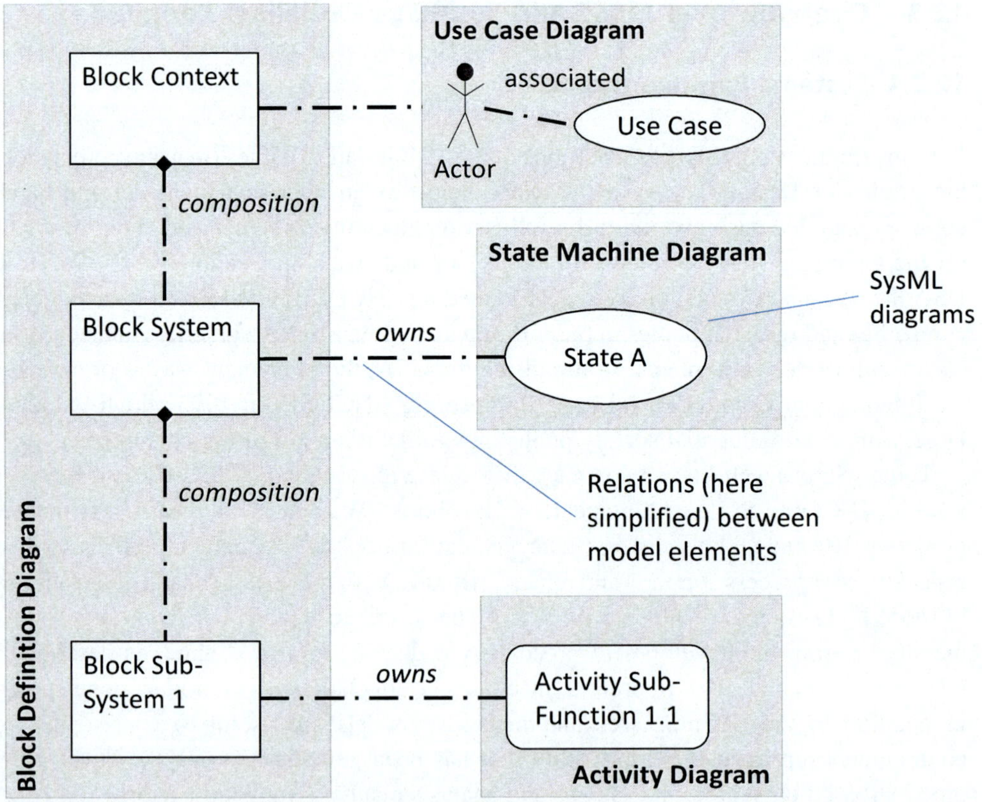

Fig. 12.2 SysML model with a few model elements and relations (simplified from standard) as well as example SysML diagrams for specific model elements

describe a system are categorised into Requirements, Behaviour and Structure Diagrams (see SysML specification (OMG 2018)).

12.3.3 Diagrams in SysML

The following explanations base on (Friedenthal et al. 2015; OMG 2018).

Requirement Diagrams are used to define the system requirements and their relationships among each other and with model elements and test cases.

Behaviour Diagrams capture functional behavioural descriptions. In particular: **Activity Diagrams** are used to define procedural aspects of the system's functional behaviour in form of control and data flows. **Sequence Diagrams** define the communication between different system elements. **State Machine Diagrams** consist of required or possible states a system or system element can have along with possible transition events to trigger these states. This helps to build event-based functional behaviour models of a system (see Sect. 12.4). The purpose of **Use Case Diagrams** is to define possible use situations of a system. They show the expectations that the relevant stakeholders have from a system in a particular use situation.

Structure Diagrams describe the composition of the system under consideration. In particular: **Block Definition Diagrams** define the structure of the system by assigning sub-systems and components to it. **Internal Block Definition Diagrams** represent the internal structure for each Block. Inside of the Internal Block Definition Diagrams, the interfaces between the sub-systems and their components is specified. **Package Diagrams** help organise the model into packages. **Parametric Diagrams** contain the parametric constraints present in between different model elements, typically consisting of mathematical equations.

12.3.4 Elements and Relations Between them in SysML

MBSE and SysML models consist of **model elements** that are related to each other. The relations and the specific semantic of the relations are described in a **metamodel.** SysML itself has a reference metamodel (OMG 2018) that can be extended according to the used methodology using so-called **profiles** (Friedenthal et al. 2015). An essential concept of SysML is the separation between the definition of model elements and their usage (as is also known in parametric CAx systems and object-oriented programming). This concept allows model elements to be defined uniformly and used for different purposes in the model—especially relevant for behaviour and structure modelling on an integral base. Here are a few relevant model elements and a few standard relations between them in the focus of this contribution (further elements and relations are described in the literature (Friedenthal et al. 2015; OMG 2018)):

- In the case of Structure Diagrams, the model elements represent, for example, the **system elements** (digital representations of the mechatronic, mechanical, electrical/ electronic, software sub-systems and components of the system as well as the context into which the system is placed). They are displayed using **Blocks** (definition of the system element) and **Part properties** (usage of the system element in the context of the composing Block). The hierarchical relation between the Blocks is a composition. The interplay between the part properties is described by means of **Ports** and **Connectors**.
- In the case of Activity Diagrams, the model elements are functions. They are represented by **Actions** and **Activities**. Activities define the Actions; the Actions themselves represent the usage of the activities. The interplay between Actions is described by means of **Pins** (represent the functional inputs and outputs of each Action) and **Object flows** (represent the flow between pins).

Depending on the used method, further model elements and relations between the model elements can be defined by inheritance, based on reference elements in the SysML metamodel using profiles.

12.3.5 MBSE and SysML in Product Development

As a summary and combination of Sects. 12.2 and 12.3, Fig. 12.3 shows the relations between real products in their respective environments, the concepts of Design Theory and Methodology (DTM), MBSE and SysML—here referring to the separation between reality and different model stages (phenomenon, information and computer models) as introduced by (Duffy and Andreasen 1995).

Conclusions for the use of MBSE and SysML in the context of product development:

- MBSE and SysML make provisions to model requirements, functional behaviour (i.e. basically functions and their realisation via solution principles) as well as structure (i.e. sub-systems and components that are part of layout and detail information of the system).
- The model elements can be related to each other with a unique semantic.
- The complete system is finally described via the relations between the model elements (e.g. the composition of the system elements). The diagrams provide context-specific views of the model elements.
- By using database systems in conjunction with the SysML modelling tools, it is also possible to version the model elements (details about the tools are not part of the contribution).
- Last but not least, MBSE/SysML offer a good starting point to realise digital twin concepts (Hausmann et al. 2021), by delivering a structure both for their definition and for the systematic capturing of information out of products' application situations. In our

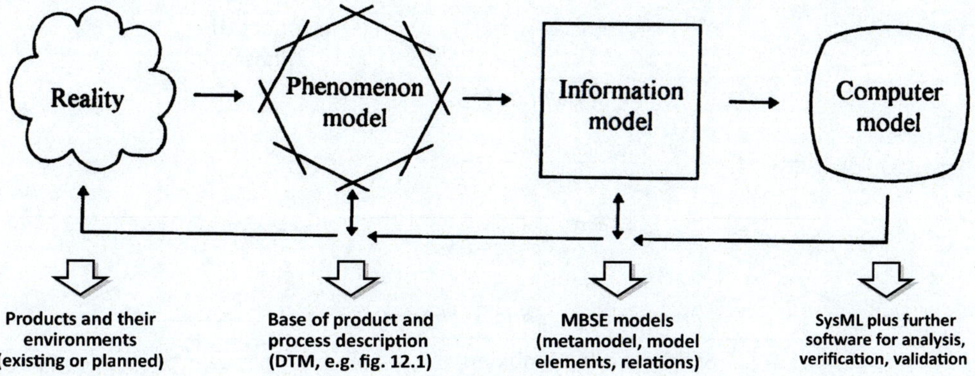

Fig. 12.3 Relations between real products in their respective environments, the concepts of Design Theory and Methodology (DTM), MBSE and SysML, based on a proposal to distinguish between reality and different stages of models according to (Duffy and Andreasen 1995)

context especially important: Collecting information about As-Is functions that can be transferred back into future development processes.

Exactly these aspects make MBSE with SysML an attractive choice as an integrated method framework, modelling concept and language for supporting product development.

12.4 Implementation of Functional Descriptions Using MBSE

12.4.1 Role of Functions in the Context of MBSE

As described in Sect. 12.2, functions play an essential role in product development. The functions, derived from the functional requirements of the system to be designed, are the basis for the decision on solution principles and for the verification of the system. Functions can change during the development, either due to changed requirements or due to changed boundary conditions during the realisation. Therefore, the functions are also needed to perform impact analyses in the context of change management. For these reasons, continuous and traceable functional description is useful, which can be achieved using MBSE methods and modelling approaches (Lamm and Weilkiens 2014). For the application of explicit functional specification methods, specific metamodels are available, such as SysML4FMArch (Drave et al. 2020) (see also Chap. 13) or MechML (Grundel et al. 2014).

The following explanation about the functional description is divided into "context" and "system" (subdivision and naming are not uniform in the literature). The context level is used to specify the demands and requirements in a more detailed form, thus delivering the base for the solution specification on the system level.

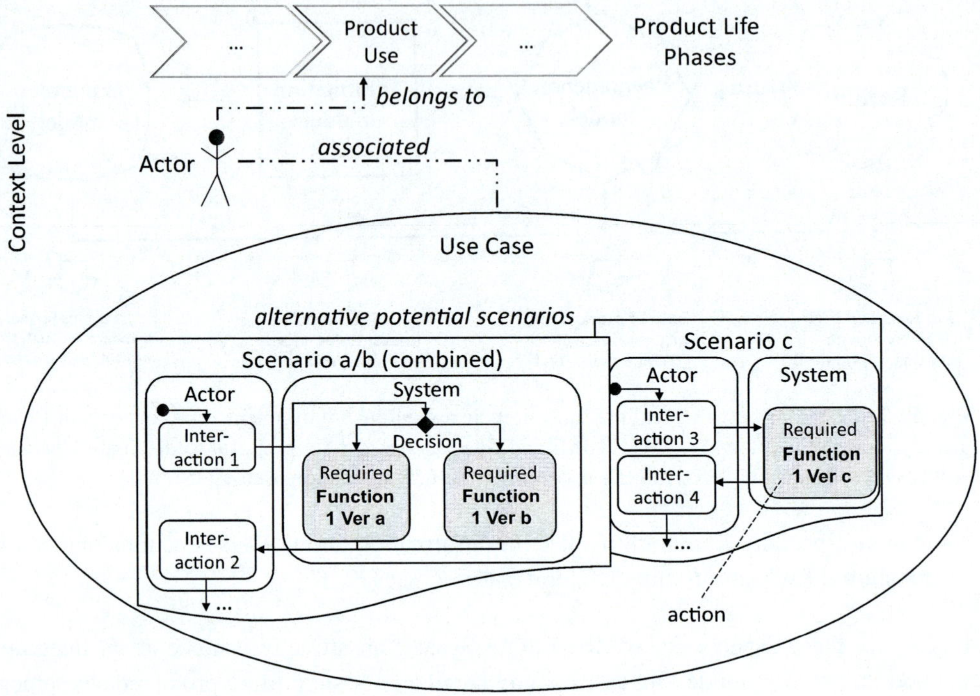

Fig. 12.4 MBSE modelling of the functional demand using Use Cases and scenarios at the context level (inspired by (Morkevicius et al. 2017))

12.4.2 Description of the Context

The following explanations focus on the description of the functional part of the context. In addition, further descriptions of the surrounding elements including the interfaces and the other requirements are necessary.

The functional demand (functional requirements) that a stakeholder in a specific product life phase expects from the system based on the needs could be described by **Use Cases** (Lamm and Weilkiens 2014) (there are also further approaches to describe the functional demand that cannot all be considered here (see also Chap. 5)). A Use Case can be further specified by describing the triggers that start it, the preconditions before the Use Case, the success end condition and the flow within the Use Case (not shown in the figures further below) (Cockburn 2006). A Use Case can be split into different **scenarios** (Fig. 12.4), depending on the interaction with the actor and the further context elements (not shown in the figures below) as well as further conditions, leading to alternative flows and results (Glinz 2000). The scenarios could be described separately or combined by use of decision model elements for the different flows.

This mechanism of creating and refining Use Cases is a decisive advantage of MBSE/SysML for functional modelling: With conventional functional modelling methods (function structures as block diagrams), even with the early computer-supported approaches, it is nearly impossible to capture different Use Cases—and within them even different scenarios—properly. In MBSE with SysML modelling, the atomic model elements are related to each other, so that possible impacts in case of changes can be analysed consistently.

Using the SysML language, scenarios can be described by several diagram types, like Activity, Sequence or State Machine Diagrams (see also (Friedenthal et al. 2015) and (Chamas and Paetzold 2018)). In this contribution, we focus on Activity Diagrams, as these are specifically appropriate for process-oriented applications in the context of product development (Friedenthal et al. 2015).

Further down (Sect. 12.5) the example of a mobile service robot is introduced. For this, Fig. 12.6 shows a scenario in the Use Case "get service information from device". The scenario displays which required functions the actor (here the user) expects from the system. These required functions are described in SysML as model elements using Actions (in product development, further sources for required functions also exist, e.g. from direct stakeholder specifications, which will not be discussed further here). The Actions (required functions) are directly assigned to the model element of the service robot (here using a Block). It is essential for the application of MBSE that the described relations between the atomic model elements (actor, Use Case, scenario, Actions inside of the scenario) can be traced according to the metamodel. In Fig. 12.4 the specific relations "associated" (Use Case is associated with the actor) and "belongs to" (actor belongs to life phase usage) are defined.

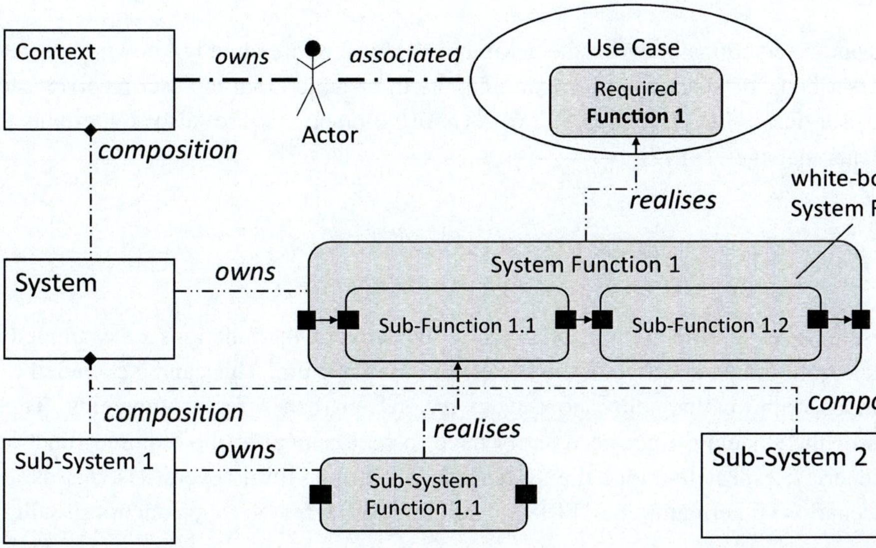

Fig. 12.5 Transition from the description of the context to the system and sub-systems (focus relevant functions), inspired by (Morkevicius et al. 2017)

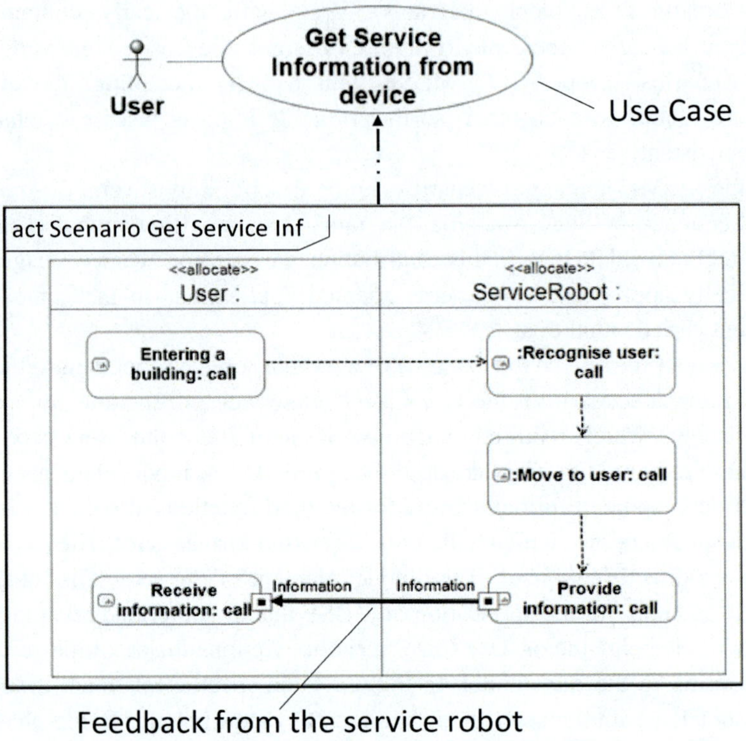

Fig. 12.6 Use Case detailing with a scenario for the example of service robot (top: Use Case (without a diagram frame), bottom: Activity diagram)

If relevant interactions between the actor and the system are already known, these can also be described, for instance in the example, the information that the user receives back from the service robot (see Fig. 12.6). These interactions are also important for requirements management.

12.4.3 Description of the System

At the system level, in order to find solutions for the demand, the functions are examined in more detail from the perspective of the system and its structure. This can be extended to a functional description that tells about what the system can offer as functions. These functions of the system do not necessarily have to be identical to the required functions in the scenarios, especially since the system usually has to fulfil several Use Cases and related scenarios (Inkermann et al. 2019). Therefore, the term "system functions" will be used here.

In SysML, the system functions could also be represented by means of Actions (usage of the functions) based on activities (definition of the functions) (Mahboob et al. 2019). Based on the model-based description, the required functions in the scenarios can be refined to the system functions (Inkermann et al. 2019), ensuring that the demands in the scenarios are met. The relations between the required functions in the scenarios and the system functions are described in the metamodel. Figure 12.5 shows this by introducing the relation "realises". Additionally, the relationship ("owns") between the Block for the system and the system function is shown, because the function belongs to the Block.

The description of the functions by means of Actions and underlying Activities enables the further decomposition of a system and its functional behaviour (Lamm and Weilkiens 2014). A black-box view of a system function (only the relation between the input and output flows is visible) becomes a white-box view of the function with the involved sub-functions by further decomposition (see Fig. 12.5 system function 1 has two sub-functions in the white-box view that have to be realized by functions of the sub-systems (Morkevicius et al. 2017)). It is important here that the flows at the inputs and outputs of the system functions can continue to be used in the white-box. Similarly, flows that are newly generated within the white-box are automatically present at the boundary of the black-box, whereby further flows are also included in the representation due to concept decisions in the white-box and at the boundary of the black-box. The decomposition could be performed as deep as necessary. Especially from the transition of mechatronic systems to the domains, it should be checked whether domain-specific descriptions are more useful (Grundel et al. 2014).

As already mentioned, existing sub-systems are often re-used in product development (see also Chap. 2). These sub-systems have already implemented functions that were usually already verified in one or more previously relevant contexts (Albers et al. 2018c)—for example: measuring functions of precision sensors (Vasilyan 2016). Therefore, the functional decomposition needs to be performed down to the black-box description of these sub-systems. For the sub-systems, an already existing model-based description of the As-Is functions (see also (Weber and Husung 2016), (Anacker et al. 2020) and (Pohl 2012)) can be used to analyse and assess these directly in relation to the top-down decomposed system functions. This also allows the consideration of variants (see also Chap. 14). An essential question is the necessary detail and abstraction of the functional description (Srinivasan et al. 2012). The functions are usually very complex, so a simplification of the description is necessary for the analysis (especially for the As-Is functions) (Anacker et al. 2020) (Mahboob 2021). For this purpose, discretizations (division of the function into several areas), linearizations or neglections of partial aspects of the functions could be applied.

12.4.4 Temporal and Logical Dependence of Functions

An essential characteristic of functions is that they represent a transformation of an input flow into an output flow. This transformation is described in the MBSE language SysML by means of the flows at the pins and supplementary descriptions of the function (e.g. for the performance, accuracy, reproducibility, . . .—here partly further necessary description languages are necessary beyond the SysML). The execution of functions has additionally a temporal and/or logical dependence. A distinction is made between (see also Mahboob 2021):

- functions executed continuously,
- functions in a temporal sequence (serial or parallel),
- selection between different functions in the sequence based on logical dependencies (e.g. based on a decision) and
- functions that are active in certain states of the system (Friedenthal et al. 2015).

It may be noted that in conventional functional modelling, the distinction between these options is not fully clear. The SysML language, however, offers modelling approaches for all these options. Especially for option 2 and 3, the so-called **control flow** (see also Fig. 12.7) is used, which is available via decision and parallelization nodes. Option 4 could be implemented by assigning functions to states. Here, the states are further model elements that are described in state machines (see also Fig. 12.2).

12.4.5 Use of the System Model for Impact Analyses

Based on the defined relations between the model elements, impact analyses can be performed in case of changes. For this purpose, all model elements (representation of the system elements, functions, etc.) that are related to the model element to be changed could be determined via analysis of the relations. For the affected model elements, the impact of the change must be checked based on the represented content of the model elements (representation of the system elements, functions, etc.) (Husung et al. 2018). This procedure can be performed iteratively for the affected model elements and is often supported by the SysML modelling tools (not a topic of this contribution). This means that the influences of changes can be analysed based on the SysML model (Morkevicius et al. 2017).

12.5 Examples

The procedure for functional descriptions by means of MBSE is to be demonstrated using the simple example of a mobile service robot. The user of the system wants to have some information (e.g. news or local information) from the mobile service robot. The mobile service robot can be located at different places in the room. Therefore, it should move to the user as soon as the robot has recognized the user (see also (Merten 2012) and (Trinh et al. 2020))—an idea based on, but strongly simplified for the article). In Fig. 12.6, the Use Case including a possible scenario is presented.

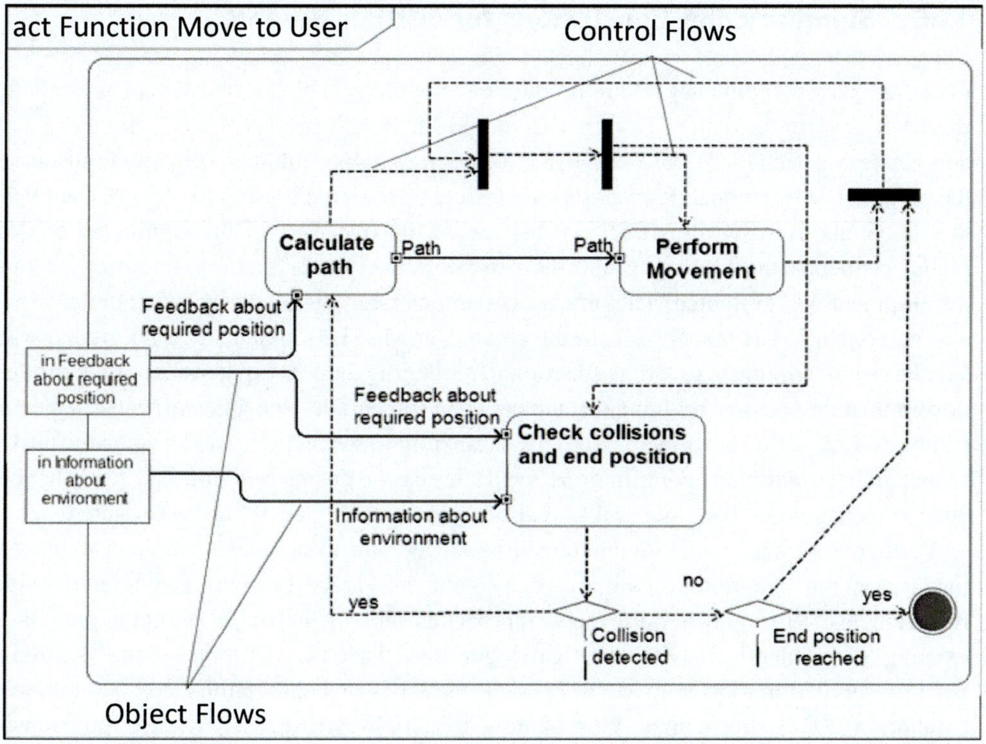

Fig. 12.7 Decomposition of the function "Move to User" by means of Actions (Activity diagram)

At the system level, the required functions from the scenario must be implemented. This is shown in Fig. 12.7 for the function "Move to User". The system function is further broken down into sub-functions. The flow of physical or other values from input to output (material, energy, information) is described by **object flows** (continuous lines in Fig. 12.7). In addition, system functions can have internal, temporal and logical dependencies. These are defined via **control flows** (dotted lines) that may contain decision nodes (represented by diamonds).

For this contribution, only the "move to user" function based on one Use Case and scenario is discussed. The functional description includes all (known) Use Cases and scenarios for the development and related functions from the robot. This ensures that the functional demand could be covered. The continuous modelling of functions (from the context to the system or sub-systems—also for several alternatives like in the scenarios) as atomic model elements with relations enables the traceability of the decision chain and enables the analysis of the interactions of the functions. The atomic model elements for the functions can be changed and concretised during the development. Based on the relations between the elements, the impact on the affected higher-level or decomposed functions can be analysed.

12.6 Summary and Conclusions for Further Research

Functions play a significant role in product development, which is shown in the article by means of work in the field of Design Theory and Methodology. The use of computer-based approaches, especially the use of MBSE, offers new opportunities. This article discusses the potential of functional descriptions in product development using the SysML language that is commonly used in MBSE. In this regard, an overview of the significant SysML model elements, their meaning and their use for functional descriptions during product development is highlighted. The process is demonstrated by means of an example.

The conclusion is that SysML in the context of MBSE is a valuable (and, by the way, standardised) approach to bring functional modelling into computer-supported product development processes, bridging the gap between stakeholder demand and domain-specific solutions (e.g. solution principles). This is an important step towards capturing design knowledge. In addition, contributions to the re-use of design solutions, to an enhanced change management handling and to realise digital twin concepts can be expected.

Further work may deal with the transition between the system levels in order to analyse the use of the sub-systems with respect to the necessary behaviour compared to the requirements. This may require new approaches and methods for building models of existing components, maybe even catalogue-like databases. Another issue is linking existing simulation tools with MBSE/SysML models: A recently published doctoral thesis (Mahboob 2021) shows first steps of how to couple MBSE/SysML descriptions with physical behaviour models (in this case represented by a physics engine) and behaviour representations in a Virtual Reality environment. Finally, with view to enhance the impact of the discussed methods in practice, further work shall deal with the efficient use of MBSE/SysML models by product developers.

References

Albers A, Bursac N, Wintergerst E (2015) Product generation development – importance and challenges from a design research perspective. In: Mastorakis NE (ed.) Proceedings of the international conference on mechanical Engineering (ME 2015), Vienna, Austria, March 15–17, pp 16–21

Albers A, Haug F, Fahl J, Hirschter T, Reinemann J, Rapp S (2018a) Customer-oriented product development: supporting the development of the complete vehicle through the systematic use of Engineering generations. In: 2018 IEEE international Systems Engineering symposium (ISSE), pp 1–8

Albers A, Mandel C, Yan S, Behrendt M (2018b): System of Systems approach for the description and characterization of validation environments. In: Proceedings of the DESIGN 2018 15th international design conference, pp 2799–2810

Albers A, Rapp S, Heitger N, Wattenberg F, Bursac, N (2018c):Reference products in PGE – product generation engineering: analyzing challenges based on the system hierarchy. In: 28th CIRP Design Conference, May 23–25, 2018, Nantes, France, Procedia CIRP 70, pp 469–474

Anacker H, Dumitrescu R, Kharatyan A, Lipsmeier A (2020) Pattern based systems engineering – application of solution patterns in the design of intelligent technical systems. In: Proceedings of the design society: DESIGN conference, vol 1. Cambridge University Press, Cambridge, pp 1195–1204

Aptiv Services US LL, AUDI AG, Bayrische Motoren Werke AG, Beijing Baidu Netcom Science Technology Co., Ltd and Continental Teves AG & Co oHG; Daimler AG; FCA US LLC; HERE Global B.V.; Infineon Technologies AG; Intel; Volkswagen AG (2019) Safety first for automated driving

Argyris JH (1955) Energy theorems and structural analysis (series of articles, pt. 1–6). In: Aircraft engineering; vol. 26

Argyris JH (1960) Energy theorems and structural analysis (revised version of Argyris 1955). Butterworths Scientific Publications, London

Bender B, Gericke K (2021) Pahl/Beitz Konstruktionslehre: Methoden und Anwendung erfolgreicher Produktentwicklung, 9th edn. Springer Vieweg Verlag, Berlin, Heidelberg

Bischoff W, Hansen F (1953) Rationelles Konstruieren. Carl Zeiss, VEB Verlag Technik, Jena and Berlin

Bracewell RH, Chaplin RV, Bradley DA (1992) Schemebuilder and layout: computer based tools to aid the Design of Mechatronic Systems. In: Proceedings of the IMechE conference on mechatronics – the integration of Engineering design (at the same time 3rd mechatronics forum conference), Dundee, Scotland, pp 1–7

Chamas MW, Paetzold K (2018) Modeling of requirement-based effect chains of mechatronic systems in conceptual stage. Int J Electr Electron Eng Telecommun 7(3):127–134

Cockburn A (2006) Writing effective use cases, 16th edn. Addison-Wesley, Boston

Drave I, Rumpe B, Wortmann A, Berroth J, Hoepfner G, Jacobs G, Spuetz K, Zerwas T, Guist C, Kohl J (2020) Modeling mechanical functional architectures in SysML. In: Proceedings of the 23rd ACM/IEEE international conference on model driven Engineering languages and Systems. ACM, Virtual Event Canada, pp 79–89

Duffy AHB, Andreasen MM (1995) Enhancing the evolution of design science. In: Proceedings of the 10th international conference on engineering design (ICED '95), Prague 22–24.08.1995, Heurista, Zurich, pp 29–35

Eigner M, Gilz T, Denger A, Fritz J (2014) Applicability of model-based system lifecycle management for cyber-physical systems. In: Proceedings of the 14th mechatronics forum international conference, pp 294–302

Eisenbart B (2014) Supporting interdisciplinary system development through integrated function modelling. Dissertation (PhD Thesis), Université du Luxembourg, Faculty of Sciences, Technology and Communication, PhD-FSTC-2014-12

Eisenbart B, Gericke K, Blessing LTM (2013) An analysis of functional modeling approaches across disciplines. II-EDAM (Artif Intell Eng Des Anal Manuf) 27(3):281–289

Friedenthal S, Moore A, Steiner R (2015) A practical guide to SysML: the systems modeling language, 3rd edn. The MK/OMG Press, Burlington

Glinz M (2000) Improving the quality of requirements with scenarios. In: Computer Science

Grabowski H, Rude S, Grein G, Meis E, El-Mejbri E (1998) Universal design theory – elements and applicability to computers. In: Grabowski H, Rude S, Grein G (eds) Universal design theory. Shaker Verlag, Aachen, Worcester, MA, pp 209–220

Grundel M, Abulawi J, Moeser G, Weilkiens T, Scheithauer A, Kleiner S, Kramer C, Neubert M, Kümpel S, Albers A (2014) FAS4M – No more: "Please mind the gap!", in: Tag des systems engineering: Bremen, 12–14 November 2014, München, Carl Hanser Verlag: 63–74

Haberfellner R, Weck OL, Fricke E (2019) Systems engineering: fundamentals and applications. Springer International Publishing, Berlin

Hanna M, Krause D (2017) Model-based support for product family design, in: understand, innovate, and manage your complex system! In: 19th international DSM conference, pp 127–136

Hansen F (1955) Konstruktionssystematik – eine Arbeitsweise für fortschrittliche Konstrukteure. VEB Verlag Technik, Berlin, 2nd ed., also 1955, online available at www.tu-ilmenau.de/konstruktionstechnik/tradition/. 3rd ed., 1968, online available at www.tu-ilmenau.de/konstruktionstechnik/tradition/

Hausmann M, Koch Y, Kirchner E (2021): Managing the uncertainty in data-acquisition by in situ measurements: A review and evaluation of sensing machine element-approaches in the context of digital twins. In: International Journal of Product Lifecycle Management

Hubka V, Eder WE (1996) Design science. Springer, London

Husung S, Lindemann G, Korobov S, Hamester M, Kleiner S (2018) Use case driven model-based systems engineering for industrial applications', EMEASEC 2018

Huth T, Vietor T (2020) Systems Engineering in der Produktentwicklung: Verständnis, Theorie und Praxis aus ingenieurswissenschaftlicher Sicht. Gruppe Interaktion Organisation Zeitschrift für Angewandte Organisationspsychologie (GIO) 51(1):125–130

Inkermann D, Huth T, Vietor T, Grewe A, Knieke C, Rausch A (2019) Model-based requirement engineering to support development of complex systems. Procedia CIRP 84:239–244

Kleiner S, Ricks M, Engelmann M, Husung S (2017) Model-based systems engineering and simulation for automotive systems development at GKN driveline; NAFEMS world congress (NWC17). Stockholm

Koch M, Meerkamm H (2002) Durchgängige Funktionsmodellierung in den frühen Konstruktionsphasen. In: Proceedings of the 13th symposium on Design for X, Neukirchen, Germany, 10-11.10.2002, pp 191–198

Lamm JG, Weilkiens T (2014) Method for deriving functional architectures from use cases. Syst Eng 17(2):225–236

Mahboob A (2021) Modelling and use of SysML behaviour models for achieving dynamic use cases of technical products in different VR-Systems. Doctoral thesis, Technische Universität Ilmenau

Mahboob A, Husung S, Weber C, Liebal A, Krömker H (2019) The reuse of SysML behaviour models for creating product use cases in virtual reality. In: Proceedings of the design society: international conference on engineering design. Cambridge University Press, Cambridge, pp 2021–2030

Merten M (2012) Design of interactive service robots applying methods of systems engineering and decision making. Doctoral thesis, Technische Universität Ilmenau

Morkevicius A, Aleksandraviciene A, Mazeika D, Bisikirskiene L, Strolia Z (2017) MBSE grid: a simplified SysML-based approach for modeling complex systems. INCOSE Int Symp 27:136–150

Moyne J, Qamsane Y, Balta EC, Kovalenko I, Faris J, Barton K, Tilbury DM (2020) A requirements driven digital twin framework: specification and opportunities. IEEE Access 8:107781–107801

Object Management Group (OMG-SysML) (2018) OMG systems modeling language (Version 1.6). October 2018, https://www.omg.org/spec/SysML/1.6/Beta1/PDF last accessed 18.06.2019

Pahl G, Beitz W (1983) Engineering design (Wallace K, Blessing LTM, trans. and eds). Springer, London

Pahl G, Beitz W (2007) Engineering design (3rd English edition, K Wallace, Blessing LTM, trans and eds). Springer, London

Pohl K (2012) Model-based engineering of embedded systems: the SPES 2020 methodology. Springer, Heidelberg

Ross DT (1956) Gestalt programming: a new concept in automatic programming. In: Proceedings of the Western joint computer conference, American Federation of Information Processing Societies (AFIPS), New York

Ross DT (1960) Computer aided design – A statement of objectives. Technical memorandum, project 8436, Massachusetts Institute of Technology (MIT), Cambridge

Schamai W, Fritzson P, Paredis C, Pop A (2009) Towards unified system modeling and simulation with ModelicaML: modeling of executable behavior using graphical notations, proceedings of the 7 international Modelica conference Como, Italy, Sep. 20–22, 2009. University Electronic Press, Linköping, pp 612–621

Schindel W, Peterson T (2013) Introduction to pattern-based Systems Engineering (PBSE): leveraging MBSE techniques, INCOSE. Int Symp 23(1):1639

Schulte M, Stark R, Weber C (1993) Feature structure for the representation of design objects as combinations of technical functions and geometry in CAD. 9th international conference on CAD/CAM, robotics and factories of the future (CARs & FOF '93), Newark/USA 18–20.08.1993. In: Sodhi R.S. (ed.) Series "manufacturing research and technology", vol. 22, Advances in manufacturing systems – design, modeling and analysis, pp 335–340

Srinivasan V, Chakrabarti A, Lindemann U (2012) A framework for describing functions in design', DS 70: proceedings of DESIGN 2012, the 12th international design conference, Dubrovnik, Croatia, pp 1111–1122

Stark R, Beier G, Figge A, Wöhler T (2010) Cross-domain dependency modelling – how to achieve consistent system models with tool support, conference: EuSEC 2010 European Systems Engineering conference. Stockholm, Schweden

Trinh TQ, Vorndran A, Schütz B, Jäschke B, Mayfarth A, Scheidig A, Gross HM (2020): Autonomous Mobile gait training robot for orthopedic rehabilitation in a clinical environment. In: nternational Symposium On robot and human interactive communication (RO-MAN), IEEE, Naples, Italy, pp 580–587

Trujillo AE, Madni AM (2020) Assessing required rework in a design reuse scenario. In: IEEE international conference on systems, man, and cybernetics (SMC), Toronto, ON, 11–14.10.2020, IEEE, pp 2946–2951

Vasilyan S (2016) High precision force measurements in horizontal direction in combination with high dead loads: non-contact flowmeter for low conducting electrolytes, Doctoral thesis, Technische Universität Ilmenau

VDI-Guideline 2221 (1987) Systematic approach to the design of technical systems and products. VDI, Düsseldorf

VDI-Guideline 2221 part 1 and 2 (2019) Design of Technical Products and Systems (latest edition of (VDI 1987)). VDI, Düsseldorf

Walden DD, Roedler GJ, Forsberg K, Hamelin RD, Shortell TM (eds) (2015) Systems engineering handbook: A guide for system life cycle processes and activities, 4th edn. Wiley, Hoboken, NJ

Weber C (2014) Modelling products and product development based on characteristics and properties. In: Chakrabarti A, Blessing LTM (eds) An anthology of theories and models of design. Springer, London, pp 327–352

Weber C, Husung S (2011) Virtualisation of product development/design – seen from design theory and methodology. In: 18th international conference on Engineering design (ICED 11). Copenhagen, 15–18.08.2011, design society, pp 226–235

Weber C, Husung S (2016) Solution patterns – their role in innovation, practice and education. In Marjanovic D, Storga M, Pavkovic N, Bojcetic N and Skec S (eds) DESIGN 2016 14th international design conference, pp 99–108

WiGeP – Wissenschaftliche Gesellschaft für Produktentwicklung (2020) Position paper "Digital Twin" (English version). http://www.wigep.de/fileadmin/Positions-_und_Impulspapiere/Positionspapier_Digital_Twin_englisch.pdf

Check for updates

Function-Oriented Model-Based Product Development

13

Georg Jacobs, Christian Konrad, Joerg Berroth, Thilo Zerwas, Gregor Höpfner, and Kathrin Spütz

Abstract

The innovative strength and competitiveness of a company depends on mastering the growing complexity of digitally networked products in an efficient way. The complexity is driven by increasing interactions among the different domains, like mechanical, electrical or software engineering on all system levels. The interdependencies require modelling approaches, that allow to explicitly and transparently reveal those interdependencies on requirements, functional architectures and solution level over all phases of the development. The increasing interdependencies and the need for more efficiency forces a change from component oriented, document-based product development to a function-oriented, model-based product development with consistently linked models across all participating domains. We propose a system architecture that describes the system in a comprehensible way across domains. The domains are able to connect their models to the architecture and link them down to the parameter level over requirements, functional architecture to the solution layer. The resulting system model allows a transparent, cross-domain mapping of functional interactions. Principle solution models close the gap between the functional and the solution layer, especially in mechanical engineering. The efficiency in development processes can be significantly increased by using model libraries to assign functions to solution models and by building ontologies to structure domain-specific models.

G. Jacobs (✉) · C. Konrad · J. Berroth · T. Zerwas · G. Höpfner · K. Spütz
RWTH Aachen University, Institute for Machine Elements and Systems Engineering (MSE), Aachen, Germany
e-mail: georg.jacobs@imse.rwth-aachen.de; gregor.hoepfner@imse.rwth-aachen.de; kathrin.spuetz@imse.rwth-aachen.de

D. Krause, E. Heyden (eds.), *Design Methodology for Future Products*,
https://doi.org/10.1007/978-3-030-78368-6_13

13.1 Introduction

In order to meet today's customer requirements, cyber-physical systems (CPS) are being developed. CPS are interconnected mechatronic systems, which in turn consist of mechanical, electrical, electronic and software components. Mastering the dependencies between the domains is one of the major challenges in product development today (Alur 2015; Eigner et al. 2012; Graessler and Hentze 2020).

The growing challenges in the development, production and operation of digitally networked technical systems can't be overcome with conventional methods of product development in mechanical engineering. Model-Based Systems Engineering (MBSE) is an approach for the cross-domain development (Eigner et al. 2012; Gausemeier et al. 2010; Broy 2010). Generic guidelines for the development of technical systems provide the methodological framework for a function-oriented product development process (VDI 2221a, b). While other domains are developing in a strongly function-oriented manner, for mechanical engineering this is more difficult due to the methodical breaks in the transfer of functions into components. Since the development of components is time-consuming compared to the small-step procedure of other domains, the mechanics have to invest a lot of effort in order to show first development results (Graessler and Hentze 2020). Thus, parallel and agile developing is more difficult. To overcome this obstacle, the mechanical domain needs methods that allow for faster functional testing and virtual validation of the system. Therefore, behavioral models must be integrated into function-oriented development processes in a standardized way. This enables model-based design decisions and collaboration with other domains in agile development processes.

The Institute for Machine Elements and Systems Engineering (MSE) at RWTH Aachen University has identified key success factors for the use of MBSE as part of research activities at the Center for Systems Engineering together with leading companies, see Fig. 13.1. The success of MBSE in industrial practice requires standards for setting up a function-oriented and model-based system architecture, the classification of existing expert models to describe the system behavior and the consistent linking of the architecture with models of the involved domains.

This chapter presents a method for function-oriented and model-based system development. The method proposed, inherits the formalization of system requirements in form of requirement models and the derivation of functional architectures based on the modelled requirements. A decomposition of the functions of the functional architecture down to so called elementary functions as described in (Koller and Kastrup 1994) is introduced as a tool to structure the solution approach. The central element of the method is the introduction of principle solution models that are linked to elementary functions via physical effects. The relation between elementary functions and function fulfilling physical effects are the basis for the clustering of principle solutions in solution libraries, see Sect. 13.2. Besides a principle solution model, the solution libraries carry additional behavior and test models that are evolved from the principle solution models. The solution models enable

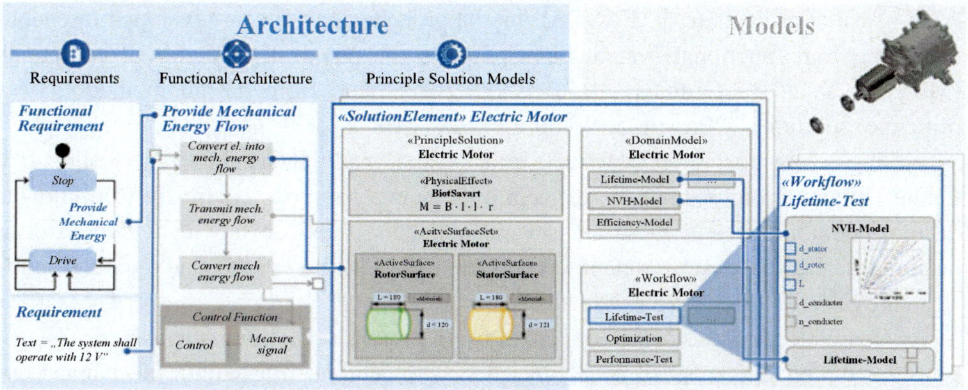

Fig. 13.1 Function oriented and model-based Architecture, classified expert models and seamless linking of architecture with expert models as key success factors

iterative functional testing and virtual validation in mechatronic system development, see Sect. 13.3.

Adding additional behavior models to the solution elements within the solution libraries with different fidelity levels for different purposes calls for necessity to structure, classify and standardize the models in organizing ontologies. On the basis of these ontologies and the SE architecture, interfaces for the corresponding model classes can then be systematically developed and used for the seamless linking of the models within the product development process. This enables the efficient reuse of expert models and forms an excellent basis for the introduction of digital seamless and more agile model-based product development processes, see Sect. 13.4.

13.2 Basic Architecture for Model-Based Systems Development

As a basis for cross-domain agile development processes, functional tests and virtual validation of system properties (e.g. dimensions, physical behavior or costs), a methodical foundation for function-oriented and model-based system architectures is needed. Thus, current methods have to be further developed, in order to enable the full description of the functional layer and its linkage to physical solution models (Eigner et al. 2012; Drave et al. 2020; Zerwas et al. 2021).

As a foundation, engineers must be able to derive a functional architecture from product requirements that is solution- and domain-neutral and can thus structure the development in a function-oriented way (Suh 1998, VDI 2221 Part 1). A function can often be realized with different solutions out of one or more domains. To support design decisions it is important for engineers to be able to define their possible solutions with little effort and test them as early as possible against requirements.

In research projects (e. g. FAS4M), first approaches have been developed to enable a transition from functional system description to the physical geometry of components (Möser et al. 2016; Grundel 2017). Another approach from design methodology are principle solutions. "Principle Solutions" describe a possible solution principle by specifying a physical effect, active surfaces and material (Roth 1994; Koller 1998; Feldhusen and Grote 2013). The concept of principle solutions is already well known in design methodology and there are several concepts based on it with regard to design catalogues and the description of components by active surfaces and guiding support structures.

As an architectural basis, the concept of using principle solutions offers an ideal starting point for the enhancement of model-based systems engineering within mechanical engineering. The approach according to Koller has been revised and further developed into a function-oriented and model-based method. MBSE modelling languages and tools are used to transfer requirements via functions into principle solutions (Drave et al. 2020; Zerwas et al. 2021; Höpfner et al. 2021).

The underlying idea of the modelling method is that properties of physical systems rely on clearly definable behavior descriptions that incorporate parameters. There are already extensive parameter classifications for technical systems in design methodology (Hubka and Eder 1988; Patzak 1982; Weber 2014). In this chapter, parameters are understood as relevant inputs and outputs of models. As models we understand representations and abstractions from real systems as 3D-representations, numerical differential equations in CAE or simple analytical equations describing physical effects. Parameters are of different types and quantifiable. However, they are not always of scalar nature but may be matrices, tuples or objects, which store scalar parameters by themselves. Examples of parameters include target values from requirements, variables and constants in physical laws, property values such as weight, volume, stiffness and post-processed outputs from models such as natural frequencies and sound pressure level maps. Therefore, an essential idea of the system architecture approach presented in this chapter is the consistent and seamless linking of parameters.

The presented modelling method ensures the function orientation for the mechanical domain similar to existing software domain approaches and thus enables the cross-domain collaboration required in today's development processes. For modeling purposes, the SysML profile SysML4FMArch is used. SysML4FMArch was developed as a basis for modelling functional architectures in all domains, with a strong focus on the improvement of the mechanical domain's particularities (Drave et al. 2020).

The modeling Method is demonstrated on an automotive cooling system, which will be briefly introduced at this point:

The main function of a vehicle is locomotion. For this purpose, the drive system provides a mechanical energy flow that is conducted to the wheels and then transferred to the road. In vehicles with combustion engines, mechanical energy is obtained from the chemical energy of a fuel. For this purpose, the physical effect of combustion is used in the cylinders of the engine, resulting in a thermal expansion of the fuel-air mixture. The sudden increase in pressure accelerates the piston, which transfers the mechanical energy to the rest of the drive system. During combustion of the fuel-air mixture, not all the chemical energy is converted into mechanical energy for propulsion: A part of the energy is conducted out of the system via the escaping exhaust gas and another part is induced into the engine components as thermal energy. Since the engine is often unable to release all of this thermal energy via its outer surfaces, its internal energy and temperature rise.

The rising component temperature is becoming increasingly critical for the component material as well as the combustion process and endangers the functional reliability of the engine. For this reason, combustion engines are usually kept within an optimum temperature window by a liquid-based cooling system. A cooling medium circulates in this cooling system, which absorbs heat from the engine and releases it to the radiator (heat exchanger). At the radiator the thermal energy flow is emitted to the environment. The cooling medium is accelerated by a pump so that it can absorb and release sufficient heat by convection and remains in motion despite the pressure losses. In our example system, the coolant pump is not operated mechanically but electrically and can thus be set to a certain rotational speed by a control unit, based on the current temperatures of the engine. In addition, the engine is simplified and consists of the components cylinder head (CH) and crankcase (CC), which both require different target temperatures.

The following subchapters each describe the methodical modelling of this example system with regard to its requirements and functions, as well as the evolvement of the solution models starting from a selection of principle solution models.

13.2.1 Modelling of Requirements

Requirements are demands and wishes that customers, users, manufacturers, legislators and many other stakeholders have for the product to be developed. A successful product must not only fulfill the wishes of the customer, but must also be able to be produced efficiently (e.g. factory standards) and comply with legal requirements. Therefore it is important to document these requirements and to continuously check their compliance during the development process (Koller 1998). There are many approaches to classifying and formulating requirements in product development (Ross and Schoman 1977; Göhlich and Fay 2021). Until today, requirements in many companies are still formulated in textual sentences and stored as a list in unlinked documents or software tools. Same applies to e.g. guidelines, norms, etc. Even if requirements engineering and -management is supported by good templates (Rupp 2014), the disadvantages are often the ambiguous formulation of requirements and the missing link to the models based on them, which are thus cut off from requirement changes (Konrad et al. 2019; Graessler et al. 2020).

There have been a wide range of suggestions on how requirements can be modeled using MBSE approaches. Many of them have two common features that make a significant difference to document-based and informal requirements. One is a clear formalization and explicit description that leaves no room for interpretation. On the other hand, the

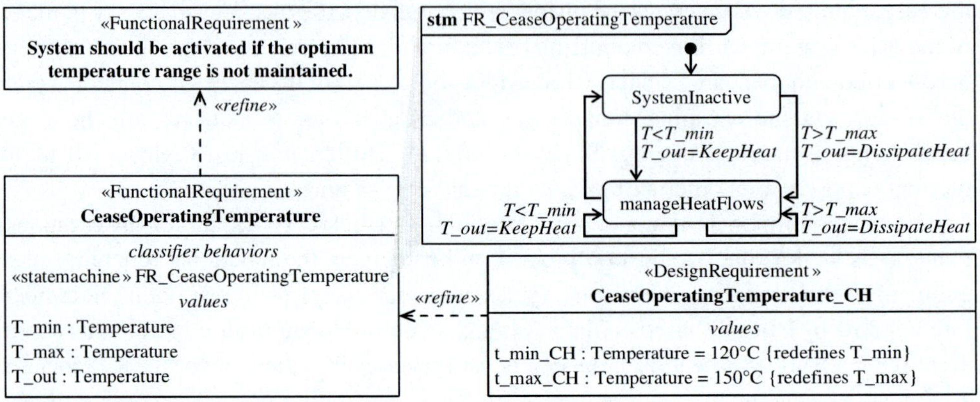

Fig. 13.2 Requirements of the example system

requirements are expressed as far as possible by (physical) quantities with concrete values. These can be linked to other development models so that changes in requirements directly reach all relevant models, see Fig. 13.2. We differentiate requirements into two categories. Functional requirements («FunctionalRequirement») specify the desired functionality of a technical system. Functional Requirements are modeled in requirement diagrams and are formalized in state machine diagrams (stm), activity diagrams (act) or sequence diagram (sd).

In the example system, the superior behavior of the complete cooling system is described as a state machine, which is always in one of two states: Either it is idle or active. The transitions between the states depend on the temperature states of cylinder head and crankcase. If, for example, the cylinder head exceeds its permissible maximum temperature, the system switches to the active state. This modelling allows for example the automated generation of test cases. This way it can be checked whether the system fulfills the prescriptive behavior and is always in the expected state (Drave et al. 2019).

The second category are restrictions or design requirements («DesignRequirement»), which limit the value range of a parameter occurring in the system. For example, the optimum range for the operating temperature of the cylinder head can be specified as 120 °C to 130 °C. Since this temperature range is decisive for the described transition in the state machine, this design requirement refines the functional requirement. Thus, the modelling of behavior and the restriction of parameter values are clearly separated, but can be used for common statements.

13.2.2 Functional Architecture

There are different types and views of functions in literature where in this paper the focus is on system environment functions according to (Srinivasan et al. 2012).

Accordingly, Functions describe the specific behavior of a product without specifying the solution, e.g. which domain, components, effects, etc. implement this behavior. The concept of functions is based on the idea that function flows enter and leave a system over a given system boundary. These function flows are quantified by parameter values and can be categorized as flows of energy, material, or signal. Functions describe not only which function flows enter and exit, but also which operation takes place (Koller 1998). The decomposition of the overall function into subfunctions results in a functional architecture (Feldhusen and Grote 2013). The SysML4FMArch profile defines that functions can be divided into decomposed functions («Architecture») and elementary functions («ElementaryFunction»).

Each elementary function describes an elementary mathematical relationship between the input and output flows (Koller 1998). Ideally, functions can directly be derived from requirements (customer functions). In any case, functions can be linked to the requirements they fulfill through function calls or satisfy relationships.

In the example system, the function *ManageHeatFlows* is decomposed into four functions, see Fig. 13.3. The function *GenerateVolumeFlow* generates a volume flow of the coolant according to the specified signal from the function *ControlHeatFlows*. This volume flow is directed to the function *DistributeHeatFlows*, where heat is absorbed (at the cylinder head and crankcase). The heated coolant flow leaves this function and releases a thermal energy flow to the environment in the elementary function *SeparateFluidAndThermalEnergy*, before it circulates back into the *GenerateVolumeFlow* function. This function can be divided into three elementary functions:

InDecreaseElectricalEnergy transforms the incoming electrical energy flow so that the following function *ConvertElectricalInMechanicalEnergy* generates mechanical power according to the specified rotational speed. This mechanical power is used in the subsequent function *ApplyMechanicalEnergyToFluid* to pressurize and accelerate the coolant flow. States of a system, as introduced in (Ponn and Lindemann 2011) are modeled via the ports of the functions.

The realization of an elementary function is often possible in multiple domains. For example, cooling circuits can be controlled by mechanical thermostats or software-based controllers. Often the flows of a function can give a hint, in which domain this function can be realized. At the latest by defining a physical effect, the further development of this elementary function is assigned to a certain domain. Therefore, the transition from functions to principle solutions also entails a shift from cross-domain to domain-specific development.

13.2.3 Principle Solution Models

While functions describe the changes of function flows, principle solutions concretize how this change is physically realized. Therefore, elementary functions and principle solution models are linked with a generalization relationship. The principle solution inherits all

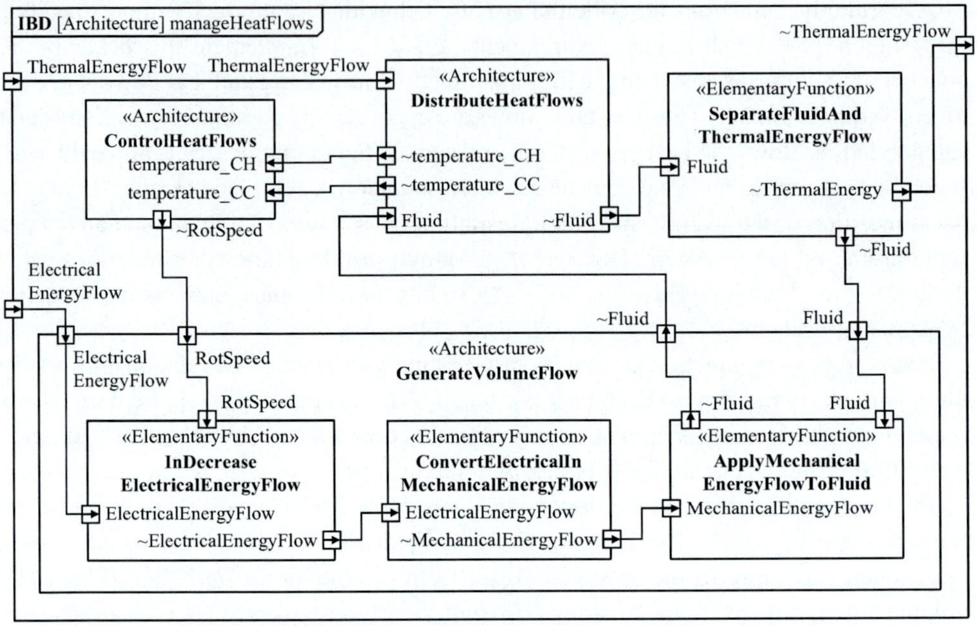

Fig. 13.3 Functional architecture of the example system

functional flows as ports from the elementary function and can use them to describe more precisely how incoming flows are transformed into outgoing flows using a physical effect, active surfaces and material. As elements of the principle solution, physical effect, active surfaces and material can be modeled with the corresponding stereotypes defined in (Drave et al. 2020) in the internal block diagram of the principle solution. Physical effects can usually be described with mathematical equations which are modeled as constraints. The parameters of such equations can depend on function flows, active surfaces and material and are linked to them accordingly. Physical quantities, which refer to function flows (e.g. volume flow), are linked to the corresponding values of incoming and outgoing function flows. Parameters that refer to geometric or material-related quantities are linked to the value properties of an «ActiveSurface» or a «Material». If the equation contains natural constants, these are modeled as value properties directly into the principle solution and linked to the constraint parameters.

The upper section of Fig. 13.4 shows the continuous modeled path from a requirement to the associated function and its principle solution for the example system. The requirement that a volume flow of 7 l/s should be generated at a rotational speed of 15 s^{-1} and against a pressure difference of 37 kPa is met by the elementary function *ApplyMechanicalEnergyToFluid*. This elementary function is now specialized by the principle solution *CentrifugalPumpWheel* that describes how mechanical energy is applied to the fluid. The lower section of Fig. 13.4 illustrates the internal block diagram of the principal solution. Several possible physical effects for the elementary function

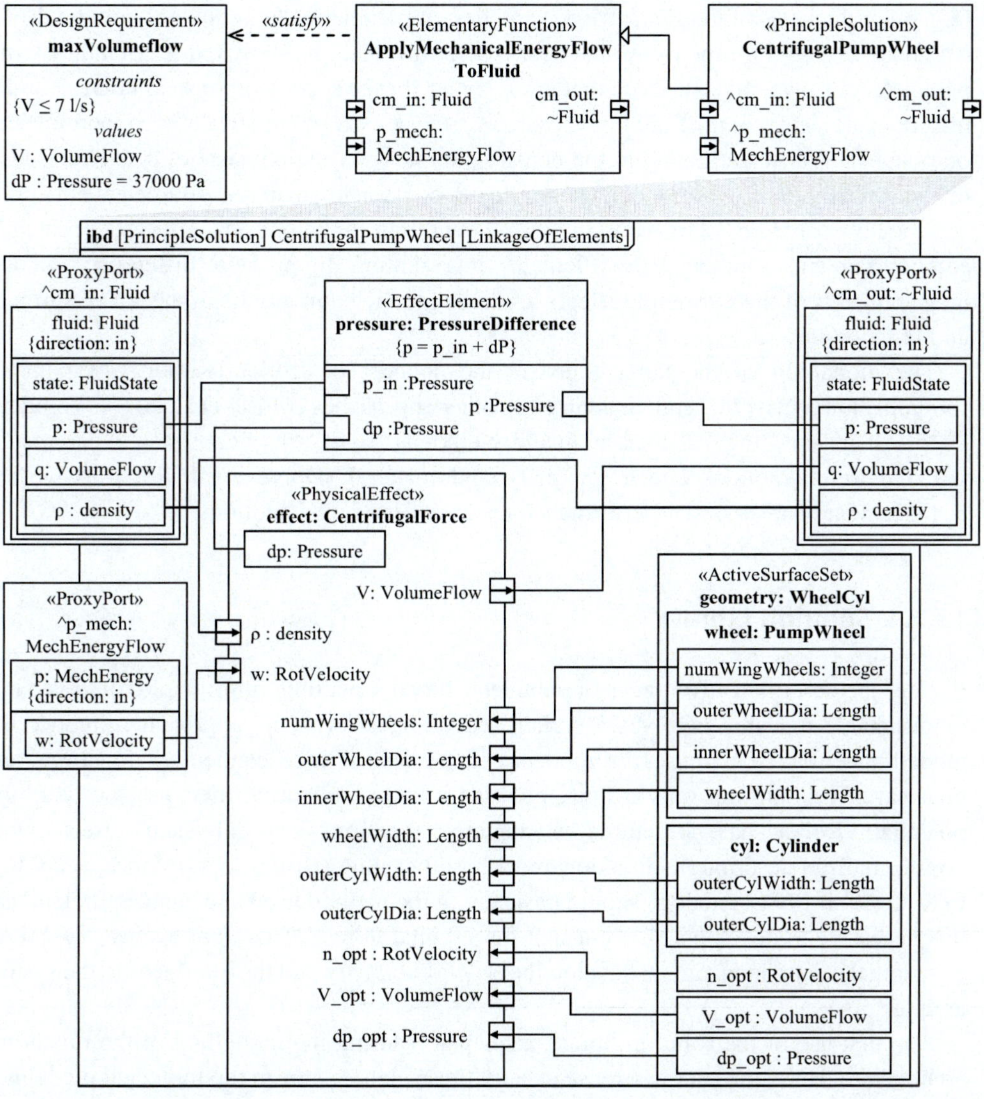

Fig. 13.4 Requirements are satisfied by elementary functions and their principle solutions whose elements, parameters and relationships are visible in the internal block diagram

ApplyMechanicalEnergyToFluid can be found in the Koller catalog (Koller and Kastrup 1994): Boyle's law, adhesion, Coulomb's law and others. Here the physical effect *CentrifugalForce* is chosen, which is modeled as «PrincipleEffect» with all relevant parameters in the principle solution model. The active surfaces are selected to match this effect: The *PumpWheel* rotates and conveys the fluid outwards against the *Cylinder*, where the induced kinetic energy is converted into static pressure and the fluid can exit through a radial opening. These active surfaces are described by a few parameters for their geometry (e.g. outer diameter of the *PumpWheel*) and design parameters (e.g. optimal volume flow), which are essential for their physical behavior. In the example shown, it is not the material parameters of the active surfaces that are relevant for the modeled physical effect, but the material parameters of the fluid flow. Therefore, the density parameters of the incoming and outgoing fluid flow are linked to the density parameter of the physical effect. This is also the reason why *PumpWheel* and *Cylinder* do not contain any material parameters here, as it is basically enabled by (Drave et al. 2020). In addition to the physical effect, the principle solution contains another «EffectElement» representing the pressure difference. Finally, the parameters of the «PrincipleEffect» are linked to the counterparts of the active surfaces and function flows.

The modelling of the latter transfers the concept of Koller (Koller 1998) into a formalization for SysML and, in contrast to other approaches (Möser et al. 2016; Weilkiens 2016; Albers and Zingel 2011; Lamm and Weilkiens 2010), uses a consistently parameter-based representation of effect, geometry and material (Drave et al. 2020). This key advancement enables initial performance testing of principle solutions.

13.2.4 Solution Library

Besides the described advantages of parameter-based modelling of principle solutions, it is unmistakable that this involves a certain modelling effort. One approach to reduce the modelling effort is to reuse the models created once. Since elementary functions and physical effects are not only a finite but also a known quantity, their reuse offers high potential. Koller has structured elementary functions and physical effects with non-formalized descriptions in a document-based catalog (Koller 1998). Since, according to Roth (Roth 1994), catalogs should basically fit the method used and enable efficient use, it is necessary to develop a new concept for a digital library. Therefore, we use SysML as conceptual design language to develop the *Solution Library* and the interfaces to the system architecture.

The first use of the solution library takes place during the modelling of the functional architecture. Here, the user can reuse exactly those elements from the finite and predefined pool of elementary functions that he needs to fulfill the requirements, see Fig. 13.5 left. By selecting the elementary function (here: *ApplyMechanicalEnergyToFluid*) the solution library is automatically filtered and only those physical effects are displayed, which can realize the chosen elementary function, see Fig. 13.5 center. After a physical effect is also selected (here: *CentrifugalForce*), the set of stored principle solutions is filtered so that only those containing the selected physical effect are listed, see Fig. 13.5 right. All these listed principle solutions are suitable as technical realization of the elementary function and are able to fulfill the initial requirement (chosen here: *CentrifugalPumpWheel*).

The described approach extends the well-known Koller catalog (Koller and Kastrup 1994) by central elements: The solution library contains not only elementary functions and physical effects (like Koller), but complete principle solutions including frequent active

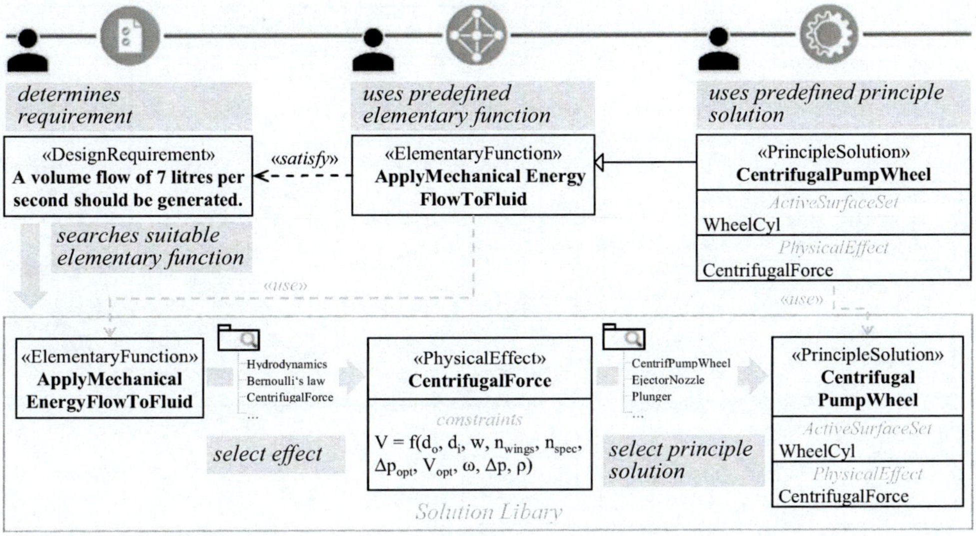

Fig. 13.5 The solution library (lower part) supports the development process (upper part) by identifying possible physical effects and principle solutions for a selected elementary function

surfaces and materials, see Fig. 13.6 left. Thus, our principle solutions can be varied not only with regard to the physical effect, but also with other active surfaces and materials. With principle solutions, initial functional tests (cf. Sect. 13.2.5) can be carried out, but more specialized models often have to be used in order to validate further requirements (e.g. service life or noise propagation). Therefore, as a second major enhancement compared to Koller, we store each principle solution together with suitable behavior models and workflows in a so-called solution element, see Fig. 13.6 right. In this way, behavior models are clearly assigned to concrete principle solutions in our solution library and can be used efficiently for virtual behavior testing of evolving solutions based on the chosen purpose (cf. Sect. 13.3).

13.2.5 Initial Performance Testing of Principle Solutions

The basis for testing a principle solution is the previously presented formalization. Due to the parameter-based representation, physical effect and active surfaces can be linked to external models, see Fig. 13.7. In our example system, the *CentrifugalForce* of the principle solution *CentrifugalPumpWheel* is linked to a MATLAB model that calculates the hydrodynamic behavior of a pump wheel. Similarly, the parameters of the «ActiveSurfaces» are linked via an Excel table to the corresponding CAD models of these active surfaces. With these links the functional behavior of the modeled principle solution (Fig. 13.4) can be calculated and validated by a software engine that handles the execution of the single solvers. In our case, we used the integrated simulation engine of a

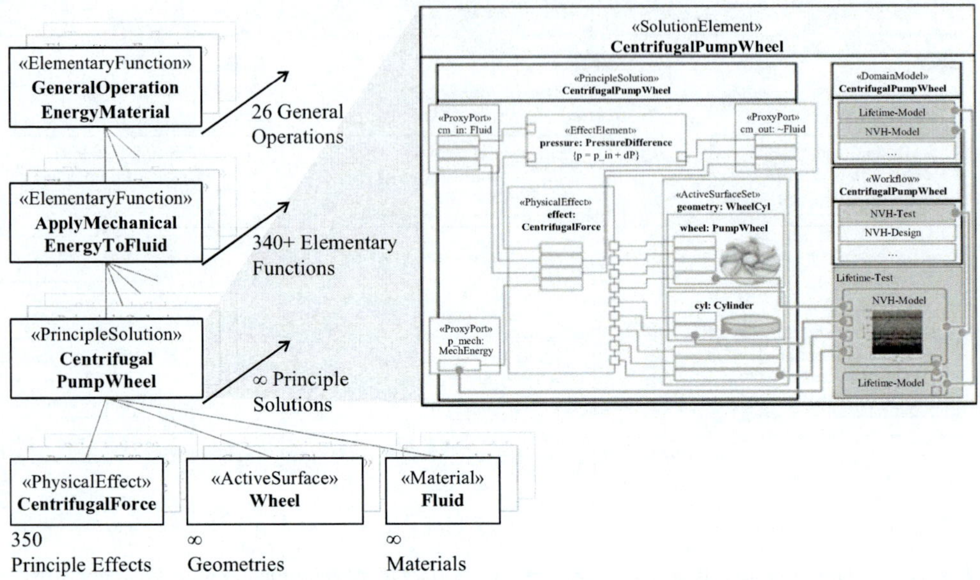

Fig. 13.6 In addition to elementary functions, physical effects, active surfaces and materials, the solution library primarily contains a set of predefined principle solutions. These principle solutions are stored together with more detailed behavior models and workflows in a so-called solution element

system modeler (e.g. Cameo Systems Modeler). For this purpose, parameter values are applied externally via function flows and read from other external models (e.g. the CAD model). All values are passed to the MATLAB function for effect calculation, which returns the calculated results. Those can be further processed in the principle solution or passed on to the subsequent principle solution via a function flow.

With the procedure described, principle solutions can be tested functionally based on the basic physical parameters that define the active surfaces without having to design complex components first. As a result, the mechanical domain can collaborate with the other domains on the function realization earlier than before and rely on objective test results on basis of a common architecture. In accordance with the described procedure, not only individual but also several principle solutions can be interconnected to create system tests. In addition, it is possible to define design processes as activities, for example to parameterize principle solutions for optimal functional fulfillment (Höpfner et al. 2021).

As the verification of the fulfillment of functional requirements using analytical equations is only a first step in the conceptual design, the discussion of the further design process using more sophisticated models and verifying further constraints is necessary. In addition, the principle solution and its parameters can be supplemented in the further course of development by more detailed models that allow to test phenomena such as cavitation or acoustics.

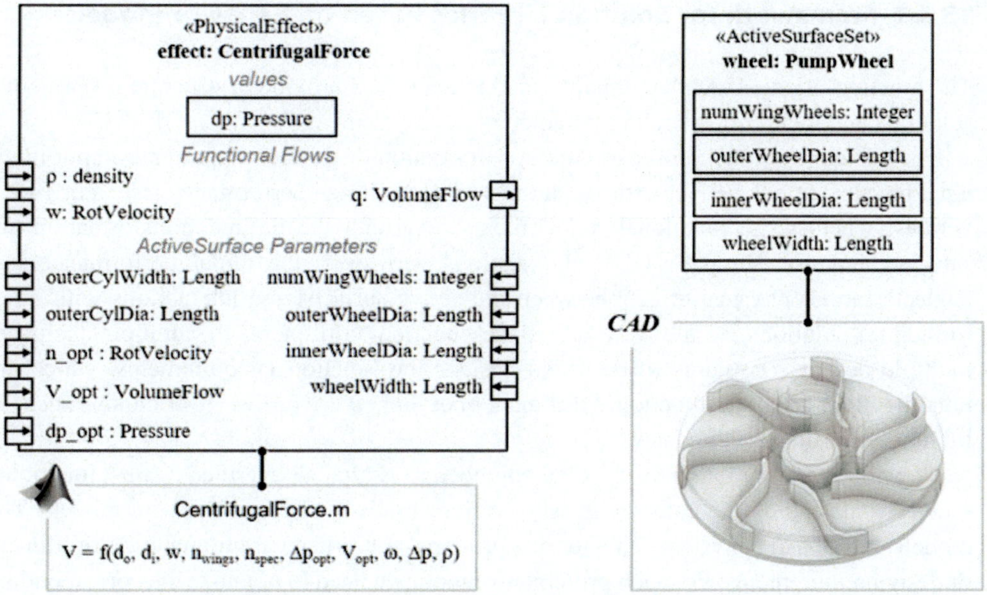

Fig. 13.7 The «PrincipleEffect» CentrifugalForce is connected to an external MATLAB model and «AcitveSurface» of the PumpWheel is connected to a CAD model (Zerwas et al. 2021)

13.3 Virtual Testing of the Behavior of Evolving Solutions

The functional testing and virtual verification of technical systems' behavior with regard to further restrictions as lifetime, efficiency or acoustics requires the use of a complex set of physical behavior models (Andary et al. 2019; Pasch et al. 2019).

These models require a specific amount of input parameters from the principle solution. Mandatory prerequisite for the efficient (re-)use of physical behavior models, and also data, economics or business models in product development, is the seamless linking of the higher fidelity behavior models with the architecture given by the principle solution model to ensure the traceability of changes down to the behavior models.

A structural framework for linking principle solutions, physical behavior models and design processes is introduced in Sect. 13.3. We propose the solution element as structuring element. The solution element contains the principle solution as described in 13.2 as architectural element. Additionally, behavior models are integrated in the solution element and connected to the principle solution parameters. By using higher-fidelity behavior models, new parameters describing the solution are added. The whole solution element evolves with new parameters and models during product development.

Furthermore, in Sect. 13.4, a suitable classification method is proposed, that can be seen as the starting point for standardization of expert models to be used to verify functions against corresponding failure modes.

13.3.1 Framework for Solution Libraries Based on Behavior Models

The function-oriented architecture described in Sect. 13.2 provides a structure for principle solutions and their corresponding parameters as key element.

The principle solution describes the transformation of different input and output flows via a physical effect, active surfaces and material. An initial performance test as in 13.2.5 verifies, whether the functional requirements regarding the desired transformation are satisfied and uses simple analytical equations. However, the initial performance test neglects further physical effects between the active surfaces and interactions with other (principle) solutions, which may affect the function fulfillment. In addition, there are multiple design constraints to be tested besides the functional requirements, which are not directly related to function fulfillment, but linked to further restrictions, such as lifetime, acoustics or efficiency.

The principle solution in mechanical engineering is typically verified against functional requirements and design constraints using various kinds of physical behavior models. The models describe the physical behavior of a principal solution regarding function fulfillment or design requirements. For each principle solution, we need to organize the corresponding behavior models and link them to the parameters of the principle solution. By connecting models to the principle solution, we can trace requirement changes down to the behavior models via function and solution. The interconnected models allow us to react to changes of requirements in later design steps, as they allow for automated repetition of simulation procedures; as well as impact analysis regarding changes. Further, the information of when to use and how to combine the physical models in order to verify requirements is needed. Therefore, we propose a solution element that combines the principle solution model, higher fidelity behavior models and testing and optimization workflows. Behavior models use principle solution parameters e.g. to predict physical behavior; verify requirements or automate design. For each solution element it is necessary to consider different models and verify different requirements. The propagated workflows allow for sequential behavior model execution in order to verify specific requirements. For virtual verification of requirements, usually a combination of multiple behavior models is necessary. Modelling the order of behavior models is done using workflows. They make the requirement verification repeatable and automatable. In workflows, we organize the model and architecture evolution as well as testing and optimization of solutions. An example for a solution element is discussed in Fig. 13.8 using the example of the coolant pump's electric motor. The motor uses the physical effect of Biot Savart's law between the two active surfaces rotor and stator. Each active surface is described via width and diameter.

The functional test for the principle solution electric motor verifies, whether the motor transforms the electrical power into mechanical power in a sufficient man-ner. However, there are further requirements related to the principle solution *Elec-tricMotor*. E.g., the principle solution has to be available for a required lifetime, efficiency might have to be as high as possible and the behavior regarding acoustics has to be pleasing. These requirements are not verified using analytical equations of physical effects. In virtual

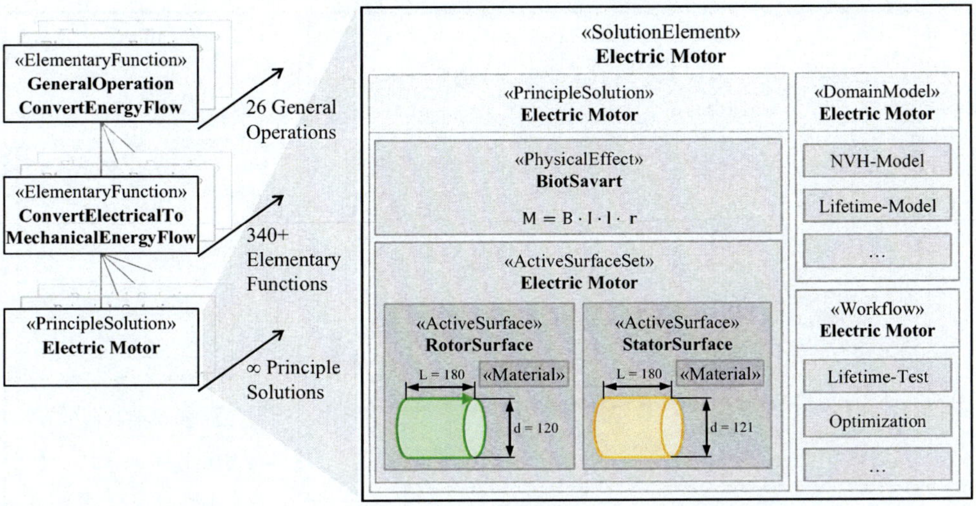

Fig. 13.8 Solution elements integrate principle solutions, their describing behavior models, as well as design and testing workflows. Workflows use the principle solution's parameters and behavior models to design and verify the principle solution

product development, specific behavior models predict the behavior for the principle solution regarding each of these requirements, and we can develop tests regarding the aforementioned requirements using combinations of these models. In the next section, we describe how to use existing models for the evolution of single solution elements as well as for the evolution of entire system models. System models consists of multiple solutions. These solutions either also consist of further solutions or refer to their solution parameters.

13.3.2 Evolving the Solution Using Physical Behavior Models

Models and workflows, which are added to the principle solution in Fig. 13.8 allow for virtual verification of the requirements, the principle solution has to fulfill. When using behavior models, we find that each model has a specific demand on parameters from the principle solution. Some models need less parameters and may be used at an early stage of design, while other models represent a high-fidelity solution requiring lot more parameters (Weber 2014). When integrating these models into our solution model we have to link the model's input parameters to the already given ones and may add new parameters. Further, models can consist of model chains that might be linked, too. In that case, outputs from one model serve as inputs of another one. Using the example of the electric engine, we derive parameter groups to be linked to the models.

As an example, Fig. 13.9 shows a still relatively simple lifetime-test workflow. The workflow combines two behavior models of the solution model electric engine, one NVH model (NVH, noise, vibration and harshness) in order to predict the torsional vibration

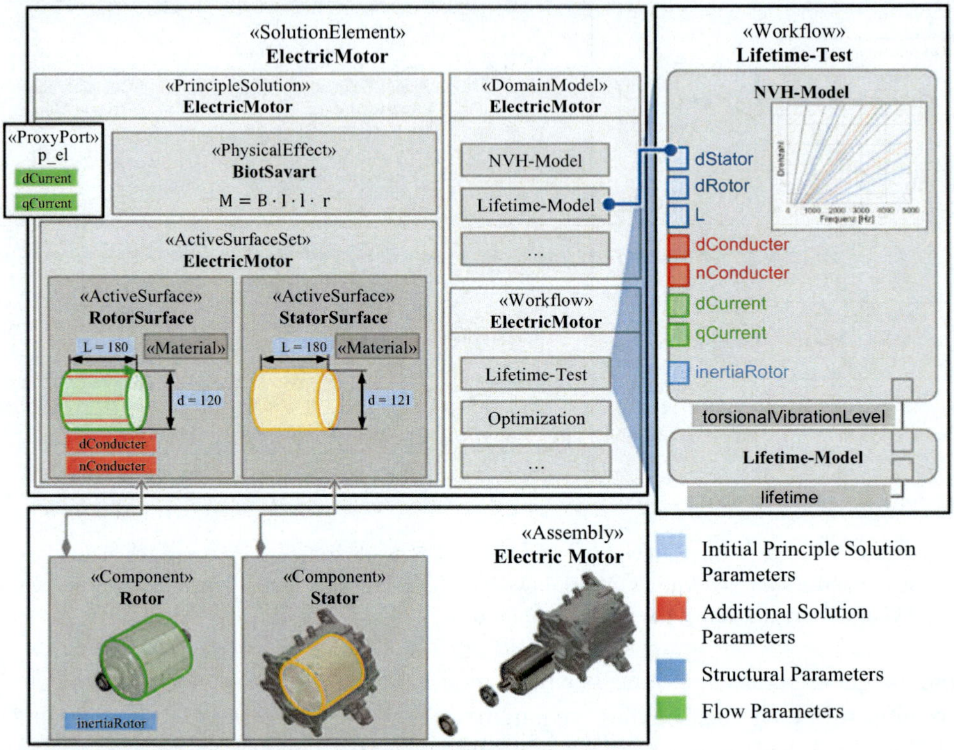

Fig. 13.9 The use of physical behavior models in test workflows during development evolves the principle solution by adding new parameters, refining geometry and physics, as well as developing components from principle solution active surfaces

level of the electric engine and one life time prediction model that uses the result in order to predict the lifetime of the electric engine. The workflow models the execution order of the behavior models and the parameter interdependencies. As the parameter need of some models is satisfied by other models, the execution order has to be modelled in the workflow to ensure parameter availability for the later ones.

In Fig. 13.9, the input parameters required for the usage of the NVH-model are shown in detail. Every model demands a specific set of parameters in order to be executable. Thereby, we differentiate four basic types of parameters:

- ***Principle solution parameters*** are already given in the principle solution in Fig. 13.8 and hence determined (in the example: *dStator*, *dRotor* and *L*, initial principal solution parameters). They can directly be linked to the model.
- For some additional parameters, we need to have more detailed information on the principle solution's active surfaces or physics. We have to continue development of the solution element, refine the active surfaces, material or physics (we add conductor elements to the rotor's active surface) and set further ***solution parameters***

(e.g. *dConductor* and *nConductor*). These parameters describe the solution itself in more detail.
- Some parameters require a further level of detail regarding the functional flow descriptions, which enter the solution via ports. For the NVH model, we need information not only about current as part of electric power, but also about the separation in d- and q-current. We add these ***flow parameters*** (e.g. *dCurrent, qCurrent*) to the functional flow, as they are inputs to the principle solution.
- Some parameters depend on the components in which the active surfaces are integrated. Components store multiple active surfaces and connect them by connecting structures. The connecting structure carries physical parameters as mass and inertia. These ***structural parameters*** can be obtained from CAD (e.g. *inertiaRotor*). As shown in Fig. 13.9, the solution element's active surfaces are the link to CAD-components, which can then be combined in assemblies and finally be aggregated to the component view of the system under development, while the system's solution elements provide a link to the functional view.

Continuing development, we are able to verify requirements using more and more sophisticated behavior models. More complex behavior models ask for more parameters and we continue evolving the solution elements and their attached components by integrating required parameters into the architecture, as the system under development increases maturity.

At a certain point of development, certain models may need a more detailed description of further, surrounding principle solutions in order to increase the detail level of behavior prediction. E.g. the stiffness of the supporting bearing system has influence on the electric motor's air gap and resulting electromagnetic forces. In this case we need a behavior model, which includes the bearing stiffness. We then increase the system scope by combining principle solution elements in a system solution element, which aggregates them. This system solution element itself does again contain behavior models and workflows. However, it does not only contain one principle solution but the solution elements of its sub solutions with connections between them. The behavior models stored in solution elements of system solutions are usually per principle solution less detailed but do have a wider system scope and consider interactions between multiple principle solutions.

Storing the whole set of principle solution, behavior models and workflows enables automated solution design of solutions from libraries. In order to set up these libraries, model classification becomes crucial. The list of behavior models needs to be classified and parameter dependencies need to be modelled. At this point, we propose model ontologies as the required next step. Therefore, an outlook on ontology development for physical behavior models is given in Sect. 13.4.

13.4 Model Frameworks and Ontologies for Efficient Model Re-Use

The solution library derived in 13.2 and 13.3 allows for a reuse of (principle) solution elements with their corresponding models in the enterprise context and classified behavior models allow the functional verification and the virtual validation of the system behavior. Expert models thus hold a key position in system development. The efficient use of expert models allows for agile development and continuous system testing. However, the solution element can only be used efficiently with a strong reusability of behavior models. In preparation for or as part of the introduction of a seamless function-oriented model-based product development, the existing behavior models have to be classified, characterized, restructured and thus prepared for efficient reuse in product development processes.

For integrating behavior models in solution elements as proposed in 13.3, behavior models have to be set up in a function oriented, modular way. In order to classify the modular behavior models, we propose a framework consisting of the three axes system scope, model purpose and model fidelity, see Fig. 13.10.

Structuring models in such a framework results in a multidimensional classification matrix for behavior models. Per model purpose, we can describe the solutions in scope and per solution identify multiple modelling fidelities with increasing detail level and parameter request. The overall behavior model for a system solution is then one possible combination of principle solution behavior models in the matrix. When using many model elements with high fidelity, we increase the total fidelity level, but also the required parameters and calculation effort, Fig. 13.11.

The classification of expert models in the coordinate system of Model Fidelity, Model Purpose and System Scope results in a classification scheme for expert models. However, models are also characterized by the input parameters they use, the output parameters they produce and the connection of their sub-models. Classification is only a first step for model reusability. By establishing standardized interfaces, expert models can be connected efficiently to handle parameter connections. Characterizing models and their corresponding parameters is required. Ontologies are a way to characterize models in that regards. Ontologies in this context describe the parameters that belong to a model and their interdependencies with each other in a formal way. Building ontologies allows for formalization and identification of dependencies and relationships between the models. That enables efficient reuse of the models to validate specific issues in development processes. Model ontologies and standardized workflows allow semi-automatic up-load into solution models and further into entire system models as discussed in 13.3.

By consistently organizing product parameters in a function-oriented architecture and linking classified behavior models to it, the presented method represents an excellent starting point for participation in highly relevant and current international research priorities such as the IEA Wind TCP Task 37, which is coordinating international research activities to analyze wind power plants as holistic systems. Within the scope of these research activities, the development of a Systems Modelling Framework and Ontology for Wind Turbines plays a key role (Dykes et al. 2017).

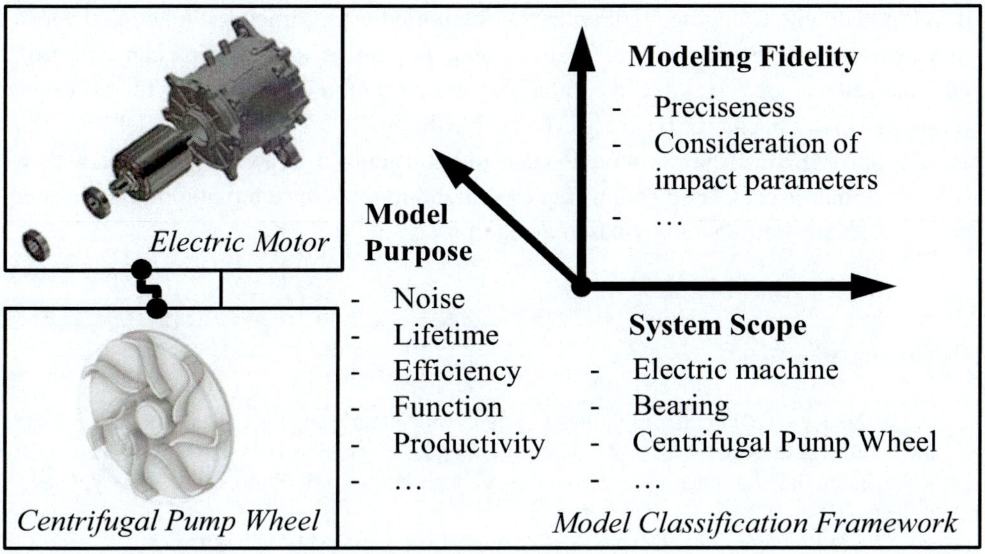

Fig. 13.10 Framework to classify expert models

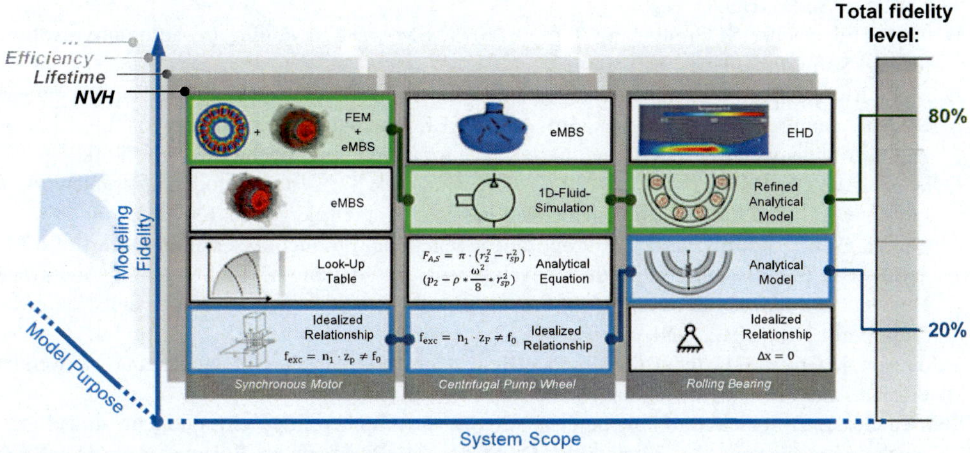

Fig. 13.11 Order matrix of expert models consisting of framework and ontologies

13.5 Summary and Conclusion

In summary, the presented method allows for consequent use of virtual behavior prediction using models throughout product development, beginning in early concept phase. It closes the gap between top level function development and expert component design and allows for reliable design decisions based on virtual models for CPS in industrial practice. Agile development with changing requirements is supported using structured functional verification processes in virtual product development. Repetitive design steps can be partially automatized and are accessible in company specific solution libraries, allowing for efficient reuse, especially in change processes. To enable the future use of expert models in solution models and MBSE, behavior models need to be organized in a clear structure with well described interfaces. Classification framework and model characterizing ontologies hence are enablers for future's model-based design process.

References

Albers A, Zingel C (2011) Interdisciplinary systems modelling using the contact and channel-model for SYSML. In: Culley SJ (ed) ICED 11: impacting society through engineering design, København, the 18th international conference on engineering design. Design Society, Glasgow, pp 196–207

Alur R (2015) Principles of cyber-physical systems. MIT-Press, Massachusetts

Andary F, Berroth J, Jacobs G (2019) An energy-based load distribution approach for the application of gear mesh stiffness on elastic bodies. J Mech Des 141(9)

Broy M (2010) Cyber-physical systems innovation Durch software-intensive Eingebettete Systeme. Springer, Berlin, Heidelberg

Drave I, Hillemacher S, Greifenberg T et al (2019) SMArDT modelling for automotive software testing. Softw Pract Exp 49(2):301–328

Drave I, Rumpe B, Wortmann A et al (2020) Modelling mechanical functional architectures in SysML. In: Proceedings of the 23rd ACM/IEEE international conference on model driven engineering languages and systems. Association for Computing Machinery, New York, pp 79–89

Dykes K L, Zahle F, Merz K, et al (2017) IEA wind task 37: systems modelling framework and ontology for wind turbines and plants. National Renewable Energy lab. (NREL), golden

Eigner M, Gilz T, Zafirov R (2012) Proposal for functional product de-scription as part of a PLM solution in interdisciplinary product development. In: Marjanović D, Storga M, Pavkovic N, Bojcetic N (eds) DESIGN 2012 proceedings of the 12th international design conference, Dubrovnik, May 2012, UNIZAG-FSB 2012, pp 1667–76

Feldhusen J, Grote K-H (eds) (2013) Pahl/Beitz Konstruktionslehre Methoden und Anwendung erfolgreicher Produktentwicklung, vol 8. Springer, Berlin

Gausemeier J, Dorociak R, Pook S, et al (2010) A computer-aided cross-domain modelling of mechatronic systems. In: Marjanović D, Storga M, Pavkovic N, Bojcetic N (eds) DESIGN 2010 proceedings of the 11th international design conference, Dubrovnik, May 2010, pp 723–32

Göhlich D, Fay TA (2021) Arbeiten mit Anforderungen: Requirements Management. In: Bender B, Gericke K (eds) Pahl/Beitz Konstruktionslehre. Springer Vieweg, Berlin, Heidelberg

Graessler I, Hentze J (2020) The new V-model of VDI 2206 and its validation. Automatisierungstechnik 68(5):312–324

Graessler I, Oleff C, Scholle P (2020) Method for systematic assessment of – requirement change risk in industrial practice. Appl Sci 10:8697–8725

Grundel M (2017) Ein Modell zur Überbrückung der derzeitigen Lücke in der modellbasierten Konzeption mechanischer Systeme. Institut für Konstruktions und Fertigungstechnik, vol 45

Höpfner G, Jacobs G, Zerwas T, et al (2021) Model-based design workflows for cyber-physical systems applied to an electric-mechanical coolant pump. In: IOP Conference Series: Materials Science and Engineering 1097, 1

Hubka V, Eder WE (1988) Theory of technical systems. A total concept theory for engineering design. Springer, Berlin, Heidelberg

Koller R (1998) Konstruktionslehre für den Maschinenbau Grundlagen zur Neu- und Weiterentwicklung technischer Produkte mit Beispielen, vol 4. Springer, Berlin

Koller R, Kastrup N (1994) Prinziplösungen zur Konstruktion technischer Produkte. Springer, Berlin

Konrad C, Riedel R, Rasor R et al (2019) Enabling complexity management through merging business process modelling with MBSE. In: Procedia CIRP. 29th CIRP design conference, vol 84. Elsevier, Amsterdam, pp 451–456

Lamm JG, Weilkiens T (2010) Functional architectures in SysML. Proceedings of the TdSE

Möser G, Kramer C, Grundel M et al (2016) Fortschrittsbericht zur modellbasierten Unterstützung der Konstrukteurstätigkeit durch FAS4M. In: Schulze S-O, Muggeo C (eds) Tag des systems engineering. Hanser, München, pp 69–78

Pasch G, Jacobs G, Höpfner G et al (2019) Multi-domain simulation for the assessment of the NVH behaviour of a tractor with hydrostatic-mechanical power Split transmission. LandTechnik AgEng:19–27

Patzak G (1982) Systemtechnik - Planung komplexer innovativer Systeme. Grundlagen, Methoden, Techniken. Springer, Berlin, Heidelberg

Ponn J, Lindemann U (2011) Funktionen. In: Konzeptentwicklung und Gestaltung technischer Produkte. VDI-Buch. Springer, Berlin, Heidelberg

Ross DT, Schoman KE (1977) Structured analysis for requirements definition. IEEE Trans Softw Eng SE-3(1):6–15

Roth K (1994) Konstruieren mit Konstruktionskatalogen Band 1: Konstruktionslehre, 2nd edn. Springer, Berlin

Rupp C (2014) Requirements-Engineering und –Management: Aus der Praxis von klassisch bis agil, 6th edn. Carl Hanser Verlag, München

Srinivasan V, Chakrabarti A, Lindemann U (2012) A framework for describing functions in design. In: Marjanović D, Storga M, Pavkovic N, Bojcetic N (eds) DESIGN 2012 proceedings of the 12th international design conference, Dubrovnik, May 2012, UNIZAG-FSB 2012, pp 1111–1122

Suh N (1998) Axiomatic design theory for systems. Res Eng Des 10:189–209

VDI 2221 Blatt 1:2019–11 Entwicklung technischer Produkte und Systeme - Modell der Produktentwicklung, 2019–11

VDI 2221 Blatt 2:2019–11 Entwicklung technischer Produkte und Systeme - Gestaltung individueller Produktentwicklungsprozesse, 2019–11

Weber C (2014) Modelling products and product development based on characteristics and properties. In: Chakrabarti A, Blessing LTM (eds) An anthology of theories and models of design. Springer, London, pp 327–352

Weilkiens T (2016) SYSMOD – the systems modelling toolbox. Pragmatic MBSE with SysML. MBSE4U

Zerwas T, Jacobs G, Spütz K et al (2021) Mechanical concept development using principle solution models. In: IOP conference series: materials science and engineering 1097, 1

14

Model-Based Systems Engineering: Discovering Potentials for Methodical Modular Product Development

Dieter Krause, Lea-Nadine Schwede, Florian M. Dambietz, and Michael Hanna

Abstract

A constantly growing market variety results in an increasing internal variety, which is reflected in increased variety costs. In order to cope with this situation, different methods for the development of modular product families and their modular product architectures were developed. During the implementation of these methods, different product data come together, which are linked in different tools. At this point, a document-based approach reaches its limits and inconsistencies occur. To counteract, the trend of Model-Based Systems Engineering (MBSE) is being integrated into methodical modular product development. Using the example of method units of the *Integrated PKT Approach for the Development of Modular Product Families*, it is shown how the deposit of a meta model of product data enables consistency. The consistent model of the method units *Design for Variety* and *Life Phases Modularization* is extended by two elements: Configuration systems and the effects of modular product architectures. A configuration system based on this enables the efficient addressing of customer requirements in sales. The linking of the effects of modular product architectures strengthens the objective of *Life Phases Modularization*. Furthermore, the resulting consistent overall model generates several analysis options and opens up new possibilities, such as the establishment of *Digital Twins*.

D. Krause (✉) · L.-N. Schwede · F. M. Dambietz · M. Hanna
Hamburg University of Technology (TUHH), Institute of Product Development and Mechanical Engineering Design (PKT), Hamburg, Germany
e-mail: krause@tuhh.de

D. Krause, E. Heyden (eds.), *Design Methodology for Future Products*,
https://doi.org/10.1007/978-3-030-78368-6_14

14.1 Introduction

Product development is faced with the great challenge of covering an ever-increasing external variety demanded by the market, which is caused by megatrends such as individualization (Krause and Gebhardt 2018). Conversely, a high level of product standardization offers the potential to achieve higher productivity and shorter throughput times with more automated processes through reduced process complexity. Modular product architecture design represents an approach from the product development point of view for overcoming this conflict of objectives. The goal is to control the variance of variants, which is regarded as a central field of application in a company (Krause et al. 2021).

Modularity in this context is a gradual property of product architectures. It can be described by the properties and characteristics of modularity (Hackl et al. 2020; Salvador 2007), where the subdivision into properties and characteristics is based on Weber (Weber 2007). The properties of modularity are *commonality* and *combinability*. *Commonality* means that a module can be used several times in different or the same product variants. This is made possible by the characteristics *interface standardization* and *oversizing* in order to be able to use modules several times. In addition to *interface standardization*, *combinability* is also made possible by *decoupling* and *function binding*. Commonality makes the reduction of the internal variety possible whereas a large external variety can be made possible by the *combinability* (Hackl et al. 2020). Modular product structuring—also called *platform design*—and variant management is the subject of current research at various research institutes. Bonvoisin et al. have focused on drivers, design principles and metrics of modularization (Bonvoisin et al. 2016). Otto et al. compared several modularization methods and generated a global view on the approach of modularization methods. The methods existing in the literature were assigned to different activities in order to clarify, which method supports in which phase of the product family development (Otto et al. 2016). Krause et al. likewise give an overview of methods for the development of modular product architectures and discuss the potentials and limits of modularization (Krause et al. 2018). Gebhardt et al. focus on the strengthening of knowledge transfer to industry. They define the platform strategy, the module strategy and the common parts strategy and provide an adapted overview of methodical approaches (Gebhardt et al. 2016).

The development of modular product architectures is increasingly holistic. The approaches often pursue a systemic view, which aims at the analysis and synthesis of technical products. According to Bender, this allows overall tasks to be broken down into subtasks, and different technical disciplines can be integrated using this approach (Bender et al. 2018). Interface consideration between different disciplines also takes place. For example, the integrated product generation model (iPeM) follows the goal of creating an interface between process management and product development (Albers et al. 2016). Further application cases for integrative approaches are product service systems (Rennpferdt et al. 2019) or the linking of structural optimization and modularization (Hanna et al. 2020).

As already indicated, many life phases and disciplines are involved in the development of modular product architectures. Module drivers from different life phases are integrated in modularization methods, in order to consider product-strategic aspects in the module formation. Impacts and effects of modularization on all life phases could also be identified (Hackl et al. 2020; Schwede et al. 2020a). Currently, approaches for the selection of suitable product architecture concepts (Richter et al. 2016) or also the selection of modularization methods (Schwede et al. 2019b) are researched in dependence of impacts.

Current research is investigating various trends in the field of modular product architecture. For example, methods are being developed in which future changes in internal and external variety are taken into account at an early stage in the development processes (Greve et al. 2020b). In this context, the scenario technique can also be used (Gausemeier et al. 2000) in an adapted way. The trend towards individualization poses a challenge for modular product development. Gräßler developed a methodology for the redevelopment of customer relevant modular kits (in German called "Baukasten"), which considers continuously variable components (Gräßler 2004). It also focuses on the extension of already known methods for the design of variable individualization options. Current research in this area deals with individual performance fulfilment through product individualization (Kuhl et al. 2021). Individualization also plays a role in sales. With the help of product configurators, individual customer requirements can be taken up and implemented (Rennpferdt et al. 2020a; Seiler et al. 2020b). The consideration of future product characteristics and also the consideration of specific customer requirements lead to necessary product changes and thus to an increasing variance-induced complexity (Lindemann 2009), which causes indirect costs in different life phases. Thus, the calculation of complexity costs plays a major role. By a cost forecast these can be considered early in the course of the concept selection (Ripperda and Krause 2015). The structured treatment of the complexity of product and process is supported by a situational use of development methods and product models (Lindemann and Ponn 2011).

The holistic approaches and the trends presented have in common that more and more knowledge is built on top of each other in the course of methodical modular product development and must be linked with each other in the most diverse ways. This leads us to another digital trend in product development: Model-Based Systems Engineering (MBSE).

In order to thematize MBSE, a distinction must first be made between it and Model-Based-Engineering (MBE). If models represent an integral part in a development process, one speaks of MBE. The goal of MBE is to increase the effectiveness of the engineering as, for example, inputs are generated from these models in individual steps of the development process (Paetzold 2017). Software plays a major role in depositing the process with models (Liebel et al. 2018). Software extends standardized 3D representations with additional product information, which is directly related to the deposited models. MBSE can be understood as a sub-discipline of MBE. This sub-discipline includes all models that support the aspects of systems engineering. MBSE describes an interdisciplinary approach to the description of technical systems. The description proceeds thereby mostly from requirements, which result from the underlying use cases. Alt describes the core of the modeling process according to the scheme of Input-Processing-Output (Alt 2012; Delligatti 2013; Walden et al. 2015). Requirement engineering represents a major activity within MBSE. Implicit requirement management must of course take place throughout the entire product development process (Göhlich and Fay 2021). Due to the increasing scope of products and company processes, MBSE is finding its way into a wide variety of industries.

Nowadays, it is used more and more to represent complex systems such as products, which contain elements from different disciplines. In addition, process steps with many participants are modeled. Through modeling, a process-accompanying system model can be developed that visualizes the dependencies of different stakeholders (Riedel et al. 2020). In order to fully exploit the potential of data linking on the company side, suppliers are increasingly demanding the use of MBSE. However, continuous process support with MBSE is very costly and not necessarily profitable for suppliers. However, as soon as MBSE is established and anchored in the processes, the modeling does not cause any additional or duplicate work, and the data and models are used and linked consistently, the benefit should increase (also for suppliers). Wilking et al. are investigating the extent to which the additional effort caused by the use of MBSE can be compensated (Wilking et al. 2020). Furthermore, the goals to be achieved through the use of MBSE should be clearly defined in order to make the success of MBSE measurable (Kößler and Paetzold 2017). In addition to the application of MBSE in industry to support processes, this approach also represents an interesting option for managing the data relationships in methodical product development. The associated potential of model-based data links fits in with the increasingly integrative approach in the field of modular product architectures. Initial solutions in this area are already present in the literature.

MBSE is used, for example, to enable a consistent representation of modular kits. This means that knowledge can be used across generations (Albers et al. 2019). MBSE can also help in the new development of modular product families to use existing knowledge for further product variants and to make changes consistently traceable during development (Küchenhof et al. 2020). Configurator systems can also be strengthened by MBSE. By storing system models, optimal variants can be configured while taking user goals into account (Wyrwich et al. 2020). In the applications presented, SysML is mainly used as the modeling language to model data relationships. In addition, the presented applications make clear that modeling with SysML in the sense of MBSE as a competence is increasingly relevant also in the education of engineers. For this purpose, e.g. the Karlsruhe SysKIT Approach was developed (Matthiesen et al. 2014).

As can be seen from the research landscape presented, there are many new research trends in the context of modular product architectures, making the topic increasingly comprehensive. First approaches show the potential of MBSE to support product development. Besides the presented application examples, especially methodical product development can benefit from MBSE. In this book chapter, the application of SysML in methodical product development is presented on the modeling of method units of the *Integrated PKT Approach for the Development of Modular Product Families* (abbr.: *Integrated PKT*

Approach). This is based on a consistent meta model of product data developed specifically for this purpose.

14.2 Integrated PKT Approach for the Development of Modular Product Families

The development of the *Integrated PKT Approach* is based on the fundamental idea of the conflict of objectives described at the beginning between desired external variety and required internal variety (Fig. 14.1). The *Integrated PKT Approach* takes into account diverse strategies and approaches of variant management (Greve et al. 2020a; Krause and Gebhardt 2018).

The current trends in the field of modular product architectures are addressed in different method units (Greve et al. 2020b). Method units are continuously being developed in the application fields of aerospace, mechanical and plant engineering, and medical technology (Rennpferdt et al. 2020b) and tested in industrial collaborations (Rennpferdt et al. 2020a). The individual method units can be combined with each other in a way that is specific to the application in order to provide customized support for companies that want to reduce their internal variety (Krause and Gebhardt 2018). This approach is characterized in particular by the fact that the targeted redesign of product architectures takes place not only at the conceptual level. The approach also provides a constructive redesign, a modification or even a redesign of components in order to reduce variant-related complexity. In addition, it is a workshop-based approach in which experts with their special product knowledge from different disciplines are integrated. In the course of this, discussions in project teams are simplified through visualizations of data relationships. The basis of the *Integrated PKT Approach* is the *Design for Variety* and *Life Phases Modularization*. *Design for Variety* can be used to achieve a better starting point for *Life Phases Modularization*. It combines technical-functional and product-strategic aspects for modularization. The aim is to achieve a modular product architecture that is geared to strategic, company-specific and product-specific benefits (Krause and Gebhardt 2018).

14.2.1 *Design for Variety* and *Life Phases Modularization*

The method unit *Design for Variety* can be understood as preliminary stage for the *Life Phases Modularization*. The implementation of both method units is supported by tools (Fig. 14.2).

The goal of *Design for Variety* is to arrange components variant-oriented, whereby only a small part of the components is to depend on the customer relevant properties (Kipp et al. 2010). In a first step, the external variety is taken up with the help of the *Tree of External Variety* (TEV) (Fig. 14.2, upper left), in which the customer relevant properties and their characteristics are recorded. For the representation of the variety of the functions, the

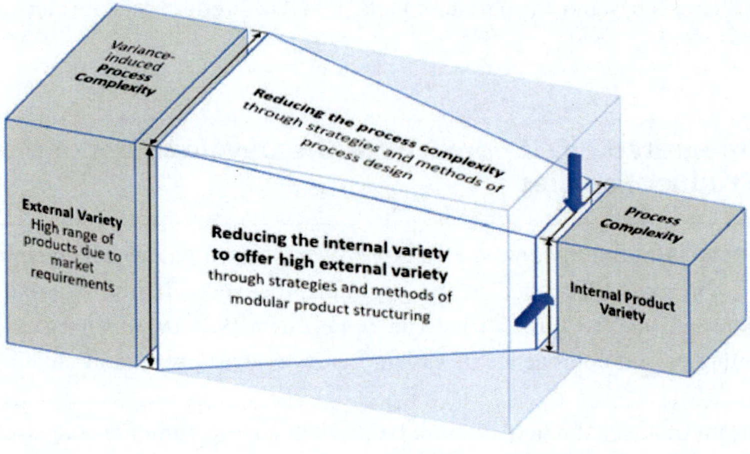

Fig. 14.1 *Integrated PKT Approach for the Development of Modular Product Families* (Krause and Gebhardt 2018)

Product Family Function Structure (PFS) (Fig. 14.2, upper right) is created (Kipp et al. 2010). The representation of the internal component variety of a product family is done in the *Module Interface Graph* (MIG) (Fig. 14.2, middle left).

The MIG shows a reduction of the design to the essentials. Like the other tools, the MIG presents the information of all product variants in the product family. The components are shown in a simplified form. The variety of the components is determined by the color of the filling in the visualizations. Standard components (white) exist in only one version within the product family, while variant components (gray) exist in several defined versions. Optional components are marked by a dashed frame. The components are connected to each other via color-coded flows (Gebhardt et al. 2014; Krause and Gebhardt 2018). The variant-oriented design of the product architecture itself is done with the help of the *Variety Allocation Model* (VAM) (Fig. 14.2, middle right). This model shows the connection between the customer relevant properties, functions, working principles and components and makes it possible to revise these by applying the four ideals of variety-oriented product structuring with the aim to reduce the variant component and to increase the standard components (Kipp et al. 2010). Within the scope of the method unit *Life Phases Modularization*, technical-functional and product-strategic views are taken into account (Greve et al. 2020a). First, the modularization of the components takes place from a technical-functional view. For this purpose, the heuristics according to Stone et al. (2000), for example, are applied to the MIG. For the product-strategic modularization of the individual life phases of a company, *Network Plans* (NP) are provided in which modules are formed on the basis of life phase-specific module drivers (Fig. 14.2, bottom left). A special case is the life phase sales, in which the customer relevant properties are used as additional module drivers. The modularization concepts of the individual life phases are then collected in the *Module Process Chart* (MPC) and subsequently harmonized (Fig. 14.2, bottom right).

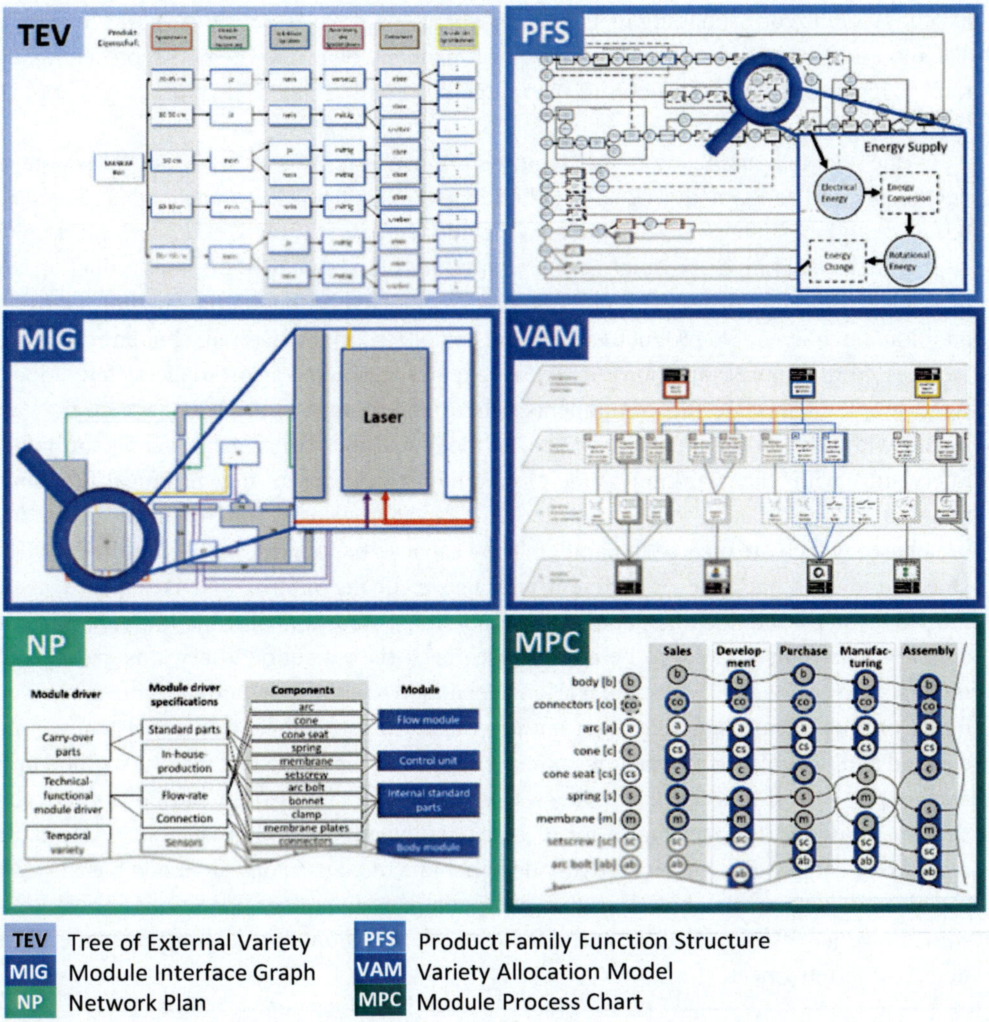

Fig. 14.2 Tools of the method unit *Design for Variety* (top and middle; according to Seiler and Krause 2020) and tools of the method unit *Life Phases Modularization* (bottom; according to Greve et al. 2020a)

14.2.2 Interim Summary—Deficits of the Document-Based Approach

For the execution of the individual method steps, different information and also different tools are needed. The tools were developed for the visualization of the data correlations. Tools in the context of methods are instruments that are intended to support the execution of method steps (Gebhardt and Krause 2016). These tools differ in terms of versioning. For example, the TEV only represents an actual state and is therefore to be regarded as static. In the VAM, among other things, data correlations are changed in the tool during the execution of the corresponding method step, whereby a new version of the VAM is created. The modification of the tools is currently done manually on printed posters representing the tools. This also makes it difficult to map large product families. Therefore, the mappable limit for the manual approach is about 80 components.

In the method application, the required information is thus only stored in paper documents and tools of individual method steps are not linked to each other. Precisely because product strategy aspects are also included in *Life Phases Modularization*, knowledge from a wide range of disciplines is required for individual method steps, which means that many different disciplines are involved in the execution of the method steps. In addition, not only single product variants are considered in the methods, but entire product families. Furthermore, some information is taken up again and again in the different tools. An example of this are the components. These appear first in the MIG, where they are linked with other components via flows. In the VAM the variant components are linked afterwards with the working principles. They reappear in the technical-functional modularization in a next version of the MIG and then also in the NPs in the individual life phases, which are created in parallel. They can also be found several times in the MPC. Due to the fact that a lot of different and also recurring information is required in the individual steps, the documentation of the individual steps and also the individual results are very important in order to be able to trace back the execution afterwards and accordingly also to be able to understand it (Hanna et al. 2018). If the method units are carried out based on individual documents that are not linked to each other, data consistency is not ensured and redundant information sources can occur. Due to the fact that the method units *Design for Variety* and the *Life Phases Modularization* can also be carried out independently and due to the fact that a lot of expert knowledge is required in some steps, the method units have a lot of subjectively designable parts. Decisions are made based on the specific products or also based on the given company boundary conditions. Thus, it can sometimes be difficult to place individual module decisions in the context of the initial objective in retrospect.

14.3 Potentials Through Model-Based Approaches

In order to maintain an overview and to address this problem, data-driven management in the form of MBSE is used. As the approach no longer has to be paper-based, the limit of the components that can be mapped can be extended, as this is no longer a limiting factor. Another important point is the strengthening of consistency. This can be increased enormously by the use of model-based approaches. Other advantages coming with the MBSE environment are an increased plausibility check possibility, increased meta model transparency, increased product architecture maintainability and versionability for product generation developments. If a system does not contain any conflicting information, it is considered to be consistent. Inconsistencies are thus often a source of contradictions in the models, which are caused by a lack of consistency management as well as knowledge that is not explicitly documented and can lead to errors (Seiler et al. 2020b). Consistency in the data will enable traceability, for example of module decisions. Traceability of subsequent changes can also take place. This facilitates and enables the maintenance of the data in case of changes and adjustments. Since changes no longer have to be tracked manually, they are more controllable. The use of MBSE also has an impact on the methodical procedure. First of all, it has to be clarified that the introduction of MBSE does not contradict the workshop-based method implementation. On the contrary: MBSE strengthens the interdisciplinarity, which plays a major role in the methodical process: Subjective aspects of the stakeholders can be integrated into the development process, enabling acting in a more objective-oriented manner. The module decision can also be supported by software support. Through additional visualizations of data correlations, these become more comprehensible and can be actively used. The follow-up to method implementations is changing. By implementing the generic data contexts, workshop results can be documented easily and quickly in a model-based manner. Also, intermediate results of individual method steps can be recorded and are thus stored in a traceable manner.

14.3.1 Ensuring Consistency Through the Development of Meta Models

As already mentioned, the potential of model-based approaches is manifold: larger amounts of data can be processed, traceability, e.g. of changes, is strengthened and method implementation is simplified. The consistency enablement in itself can be understood as the basis for this. The consistency is thus made possible by a model-based approach in which the methodical modular product development is deposited with a consistent meta model for product data (Seiler et al. 2020b). To exploit this potential for the *Integrated PKT Approach*, it is backed by a meta model of product data developed for this purpose (Fig. 14.3). The meta model of product data defines the scope of the approach and provides a basis for software-supported implementation in modeling languages such as SysML.

In the meta data model, the individual elements of the tools are linked to each other via connections, which are not specified in more detail here in order to remain solution-neutral. The scope shown in Fig. 14.3 contains the tools presented in Sect. 14.2.1. It should be noted that this model is extensible due to its clear representation (Hanna et al. 2018).

14.3.2 Consistent Model-Based Implementation in SysML

The data relationships defined in the meta data model can now be implemented using MBSE. In the context of MBSE, SysML is a modeling language for object-oriented modeling of system models. The SysML contains nine different diagrams, from which four are structure diagrams, four are behavior diagrams and one is a requirement diagram (Alt 2012). They are used to model a functional system architecture at the system level. For example, Cameo Systems Modeler or Papyrus can be used as a modeling environment. The

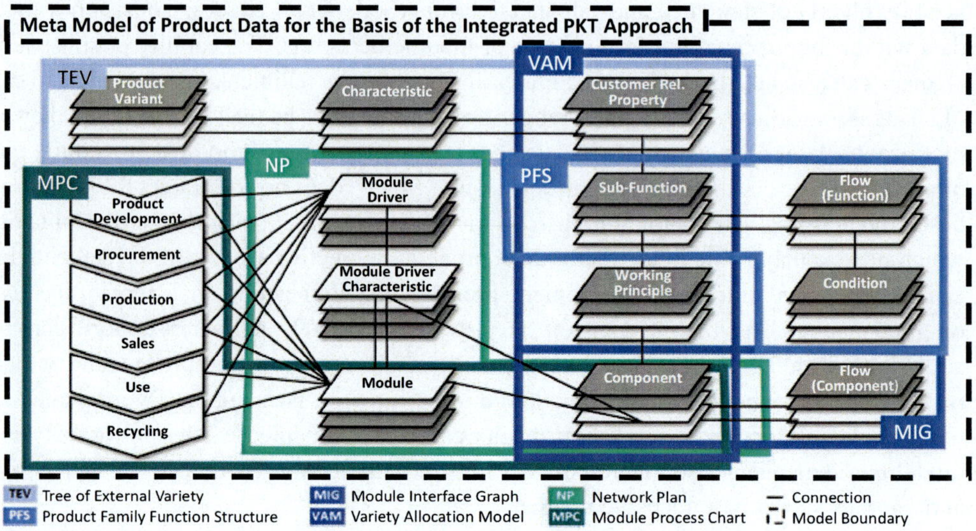

Fig. 14.3 Meta model of product data for the basis of the *Integrated PKT-Approach* (Hanna et al. 2018)

Cameo Systems Modeler has the special feature of providing a unique data store. In addition, it has more diagrams, tables and matrices, which allow the analysis of data relationships in the diagrams. With the help of consistent modeling, which is made possible by MBSE, horizontal and vertical traceability of individual elements can be enabled (Gilz 2014). A unique database, i.e., each element is contained only once in a model, and a generic description of the data relationships, for example in a meta data model, ensure consistency and the continuity based on it (Hanna et al. 2018). Thus, the basis of a model is set. In the following, such a model is presented, using a laser system as an example. The model is based on the meta data model presented in Sect. 14.3.1. In addition, the Cameo Systems Modeler is used to generate a uniform data base (Fig. 14.4). With the help of the elements contained therein, the tools from Sect. 14.2.1 could be built.

In the center of the figure, a containment tree can be seen, which illustrates the consistent and unique data base. On the sides, selected tools of the *Integrated PKT Approach* are presented, which, as can also be seen in the meta data model in Fig. 14.3, all access elements of the "components" group. By storing the data at a central place we have a consistent data management. Due to the fact that there is a meta data model, other product application examples can now be created quickly and according to the same schema.

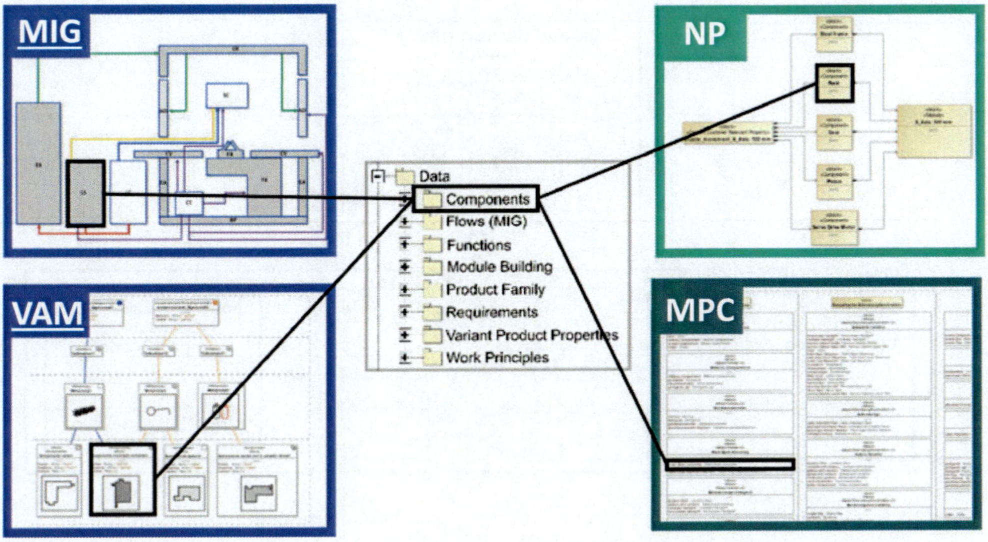

Fig. 14.4 Implementation of tools on a consistent data base in SysML (adapted from Eichmann et al. 2018)

14.4 Extension of the Model-Based Implementation on the Basis of Two Application Examples

Methodical product development can be strongly supported by MBSE. For example, different tools can be demonstrated based on a consistent meta data model. However, the storage of a consistent meta data model also has many other advantages. In this section, two applications are presented, which could be developed on the basis of the model provided in Sect. 14.3.2 (Fig. 14.5). This results in an integrated holistic model.

These are on the one hand a linked configurator system and on the other hand the linking of concepts for modular product families with their effects on economic target values. In the following two sections, the two application examples for extensions are presented. The links, which are shown as generic plus signs in Fig. 14.5, will be detailed and written more precisely.

14.4.1 Configuration Systems for Laser Processing Systems

Based on a modular product architecture, configuration systems are described as indispensable for mastering the increasingly comprehensive, variant-rich product architectures (Seiler et al. 2019). The configuration systems represent an instrument to optimally map concrete requirements of a customer to the most suitable product variant. However, a

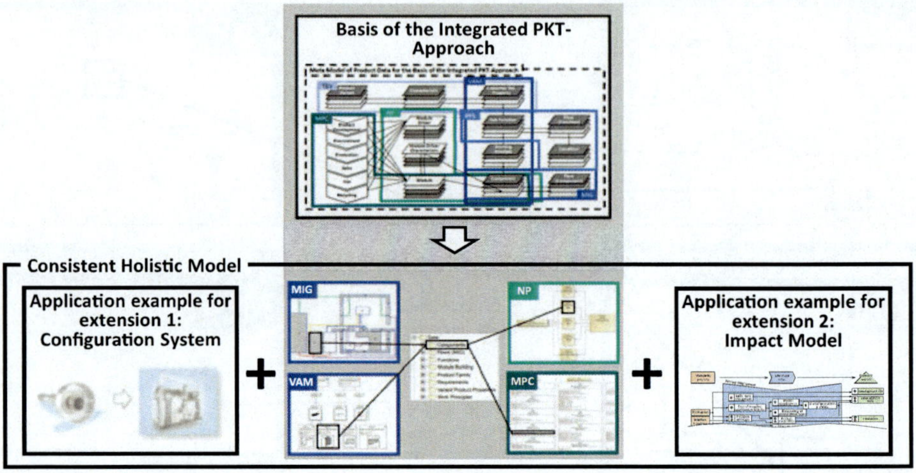

Fig. 14.5 Concept for the development on an integrated holistic model

general limitation is that the configuration options usually cannot cover the complete range of customer requirements exactly (Liebisch 2014).

Since customers do not buy customer-relevant features, but rather decide on the basis of these features for variant module versions, a configurator generally has the task of processing all the information between customer requirements and bill of materials. This process is also often referred to as "translation", representing one of the main tasks of a configuration system by translating customer requirements into discrete modules and module variants while consistently checking the determined product variant's plausibility. The interaction platform towards the customer such a configuration system corresponds here to the front-end, by means of which the modular product architecture incorporated in the data structure is made accessible to the user. In this context, the use of MBSE offers a way of managing this modular product architecture with all its dependencies and limitations with regard to the configuration of the modules, which is intended to make verifiability for consistency traceable. In addition, it is not only necessary to ensure that the configurability function is visible to the user, but also to document it in conjunction with all decisions made on the basis of the product architecture and the set of dependencies and constraints linking the individual modules. Support for the maintenance of the configuration system in the course of changes or versioning of the underlying product architecture must also be guaranteed. MBSE opens up the potential to efficiently use structural and behavioral information as well as abstracted links in order to merge individual models into a coherent meta data model. In addition, the use of an appropriate MBSE tool, such as Cameo Systems Modeler, makes the individual configuration options traceable and verifiable. Accordingly, the database of such a configuration system must be able to map qualitative data, such as requirement links or customer relevant properties. This becomes all the clearer when the most important requirements for configuration systems are considered: These are to enable a consistent product configuration on the basis of a complete information system as well as the possibility of plausibility checks.

Above all, the forward and backward integration of all systems is one of the greatest advantages created by MBSE. This database is also used when using product configurators, which are described below using the example of customized laser machines. At this point, one majorly important fact to be stated considers the relation of the customer and the company product. As for the example of customer individual laser processing machines, customers tend to consider the individual machines' (the company product) as black boxes, forming complex systems which enable the processing of the customer product. The customer product, for example a part of a car's gear train, imposes a set of customer requirements towards the machine they are processed on. As the customer itself usually is not able to express these requirements in the perspective of the machine building company, a translation process is required. The customer requirements—and therefore the customer perspective—is translated into the company perspective, leading to a set of customer relevant properties. These customer relevant properties can then be used as a baseline for the subsequent configuration process. In order to realise the aforementioned translation of customer relevant properties into suitable module variants, selected tools of the *Integrated PKT Approach* are examined. Here, the use of a NP adapted to the special requirements of a product configuration system appears to be the most suitable from a sales perspective (Fig. 14.6).

Figure 14.6 shows both the generic structure of this NP (here: *Configuration Network Plan*) with its relation to customer and company product and the implementation of such a possible NP for individually configurable laser welding systems in the MBSE environment of the Cameo Systems Modeler (Seiler et al. 2020a). As the figure displays, the customer product imposes a set of customer requirements towards the company. These customer requirements are then used in order to determine the matching modules as well as the components they consist of.

By directly assigning customer relevant properties to components clustered into individual modules, the underlying product architecture can be modelled in its entirety. By using corresponding dependency matrices for the definition and representation of the individual object links, the set of rules for completing the modular kit can also be modelled semantically. The extraction of this modular product architecture as well as the set of rules via a corresponding user interface then in turn provides the continuous and consistent data basis for the configuration system. With reference to this database, the appropriate product variant is determined for each data set of customer requirements and their individual characteristics are recorded via the user interface of the configuration system (front-end) (Laukotka et al. 2020b).

In the case of the exemplary gear train part, one of the customer requirements describes the ability of the machine to process the part in its geometric dimensions. This customer requirement is then translated by using the configuration system's frontend into a customer relevant property. By applying the configuration hyperspace algorithm as described in Seiler and Krause (2020), the configuration system determines the corresponding module

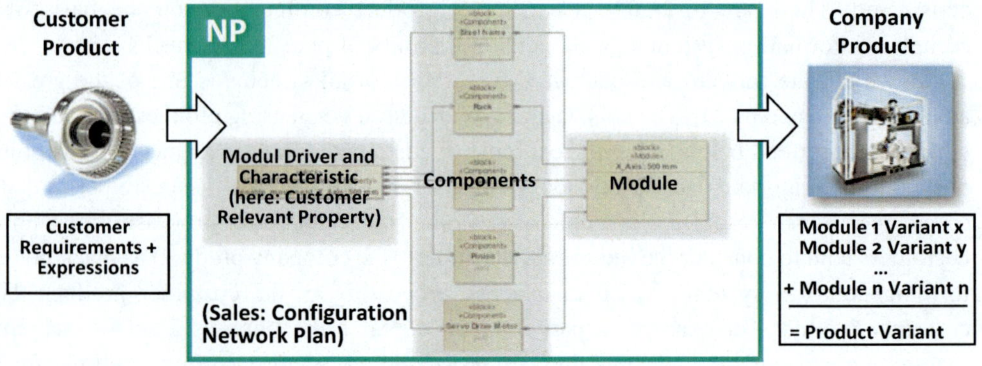

Fig. 14.6 *Configuration Network Plan* of laser systems

to this customer relevant property according to the underlying NP. Furthermore, the components and therefore, specific article numbers forming individual models are linked within the MBSE structure as well, enabling an automated generation of a final product variant bill of material (BOM). All other customer requirements are linked to corresponding modules analogically, leading to a final configured product variant.

14.4.2 Model-Based Representation of the Effects of Modular Product Families

Another application of the procedure of modeling based on the meta model of product data in Fig. 14.3 is the representation of the effects of modular product families, which have been recorded in the past years and visualized in an *Impact Model of Modular Product Families* (abbr.: *Impact Model*) (Hackl and Krause 2017; Hackl et al. 2020). The *Impact Model* shows impact chains and is a cause-effect model. Based on the properties and characteristics of modularization, effects are presented in the form of impact chains, which are assigned to the different product life phases. Finally, these effects lead to economic target values time, costs, quality or flexibility. The initial literature-based content of the *Impact Model* was validated in industry surveys and interviews (Schwede et al. 2020c; Greve et al. 2020a). The *Impact Model* is a visualization that contains many different elements (Fig. 14.7, top).

In addition to the main visualization, additional information such as boundary conditions for individual effects are available. An example therefore is the boundary condition 'production type'. This boundary condition has an impact on the effect 'lowering procurement costs': Due to the strengthening of the *commonality*, the effect of 'lowering procurement costs' is more pronounced for mass producers than for individual producers (Hackl and Krause 2017). Other additional information may include, for example, key figures or validation results for individual effects.

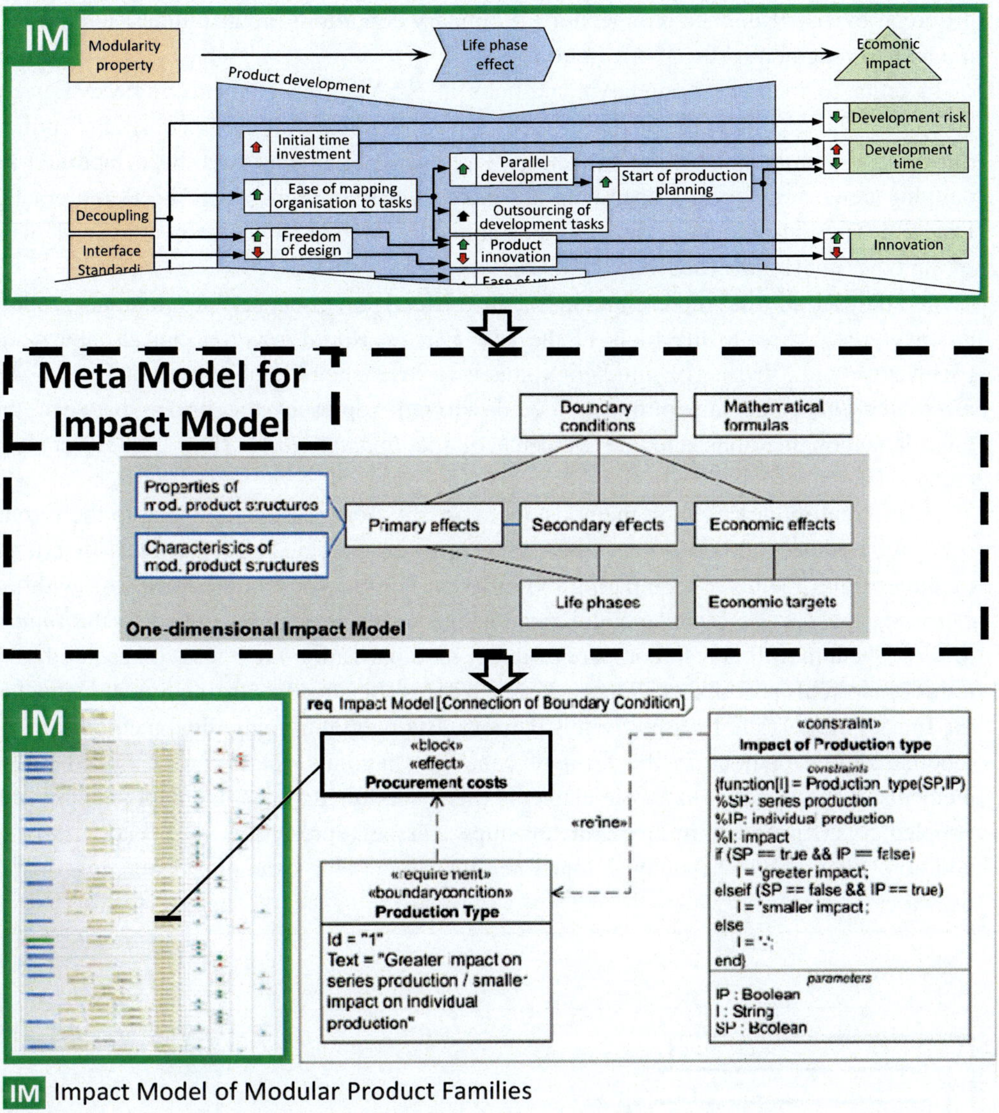

Fig. 14.7 Procedure for model-based *Impact Model*

For configuration systems, the model is based on the meta model of product data. As described, there are a large number of different elements for the *Impact Model*. In order to obtain an initial overview, similar to the development of the meta model of product data, a meta model for the *Impact Model* was created according to the same schema (Fig. 14.7, middle). The *Impact Model* itself is represented in a Block Definition Diagram, with the various visualization elements integrated as Blocks. The Block Definition Diagram is a good choice in this case, as it can be used to represent a wide variety of elements with different, even self-defined, connections. Boundary conditions are also integrated into the model to strengthen it (Fig. 14.7, bottom).

In order to map the boundary conditions in SysML, the elements in SysML were analyzed. The combination of the element types "requirement" and "constraint" turned out to be the most suitable for modeling. The SysML-elements have been modified by defining new stereotypes. Constraints are linked via Requirements in the Requirements Diagram. The causal relationship described above is stored in an if-statement in an element of the type 'constraint' (Schwede et al. 2019a).

In addition to the representation of the effects of concepts of modular product architectures, a link to the tools of the *Life Phases Modularization* can also be made (Schwede et al. 2019b). The aim behind this is to strengthen the objectives of *Life Phases Modularization*. For this purpose, module drivers represent the bridge between the modularization methods and the elements of the *Impact Model* (Fig. 14.8) (Schwede et al. 2020a).

In different modularization methods, there are different reasons that lead to the formation of modules; this is the case with product-strategic and technical-functional modularization methods. Comparing the different reasons for module formation with the effects in the *Impact Model*, modularization methods can also be linked to the *Impact Model*. The aim of this is to compare different modularization methods with each other in terms of the goals and effects addressed in each case. The already existing SysML-model of the Impact Model is to be used for this purpose. By integrating individual method steps of modularization methods in the form of behavior diagrams, these can be linked to the elements of the *Impact Model* via elements (here: module drivers). By implementing this coupled SysML-model, indirect relationships can be represented in special views. In addition, queries can be defined that filter the model, for example, depending on the use case.

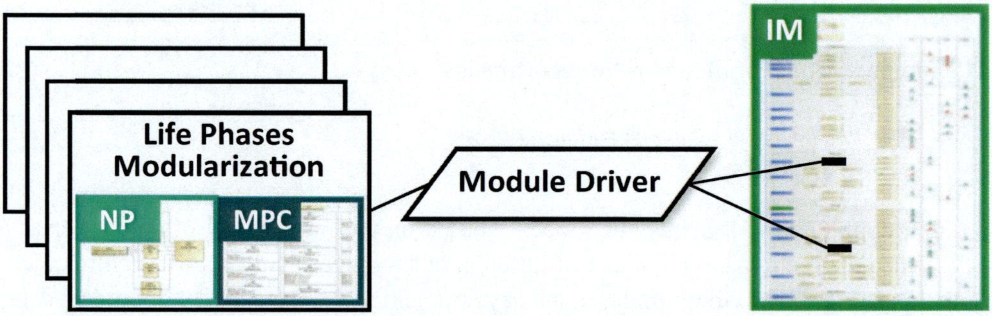

Fig. 14.8 Linking *Life Phases Modularization* to *Impact Model* via module drivers (Schwede et al. 2020a)

14.4.3 Derivation of the Potentials of the Model-Based Approach Using the Application Examples

By using a meta model of product data, an extensible consistent holistic model could be derived, which in turn can strengthen the initial meta model of product data (Fig. 14.3. By linking configuration systems, customer requirements can be addressed better and more efficiently. By linking the effects of modular product families from the *Impact Model*, the objective of *Life Phases Modularization* can also be strengthened and concretized.

Intra sub model analyses are made possible in a next step. For example, based on customer relevant properties, product variants can be easily configured. If now, for example, there are two product variants that match the customer relevant properties, further effects on different life phases can be illustrated with the help of the linked *Impact Model*. In addition to using the customer relevant properties as module drivers in the NP, it would also be possible to identify other module drivers that bring along desired effects. All in all, it can be said that the technical requirements can be satisfied by the configurator and that the economic effects or objectives can be kept in mind by linking the *Impact Model*.

Furthermore, as modular product architectures come with a higher degree of complexity than classically structured product architectures, the data linking between different models can be performed in a more efficient, traceable and versionable way. Additionally, the modular product architecture's maintainability can be increased as changes and their effects on linked components and models can be perceived directly. This allows new findings to be integrated on an ongoing basis. The open formulation of the data relationships in SysML makes it easy to add these new insights and to involve experts. As another aspect, plausibility checks for changes and adaption can directly be implemented into the MBSE environment. These specific advantages can directly be transferred towards the connected applications, such as the product configuration system or the described *Impact Model*.

14.5 Conclusion and Outlook

In addition to the applications of MBSE in the context of product development listed in Sect. 14.1, this book chapter showed that methodical modular product development can also benefit from MBSE. By implementing tools from different method units of the *Integrated PKT Approach*, a holistic model is developed. Due to the open design of interfaces, this model is expandable, thus, other tools of method units for current trends in product development can also be integrated. A planned enhancement is the key figure background of the *Impact Model* in order to strengthen it further. Furthermore, an interface to the modeling of production systems is to be created, which will enable a coordinated design of product architecture and production system (Schwede et al. 2020b).

The data correlations that arise during the development of modular product families can be used consistently, which opens up completely new possibilities in development. One of these possibilities is the use of this holistic model as a reference model for *Digital Twins*.

Among other things, the holistic model provides a generic description of product families. From this holistic model individual master models of product variants, then used as masters for *Digital Twins,* can be derived (Laukotka et al. 2020a, b). This procedure represents an extension to the assumption that frequently only individual products are focused in the context of *Digital Twins* (Stark et al. 2020). By using the holistic model, which represents entire product families, the approach of Laukotka et al. is set one level higher. This takes into account the current trend of developing product families instead of individual products (Laukotka et al. 2020b).

References

Albers A, Reiss N, Bursac N, Richter T (2016) iPeM—integrated product engineering model in context of product generation engineering. In: Wang L, Kjellberg T (eds) Procedia CIRP design conference 2016. Elsevier, pp 100–105. https://doi.org/10.1016/j.procir.2016.04.168

Albers A, Bursac N, Scherer H, Birk C, Powelske J, Muschik S (2019) Model-based systems engineering in modular design. Des Sci 5:183. https://doi.org/10.1017/dsj.2019.15

Alt O (2012) Modellbasierte Systementwicklung mit SysML. Carl Hanser, München. https://doi.org/10.3139/9783446431270

Bender B, Feldhusen J, Krause D, Beckmann G, Paetzold K, Hövel A (2018) Grundlagen technischer Systeme und des methodischen Vorgehens. In: Grote K-H, Bender B, Göhlich D (eds) Dubbel. Springer, Berlin, pp 372–411. https://doi.org/10.1007/978-3-662-54805-9_41

Bonvoisin J, Halstenberg F, Buchert T, Stark R (2016) A systematic literature review on modular product design. J Eng Des 27:1–27. https://doi.org/10.1080/09544828.2016.1166482

Delligatti L (2013) SysML distilled: A brief guide to the systems modeling language. Addison-Wesley

Eichmann O, Melzer S, Hanna M, God R, Krause D (2018) A model-based approach for the development of modular product families considering different life phases. In: Proceedings of the EMEA sector systems engineering conference 2018, pp 269–310

Gausemeier J, Lindemann U, Reinhart G, Wiendahl H-P (2000) Kooperatives Produktenginee-ring: Ein neues Selbstverständnis des ingenieurmäßigen Wirkens. HNI, Bd. 79

Gebhardt N, Krause D (2016) A method for designing visualisations as product development tools. In: Marjanovic D, Storga M Pavkovic N, Bojcetic N, Skec S (eds) Proceedings of the DESIGN 2016 14th international design conference, pp 611–620

Gebhardt N, Bahns T, Krause D (2014) An example of visually supported design of modular product families. In: Morini G, Tolio T (eds) Procedia CIRP design conference 2014, Elsevier, pp 75–80. https://doi.org/10.1016/j.procir.2014.03.136

Gebhardt N, Kruse M, Krause D (2016) Gleichteile-, Modul- und Plattformstrategie. In: Linde-mann U (ed) Handbuch Produktentwicklung. Hanser, München, pp 111–149. https://doi.org/10.3139/9783446445819.006

Gilz, T (2014) PLM-integrated interdisciplinary system models in the conceptual design phase based on model-based systems engineering. Dissertation, Tech. University Kaiserslau-tern

Göhlich D, Fay T-A (2021) Arbeiten mit Anforderungen: Requirements Management. In: Bender B, Gericke K (eds) Pahl/Beitz Konstruktionslehre. Springer, Heidelberg, pp 211–229. https://doi.org/10.1007/978-3-662-57303-7_8

Gräßler I (2004) Kundenindividuelle Massenproduktion: Entwicklung, Vorbereitung der Herstellung, Veränderungsmanagement. Springer, Berlin. https://doi.org/10.1007/978-3-642-18681-3

Greve E, Fuchs C, Hamraz B, Windheim M, Schwede L-N, Krause D (2020a) Investigating the effects of modular product structures to support design decisions in modularization projects. In: 2020 IEEE international conference on industrial engineering & engineering management, pp 295–299. https://doi.org/10.1109/IEEM45057.2020.9309820

Greve E, Spallek J, Kuhl J, Küchenhof J, Rennpferdt C, Krause D (2020b) Neue Trends in der methodischen Entwicklung modularer Produktfamilien. In: Krause D, Hartwich TS, Renn-pferdt C (eds) Produktentwicklung und Konstruktionstechnik, Forschungsergebnisse und -projekte der Jahre 2016 bis 2020. Springer, Berlin, pp 65–89. https://doi.org/10.1007/978-3-662-62393-0_3

Hackl J, Krause D (2017) Towards an impact model of modular product structures. In: Maier A, Skec S, Kim H, Kokkolaras M, Oehmen J, Fadel G, Salustri F, VDLos M (eds) Proceedings of the 21st international conference on engineering design (ICED 17), pp 151–160

Hackl J, Krause D, Otto K, Windheim M, Moon SK, Bursac N, Lachmayer R (2020) Impact of modularity decisions on a firm's economic objectives. J Mech Des 142:5. https://doi.org/10.1115/1.4044914

Hanna M, Schwede L-N, Krause D (2018) Model-based consistency for design for variety and modularization. In: Carlo L, Browning TR, Eppinger SD, Becerril L (eds) The 20th international DSM conference, pp 239–248

Hanna M, Schwenke J, Krause D (2020) Inconsistency management for product families with many variants through a model-bases approach in modular lightweight design. In: Proceedings of the design society: design conference, pp 917–926. https://doi.org/10.1017/dsd.2020.309

Kipp T, Blees C, Krause D (2010) Anwendung einer integrierten Methode zur Entwicklung mo-dularer Produktfamilien. In: Krause D, Paetzold K, Wartzack S (eds) Proceedings of the 21st symposium on design for X, pp 157–168

Kößler J, Paetzold K (2017) Integration of MBSE into existing development processes—expectations and challenges. In: Maier A, Skec S, Kim H, Kokkolaras M, Oehmen J, Fadel G, Salustri F, VDLos M (eds) Proceedings of the 21st international conference on engineering design (ICED 17), pp 51–60

Krause D, Gebhardt N (2018) Methodische Entwicklung modularer Produktfamilien: Hohe Pro-duktvielfalt beherrschbar entwickeln. Springer, Berlin. https://doi.org/10.1007/978-3-662-53040-5

Krause D, Spallek J, Blees C, Kipp T (2018) Modulare Produktstrukturierung. In: Rieg F, Steinhilper R (eds) Handbuch Konstruktion, 2nd edn. Hanser, München. https://doi.org/10.3139/9783446456198.025

Krause D, Vietor T, Inkermann D, Hanna M, Richter T, Wortmann N (2021) Produktarchitektur. In: Bender B, Gericke K (eds) Pahl/Beitz Konstruktionslehre. Springer, Berlin, pp 335–393. https://doi.org/10.1007/978-3-662-57303-7_12

Küchenhof J, Schwede L-N, Krause D (2020) Planning & tracking the changes—matrix mapping of modular product family generations. In: Mortensen NH, Hansen CT, Deininger M (eds) Proceedings of NordDesign 2020. https://doi.org/10.35199/NORDDESIGN2020.11

Kuhl J, Ding A, Ngo NT, Braschkat A, Fiehler J, Krause D (2021) Design of personalized devices—the tradeoff between individual value and personalization workload. Appl Sci 11:241. https://doi.org/10.3390/app11010241

Laukotka F, Hanna M, Schwede L-N, Krause D (2020a) Lebensphasenübergreifende Nutzung Digitaler Zwillinge. Zeitschrift für wirtschaftlichen Fabrikbetrieb 115:101–104. https://doi.org/10.3139/104.112332

Laukotka F, Seiler F, Krause D (2020b) MBSE als Datenbasis zur Unterstützung von Konfigura-toren und Digitalen Zwillingen modularer Produktfamilien. In: Krause D, Paetzold K, Wartzack S (eds) Proceedings of the 31th symposium design for X, pp 61–70. https://doi.org/10.35199/dfx2020.7

Liebel G, Marko N, Tichy M (2018) Model-based engineering in the embedded systems domain: an industrial survey on the state-of-practice. Software Syst Model 17:91–113. https://doi.org/10.1007/s10270-016-0523-3

Liebisch M (2014) Aspektorientierte Datenhaltung in Produktkonfiguratoren—Anforderungen, Konzepte und Realisierung, Dissertation, Friedrich Schiller University

Lindemann U (2009) Methodische Entwicklung technischer Produkte: Methoden flexibel und si-tuationsgerecht anwenden, 3rd edn. VDI-Buch. Springer, Berlin

Lindemann U, Ponn J (2011) Konzeptentwicklung und Gestaltung technischer Produkte: Systematisch von Anforderungen zu Konzepten. VDI-Buch. Springer, Berlin

Matthiesen S, Schmidt S, Moeser G, Munker F (2014) The Karlsruhe SysKIT approach—a three-step SysML teaching approach for mechatronic students. In: Morini G, Tolio T (eds) Procedia CIRP design conference 2014. Elsevier, pp 385–390. https://doi.org/10.1016/j.procir.2014.03.136

Otto K, Hölttä-Otto K, Simpson TW, Krause D, Ripperda S, Moon SK (2016) Global views on modular design research: linking alternative methods to support modular product family concept development. J Mech Des 138. https://doi.org/10.1115/1.4033654

Paetzold K (2017) Product and systems engineering/CA tool chains. In: Biffl S, Lüder A, Gerhard D (eds) Multi-disciplinary engineering for cyber-physical production systems. Springer, Cham. https://doi.org/10.1007/978-3-319-56345-9_2

Rennpferdt C, Greve E, Krause D (2019) The Impact of modular product architectures in PSS design: a systematic literature review. In: Putnik GD (ed) Procedia CIRP design conference 2019. Elsevier, pp 290–295. https://doi.org/10.1016/j.procir.2019.04.197

Rennpferdt C, Greve E, Kuhl J, Küchenhof J, Seiler F, Krause D (2020a) Modularisierung in der industriellen Anwendung. In: Krause D, Hartwich TS, Rennpferdt C (eds) Produktentwick-lung und Konstruktionstechnik, Forschungsergebnisse und -projekte der Jahre 2016 bis 2020. Springer, Berlin, pp 229–254. https://doi.org/10.1007/978-3-662-62393-0_10

Rennpferdt C, Hartwich TS, Krause D (2020b) PKT—Kontinuierliche Weiterentwicklung in Forschung und Lehre. In: Krause D, Hartwich TS, Rennpferdt C (eds) Produktentwicklung und Konstruktionstechnik, Forschungsergebnisse und -projekte der Jahre 2016 bis 2020. Springer, Berlin, pp 1–30. https://doi.org/10.1007/978-3-662-62393-0_1

Richter T, Inkermann D, Vietor T (2016) A framework for integrated product architecture design. In: Boks C, Sigurjosson J, Steinert M, Vis C, Wilvik A (eds) Proceedings of NordDesign 2016, pp 310–320

Riedel R, Jacobs G, Konrad C, Singh R, Sprehe J (2020) Managing knowledge and parameter dependencies with MBSE in textile product development processes. In: Mpofu K, Butala P (eds) Procedia CIRP design conference 2020. Elsevier, pp 170–175. https://doi.org/10.1016/j.procir.2020.01.138

Ripperda S, Krause D (2015) Cost prognosis of modular product structure concepts. In: Weber C, Husung S, Cascini G, CantaMessa M, Marjanovic D, Rotini F (eds) Proceedings of the 20th international conference on engineering design (ICED 15), pp 13–22

Salvador F (2007) Toward a product system modularity construct: literature review and re-conceptualization. IEEE Trans Eng Manag 54:219–240. https://doi.org/10.1109/TEM.2007.893996

Schwede L-N, Hanna M, Wortmann N, Krause D (2019a) Consistent modelling of the impact model of modular product structures. In: Proceedings of the design society: 22nd international conference on engineering design, pp 3601–3610. https://doi.org/10.1017/dsi.2019.367

Schwede L-N, Seiler F, Krause D (2019b) Anknüpfung von Modularisierungsmethoden an ein Wirkmodell modularer Produktstrukturen. In: Krause D, Paetzold K, Wartzack S (eds) Proceedings of the 30th symposium design for X, pp 159–170. https://doi.org/10.35199/dfx2019.14

Schwede L-N, Greve E, Rennpferdt C, Hanna M, Krause D (2020a) Auswirkungen modularer Produktstrukturen bei der Bewertung von Produktstrukturkonzepten. In: Krause D, Hartwich TS, Rennpferdt C (eds) Produktentwicklung und Konstruktionstechnik, Forschungsergebnisse und -projekte der Jahre 2016 bis 2020. Springer, Berlin, pp 91–109. https://doi.org/10.1007/978-3-662-62393-0_4

Schwede L-N, Winter M, Lödding H, Krause D (2020b) Darstellung des Zusammenhangs von Produktarchitektur- und Produktionssystemgestaltung in SysML. In: Krause D, Paetzold K, Wartzack S (eds) Proceedings of the 31th symposium design for X, pp 41–50. https://doi.org/10.35199/dfx2020.5

Schwede L-N, Greve E, Krause D (2020c) Validation concept for the investigation of effects of modular product families. In: Proceedings of the design society: design conference, pp 2395–2404. https://doi.org/10.1017/dsd.2020.114

Seiler FM, Krause D (2020) A multi-dimensional configuration algorithm for modular product architectures. In: Proceedings of the design society: design conference, pp 2405–2414. https://doi.org/10.1017/dsd.2020.283

Seiler FM, Greve E, Krause D (2019) Development of a configure-to-order-based process for the implementation of modular product architectures: a case study. In: Proceedings of the design society: international conference on engineering design, pp 2971–2980. https://doi.org/10.1017/dsi.2019.304

Seiler FM, Kuhl J, Krause D (2020a) A simulation-based decision support method for modular product architecture alternatives. In: Stowe H, Browning TR, Eppinger SD, Trauer J (eds) Proceedings of the 22nd international DSM conference (DSM 2020). https://doi.org/10.35199/dsm2020.9

Seiler FM, Hanna M, Schwede L-N, Laukotka F, Krause D (2020b) MBSE zur Unterstützung der Produktentwicklung von modularen Produktarchitekturen. In: Krause D, Hartwich TS, Rennpferdt C (eds) Produktentwicklung und Konstruktionstechnik, Forschungsergebnisse und -projekte der Jahre 2016 bis 2020. Springer, Berlin, pp 111–134. https://doi.org/10.1007/978-3-662-62393-0_5

Stark R, Anderl R, Thoben K-D, Wartzack S (2020) WiGeP-Positionspapier: "Digitaler Zwil-ling". Zeitschrift für wirtschaftlichen Fabrikbetrieb 115:47–50. https://doi.org/10.3139/104.112311

Stone RB, Wood KL, Crawford RH (2000) A heuristic method to identify modules from a functional description of a product. Des Stud 5–31. https://doi.org/10.1016/S0142-694X(99)00003-4

Walden DD, Roedler GJ, Forsberg K, Hamelin RD, Shortell TM (2015) Systems engineering handbook: A guide for system life cycle processes and activities. Wiley, Hoboken, NJ

Weber C (2007) Looking at DFX and product maturity from the perspective of a new approach to modelling product and product development processes. In: Krause F-L (ed) Procedia CIRP design conference 2007. Springer, pp 85–104

Wilking F, Schleich B, Wartzack S (2020) MBSE along the value chain—an approach for the compensation of additional effort. In: 2020 IEEE 15th international conference of system of systems engineering, pp 61–66. https://doi.org/10.1109/SoSE50414.2020.9130497

Wyrwich C, Jacobs G, Siebrecht J, Konrad C (2020) Model-based product configuration of high variety product portfolios. In: Proceedings of the design society: design conference, pp 2435–2444. https://doi.org/10.1017/dsd.2020.287fgddfsg

Check for updates

Methodical Approaches for Engineering Complex Systems

15

Thomas Vietor and Tobias Huth

Abstract

Technical systems are becoming increasingly complex due to technological progress, changing customer needs and (market-specific) constraints, e.g. addressing the eco-friendliness of new products. This leads to an increasing complexity of the overall development of these systems. One way of dealing with this complexity during the development is the use of design methodologies in general and especially the methods of systems engineering. A systemic view of the systems as well as of the processes in the sense that the system theory is of central importance for developing complex eco-friendly multidisciplinary systems. This article presents an overview of current research findings to support the development of the aforementioned systems and provides an outlook on how such design methods can contribute to the idea of a Circular Economy.

15.1 Introduction

Today's products are used in an environment that is subject to accelerated change. This change in general is characterized by rapid changes in consumer needs, e.g. regarding improved product functions, higher quality standards, increased product variety or customized products, and frequent technological advances (Reik et al. 2006; Wichmann et al. 2019). These conditions lead to shorter product life cycles, which make it difficult for products to be present on the market in the long term. Thus, there is an increasing trend to decommission products before they reach the actual end of their physical life (Prakash et al.

T. Vietor (✉) · T. Huth
Technische Universität Braunschweig, Institute for Engineering Design, Braunschweig, Germany
e-mail: t.vietor@tu-braunschweig.de

D. Krause, E. Heyden (eds.), *Design Methodology for Future Products*,
https://doi.org/10.1007/978-3-030-78368-6_15

2016). A common reason for the elimination or replacement of products is that the perceived value of products is increasingly based on enthusiast features such as convenience, entertainment or connectivity. These types of features are often based on functions that are realized through software components and have relatively short innovation and technology life cycles due to rapid technological developments in the fields of software and electronics (Umeda et al. 2007).

This shift from mainly mechanical based to software-intensive systems is not only a "curse" but also a "blessing", as it also offers an enormous expansion of modernization/upgrade possibilities during the use phase (Inkermann et al. 2018; Şahin et al. 2020). As a result, even the current but especially the future products will be more upgradable during their lifetime and will be more frequently provided with new features that keeps the customer value high. It can be assumed that these products will then have a generally longer use phase and will thus not only have an increased sustainability but will also be an enabler for the establishment of the concept of the Circular Economy (European Environment Agency 2017).

However, in order to be able to use these potentials in the future, methodical approaches are needed to support among others planning, development and production of such systems. These must address different challenges in the development of sustainable complex systems. For example, new approaches are needed to support the interdisciplinary development of software-intensive complex systems, but also to plan and ensure the continuous upgrading of these systems during their use phase. Besides the mere technological challenges, there are organizational challenges to overcome within the associated development projects and team processes. These may result e.g. from the high degree of individuality of the projects and the involvement of locally distributed teams composed of different engineering domains such as software, electrics/electronics or mechanics (Vietor et al. 2015). Other constraints that may change over the course of the project, such as unforeseeable necessary technical changes, budget cuts or the number of project members, also influence the individuality of a product development project and may require agile changes during the course of the project (Baschin et al. 2020).

In summary, a high demand for methodological support in the development of modern complex systems can be derived at different levels. In this chapter, we will first briefly present an understanding of circular system life cycles and the importance of product development within them. Then, four methodological approaches from current research at the Institute for Engineering Design (IK) will be presented, which support the development of sustainable complex systems and the implementation of corresponding development projects. Finally, a short summary and an outlook on future research follows.

15.2 Development of Complex Systems for Supporting the Concept of the Circular Economy

The concept of the Circular Economy describes a cycle of products and thus ultimately of resources that are bound up in these products. In the Circular Economy, products are converted at the end of their use phase into resources for the maintenance of existing products or the manufacture of new ones. The concept of the Circular Economy aims to achieve closed cycles in industrial ecosystems as far as possible. It thus describes a changed economic logic in which pure production is supplemented by sustainability. The aspects of reuse, recycling, repair and reprocessing play an important role here. (Stahel 2016; Webster 2017).

To achieve a Circular Economy, it will be crucial to design products more intelligently, to understand and influence their role in the product environment and to extend their use phase. Therefore, strategies for reuse, repair, remanufacturing and refurbishment are key enablers for a Circular Economy (Pigosso et al. 2010). Consideration of such strategies during the design phase has the potential to leverage environmental and economic benefits in products and reduce their environmental impact.

Linear resource use as the dominant economic model, is based on the cost-efficient production of goods to be sold to consumers. In this way, the majority of societal needs such as mobility, communication and housing are satisfied. Current trends suggest that the role of products in society is changing. For example, the development of additive manufacturing technologies can encourage the repair rather than replacement of products, if spare parts can be delivered on demand at short notice or even manufactured by users themselves (Wits et al. 2016). The transition to a Circular Economy requires insights into the current drivers of product development as well as better knowledge of the interrelationships between products, their use scenarios, their integration into the product environment and societal trends that determine their life cycle. (European Environment Agency 2017).

The basic understanding of the Circular Economy (context see also Ellen MacArthur Foundation 2017; European Commission 2018), in which the life cycle of a system is assumed to be a (mostly) closed loop of life cycle phases is shown in Fig. 15.1. Several variants of this model exist, showing more or less detailed the same principle: The phases of planning and design are followed by production and re-manufacturing, distribution, the phase of use, reuse and repair, collection, recycling in combination with the residual waste, and the addition of raw materials. In this picture, a hint concerning planning shows that new information (coming e.g. from the life cycle itself, from other strategic decisions within a company or network) must be considered during development to affect the cycle. This understanding of the life cycle of a system is highly relevant for developers. The planned future life cycle of a system, including many possibilities for later upgrades or, for example, recycling options of subsystems, is determined to a large extent already in the planning and design phase of the system. For this reason, information from all lifecycle phases (e.g.,

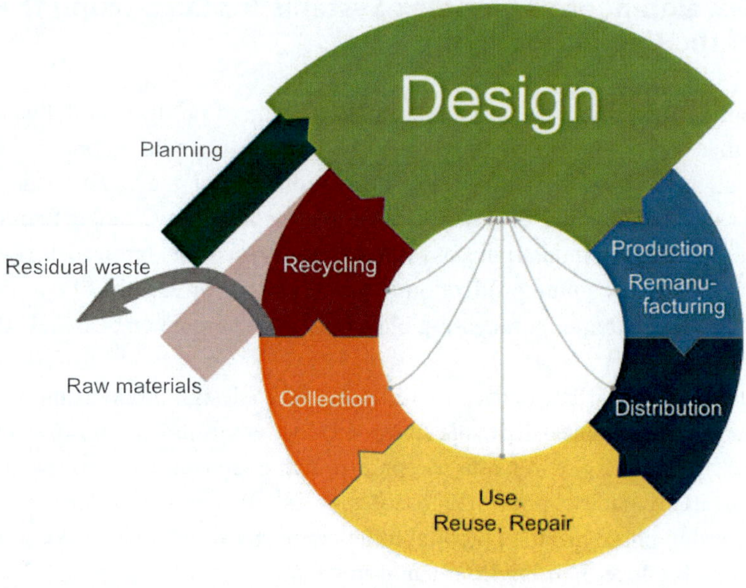

Fig. 15.1 Basic concept of the Circular Economy with emphasis on the design phase as an important enabler for new sustainable products

based on previous product generations) must be taken into account during development. This is symbolized by the arrows in Fig. 15.1.

As mentioned before, within product development a huge part of the responsibility for the systems and their lifecycles is being determined here. Against the background of the increasing environmental impact of greenhouse gas emissions, especially from burning fossil fuels within industry and the mobility sector, it is steadily becoming more important to anticipate a system's whole life cycle and its surrounding services and processes. Current research projects, e.g. RePARE (Brinker et al. 2020), ReLIFE (Werkzeugmaschinenlabor WZL der RWTH Aachen 2020) or MoDeSt (Fraunhofer-Institut für Zuverlässigkeit und Mikrointegration IZM 2019), aim to gather and further develop knowledge about strategies and methods for product development in order to address both the extension of the use phase of systems and their components as well as the intensification of the usage of these systems. Furthermore, the other life cycle phases are addressed as well to be able to contribute to a Circular Economy, increase resource efficiency and, herewith, reduce the massive consequences for the environment, humans and animals.

One of these projects is LifeCycling^2. [1] The project will investigate electric cargo bikes, their components, operating software, apps, environmental services and underlying

[1]The project "LifeCycling^2—Reconfigurable design concepts and services for resource-efficient and (re-)use of e-cargobikes" (German title: "LifeCycling^2—Rekonfigurierbare Designkonzepte und Services für die ressourceneffiziente (Weiter-)Nutzung von E-Cargobikes") is funded by the Federal Ministry of Education and Research (BMBF) in the funding measure "Resource-efficient Circular

business models as a research subject. Opportunities will be sought to develop cargo bikes and their ecosystem in a way that increases resource efficiency by preserving and circulating systems, components and materials. Updates, upgrades and modular products as well as second life strategies and business models such as leasing and sharing—i.e. implying services beyond the product—are used to achieve these goals. At the end of the project, we will be able to give general recommendations for the development of eco-friendly product-service systems.

15.3 Methodical Approaches for Developing Complex Systems

In the following sections, we briefly present selected approaches and research results that aim to support the development of complex systems from different aspects. These approaches may be first methodological building blocks for overcoming some of the challenges when establishing the Circular Economy.

15.3.1 Planning of Product Upgrading and Evolution by Release Planning

Today's products are undergoing an accelerated change of their environment, characterized by rapid changes of consumer needs and frequent advances in technology. These conditions result in shorter product life cycles, hindering the long-term market presence of products (Prakash et al. 2016). The key for high and sustaining value in such conditions are products that are evolving continuously to meet or anticipate changes. For this purpose, products are continuously upgraded by new features and introduced as new releases to the markets. In addition to the increasing dynamics of environmental conditions, the transformation of mechanically-based products to software-intensive and/or service-intensive systems results in an enormous expansion of upgrading possibilities at different product levels (Inkermann et al. 2018; Şahin et al. 2020). As a result, future products will have to be upgraded more frequently with short reaction times on different levels (e.g. the system with its components or higher-level modular kits / platforms) during development and usage.

Over time, varieties of approaches have been developed to address the issue of continuous product upgrading and evolving. Approaches such as road mapping (Groenveld 1997), product modularisation (Schuh and Riesener 2018) or product generation engineering (Albers et al. 2015) address the mapping of a product's long-term vision, technical evolution principles or the design of upgradeable systems. Here, the approach of release planning proposes the proactive planning of a product's evolution and iterative development (Ruhe 2010). This is achieved by strategically planning of future product versions and

Economy—Innovative Product Cycles (ReziProK)" (German title: "Ressourceneffiziente Kreislaufwirtschaft—Innovative Produktkreisläufe (ReziProK)")—Funding reference: 033R232A.

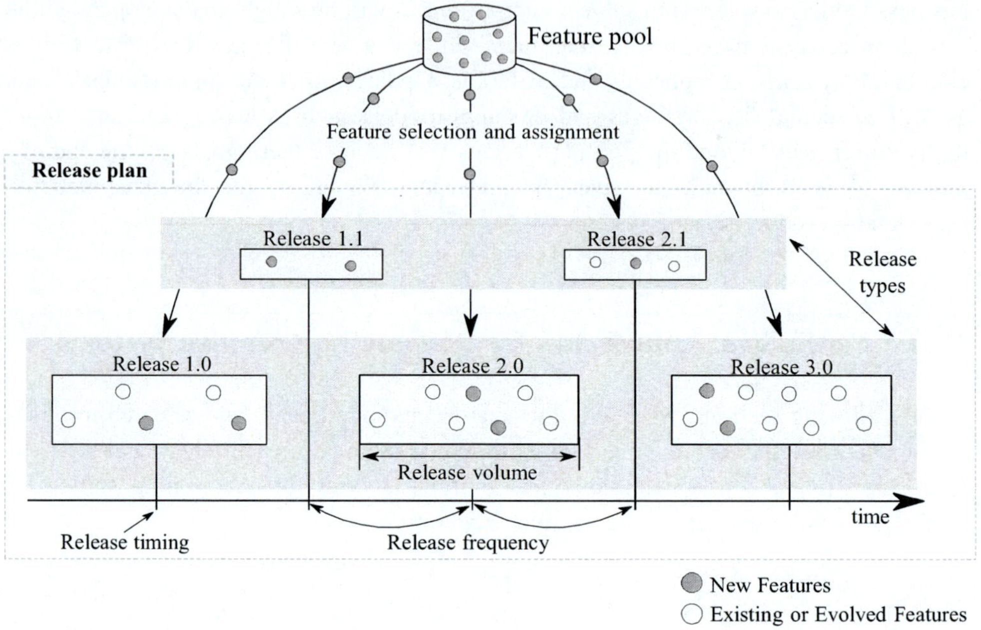

Fig. 15.2 Decisions and structure of release planning

upgrades to improve or sustain the product value over time. The key decisions in release planning are therefore the definition of appropriate release types and their timing as well as the selection and assignment of valuable features to releases (Şahin et al. 2019) (Fig. 15.2).

Thus, release planning provides an approach to design a product's life cycle by determining which features are delivered in which releases at which point in time to customers and stakeholders (Carlshamre 2002). That is why release planning plays a translating and balancing role of stakeholder objectives (e.g. customer needs) to development projects. Several approaches and concepts in literature present release planning as an approach for the systematic planning of product evolution and upgrades (Svahnberg et al. 2010). However, a minor part of the current approaches provides actual operatively applicable methodologies. On the one hand, punctual and domain-specific solutions are offered which are missing essential steps and views for release planning. On the other hand, solutions are provided which are offering isolated views, e.g. a strict effort orientation during planning of features and releases, thereby disregarding customer and business value. Moreover, most existing approaches focus on operative release planning which is used for the organisation of development activities to realise the planned releases by strategic release planning. (Inkermann et al. 2018; Şahin et al. 2019; Svahnberg et al. 2010).

Therefore, approaches are required that are offering methodological support in strategic release planning. In particular there is a need for operatively applicable methods that can be implemented in industrial processes. There is also a need for concepts addressing a consistent value orientation during the planning of product upgrades, considering not only the realisation view but also the value for customers, strategy, and business. Ultimately, future approaches—for release planning—should be applicable to today's mechatronic products.

Therefore, we analysed fundamental challenges and identified potentials of release planning for future cross-domain products (Inkermann et al. 2018). In addition, the needs of current practice were analysed, using the example of different industry sectors such as the automotive, smartphone or consumer software industry where also successful product upgrading strategies were identified (Şahin et al. 2021). Furthermore, we were able to define key requirements, criteria and concepts for a consistent value-orientation in the planning of future releases and features (Şahin et al. 2019). Based on these foundations, initial operatively applicable approaches were developed that support value-oriented strategic release planning by defined procedures and allocated methods that are applicable for today's cross-domain products. For a detailed description and explanation of the approaches developed, reference is made here to Şahin et al. (2020).

Future activities will focus on further development of the methods as well supporting intelligent tools and algorithms to improve decision-making in release planning. The developed methods need to be applied and evaluated in industrial sectors, such as the automotive industry, where the increasing digitisation of products requires new ways of product upgrading. In addition, new concepts will be developed and integrated into engineering design education to address the issue of release planning and to raise awareness about the rising relevance of continuous product upgrading.

15.3.2 An Approach for Modelling Requirements and Systems at Different Hierarchical Levels

The change in the "composition" of today's systems, already described above, leads to increased complexity, which must be dealt with in the course of development through the use of appropriate methods. Two classes of methodological approaches that have been developed specifically for the interdisciplinary development of complex multidisciplinary systems are the methods of models of systems engineering (SE) and model-based systems engineering (MBSE) (Eigner et al. 2014; Gausemeier et al. 2015; Weilkiens 2014). They are built on the basic understanding of the systems theory according to Ropohl (1975) and thereby distinguish hierarchical, structural and functional concepts of the system to be developed.

This class of methods can equally support component-oriented development as well as function-oriented development, since they contain description alternatives for both "worlds". In MBSE, computer-interpretable models built with generic object-oriented modelling languages such as UML (Object Management Group 2017b) or SysML (Object Management Group 2017a) are used to describe the systems to be developed. The modelling languages have different language elements with which the functional, structural and requirements related aspects of the system to be developed can be represented as aspect-specific "views". In an abstract way, the models can be interpreted as a system of relationships similar to a graph composed of nodes and edges, thus documenting the dependencies between different development artefacts such as requirements, functional descriptions and system structures. Hierarchisations within the models and individual views are possible, enabling decomposition across several system levels (from the overall system through levels of subsystems to the components) and thus also their traceability.

MBSE methods are usually closely linked to a systematic approach and are based on process models such as the V-model (Estefan 2008; Graessler and Hentze 2020; Kleiner and Kramer 2013). An overview of SE and MBSE methods addressing a general applicable systems engineering perspective is given for example by Estefan (2008). In the context of MBSE, additional reference is made to Chaps. 12, 13 and 14 in this book.

Requirements engineering (RE) is the first step of various process models and industrially used development processes. The requirements elicited during RE activities serve, among other things, to document the wishes and needs of various stakeholders for the system under development, but also to specify boundary conditions, capabilities and properties the future system must take into account or exhibit (ISO/IEC/IEEE 24765 2017; Rupp 2014). Requirements are often subject to change, especially at the beginning of development. Cascading effects in the propagation of changes are major sources of inadequacies in product development and increase the risk of project failure (Graessler et al. 2020). Requirement engineering, especially in the automotive sector, is more than ever a key factor due to the aforementioned increase of multidisciplinary of the overall vehicle system. Previous requirements engineering processes are usually oriented towards the creation of document-based requirement specifications for the upper (e.g. a complete vehicle) and lower (e.g. components / modules) levels of the system composition (Weber and Weisbrod 2003). They are no longer able to cope with the increased complexity. An effective description of all composition levels cannot be realized with them either (Weber and Weisbrod 2003).

Proper implementation of artefact traceability can provide insight into system development and support overall understanding of the system, impact analysis and even reuse of existing artefacts (Dömges and Pohl 1998). Traceability of artefacts can be divided into two dimensions. The first dimension concerns vertical and horizontal relations. Horizontal relations refer to the traceability of relations between elements of the same type of artefact (e.g. relations between requirements). Vertical relations, on the other hand, refer to relations of an artefact to various other types of artefacts (e.g. relation between a requirement and a system component) (Smit et al. 2016). The second dimension includes pre- and post-traceability, also known as forward and backward traceability (Gotel and Finkelstein 1994). Pre-Traceability describes the relationships between requirements and their sources (e.g. stakeholders or use cases), while post-traceability describes the relationships between requirements and artefacts created in later development phases (e.g. component specifications or interface definitions).

In automotive development, the system under development (the vehicle) is hierarchically subdivided over several abstraction levels into subsystems (e.g. the powertrain) and finally into components (e.g. the electric motor or the pulse inverter) as the smallest unit in order to define individual completed development scopes. According to this structure, the requirements are also cascaded and decomposed through the abstraction levels, starting from the overall vehicle, whose requirements are defined at the beginning of a vehicle development project, through its subsystems to the component level. Within the framework of this cascading, it is necessary to assign requirements from the upper levels accordingly only to subsystems of the next lower level for which these requirements are also relevant. The assigned requirements are decomposed into additional requirements in the further course of development and thus further detailed at the lower levels of abstraction. The same applies to the subsequent abstraction levels. In this component-oriented decomposition of the system under development, the "functions" that have become innovation drivers in recent years, such as the driver assistance function or the start-stop function, are not considered. For this reason, it is a current challenge for OEMs to define corresponding requirement collectives for the individual development scopes, which contain both the requirements induced by the component-oriented view of the system as well as the requirements to be fulfilled from the function-oriented view, cf. (Pohl 2012).

The developed approach addresses the description of systems under development from a functional and structural perspective and builds on the basic understanding of product architecture according to Krause and Vietor (Krause et al. 2021). The aim is to support requirements engineering in the context of the interdisciplinary development of multidisciplinary (vehicle) systems. For this purpose, a three-part description of the system under development is used (see Fig. 15.3). The functional description represents in a certain way the description of the required behaviour of the system, and the structural description represents the desired physical realization of the system under development. In order to relate these two descriptions to each other and to establish an assignment of the required behaviour to physical components, the description of the functional realization is introduced as a third part of the system description. It represents a mapping between function and structure description and maps the distribution of functions to implementing components and systems. Figure 15.3 shows the three parts of the system description in combination.

Within this model-based approach, five partial models are used to specify the system under development using the three-part system description introduced above. The functional description is provided by the two partial models "Capabilities" and "Functions". The partial model "Capabilities" first describes the required capabilities of the system under development in a solution-neutral way. The "Functions" partial model is used to define the functions of the system under development that implements the defined capabilities. Functional and logical structures are modelled for the individual functions based on activities and states. The structural description takes place via the partial models "Product Structure" and "System Structure". The partial model "Product Structure" structures the system in a non-functional way and rather represents the company organization. This takes into account, the fact that development scopes are the responsibility of organizational units and that closed requirement collectives must be created for the individual development

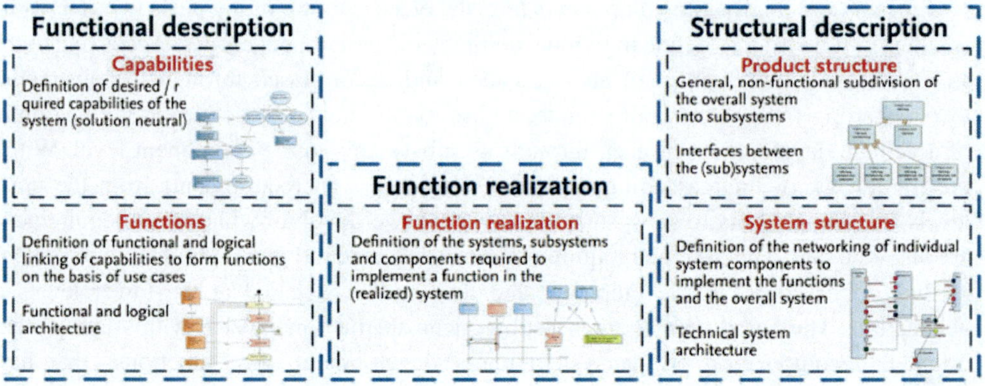

Fig. 15.3 Five partial models for describing automotive systems from an requirements point of view

scope. The partial model "System Structure" contains all physical components of the system under development as elements. Within this partial model there is no hierarchical structuring applied. The formation of subsystems is done by linking system elements with an element of the partial model "Product Structure" representing the assembly or subsystem. Each element of the "System Structure" owns a set of different interfaces, via which they can be connected to neighbouring elements by means of connectors. The result is the topological structure of the system under development based on the interfaces. The fifth partial model, the "Function Realization", basically links the functional and structural descriptions of the system to be developed.

Requirements play an important role for each of the partial models mentioned above. On the one hand, the contents of these models are defined by requirements, but on the other hand, new requirements also arise during the development of the respective "view". These in turn must also be documented in dependence on the respective partial model. For this reason, there is no dedicated requirements model, but each partial model contains, in addition to its specific elements and their relationships, the requirements that have arisen based on the individual partial model. The mapping between the functions and the realizing components within the "function realization" enables traceability through all five partial models and thus from the structural to the functional description. In this way, requirement specifications can be created that are organizationally required for scopes within the "product structure", such as assemblies or subsystems. These requirement collectives can be generated starting from the elements of the product structure, taking into account the previously mentioned traceability dimensions. Figure 15.4 illustrates the entire model as an example. The red dot-dash line visualizes how the graph can be traversed starting from the start element "Vehicle::Drive train::Assy-Axis (FA)" to the end of each path (the requirement of the capabilities). During this traversal of the graph, all requirements on the respective elements are added to the formed requirements collective.

The presented approach aims to support requirement management in the development of complex multidisciplinary systems and was developed in collaboration with an OEM. This

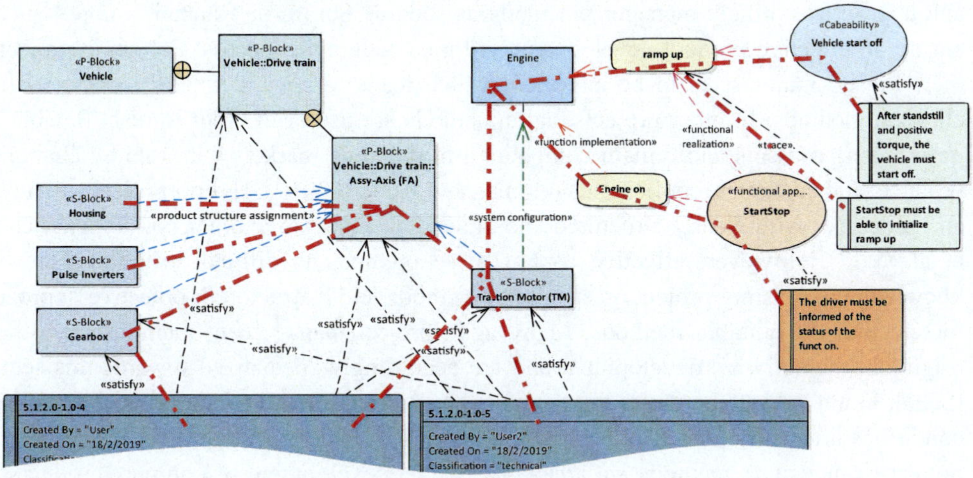

Fig. 15.4 Example of the model with elements of each partial model and requirements

resulted in many boundary conditions that were considered during development and shaped the approach accordingly. For a first pilot application within an industrial scope the system description including its partial models were implemented with the modelling languages UML (Object Management Group 2017b) and SysML (Object Management Group 2017a), which were partially extended by additional stereotypes using the profiling mechanism of UML. For collecting the requirements throughout the model, a prototypical software was implemented. Building on the experience and results gained through the development of the presented approach, further work is being done on approaches, to requirements and system modelling in various research projects in the field of vehicle development.

15.3.3 Hybrid Concepts for Project Management in Product Design Processes

Product development projects have a high degree of individuality. While some products have a simple and primarily mechanical structure, other products consist of a multitude of components. (Albers et al. 2019; Hüsselmann et al. 2019; Riesener et al. 2019) These components are generally developed by several local distributed teams, involving different engineering domains such as software, electronic or mechanic (Vietor et al. 2015). More boundary conditions, for example, unforeseeable required technical changes, the size of the budget or the number of project members, also affect the individuality of a product development project (Baschin et al. 2020).

To structure and plan these projects and related development processes, the literature suggests a variety of classical and agile project management methods. The use of suitable and appropriate project management methods such as Scrum or Kanban is important to ensure an effective product development and thus economic success. For some projects, there are no major risks to be expected, which leads to clear defined processes using classical methods. In other projects, agile methods are useful in order to react flexible to unforeseen problems and remain competitive in dynamic markets with rapidly changing products and customer requirements (Komus and Kuberg 2019). Hybrid project structures and processes with clearly structured and agile sequences are another possibility (Schuh et al. 2017). However, effective use of these methods is difficult without extensive knowledge of existing project management methods and a structured, objective approach to selecting the suitable methods. Many agile project management methods have their origins in the software development and are accordingly specialized towards this sector (Schuh et al. 2018). Therefore, the methods have to be tailored to the projects boundary conditions after selection. It is generally difficult to decide for the project leader, which project management methods are adaptable for the development of complex products in mechanical engineering. In addition, the methods have to be integrated into the product development process to ensure a problem-free execution of the project. There are already some approaches or decision-making concepts for the selection and tailoring of project management methods in practice. However, these are difficult to apply or not directly transferable to the development of mechanical and mechatronic systems (Hüsselmann et al. 2019). A systematic methodology for selecting, tailoring, and integrating project management methods is rarely established in enterprises of mechanical engineering. Therefore, the aim of the research is to develop hybrid concepts for project management, to reach a suitable degree of agility in product design processes.

To handle this challenge, a first concept for a needs-orientated selection, tailoring and integration of agile project management methods in traditional product design processes has been worked out during the KAMiiSo[2] project. The resulting concept with its six steps to conduct is shown in Fig. 15.5. At the beginning, the context of the project is analysed (step 1). The ZOPH[3] model, which was extended by Browning et al. to include the thematic area "tool", serves this purpose (Browning et al. 2006; Negele et al. 1997). Subsequently, more than 400 context factors (e.g. describing general or organizational boundary conditions, aspects of project management or project type) were collected, selecting those factors that best represent the context of product development in mechanical and

[2]The research and development project "KAMiiSo—Digital tools for communication and methods deployment in multi-site product development" was funded by the German Federal Ministry of Education and Research (BMBF) within the Program "Innovations for Tomorrow's Production, Services, and Work" (02L15A250) and managed by the Project Management Agency Karlsruhe (PTKA).

[3]ZOPH is the German abbreviation for Zielsystem (goal system), Objektsystem (product system), Prozesssystem (process system) and Handlungssystem (agent system) according to the approach of Negele et al. (1997).

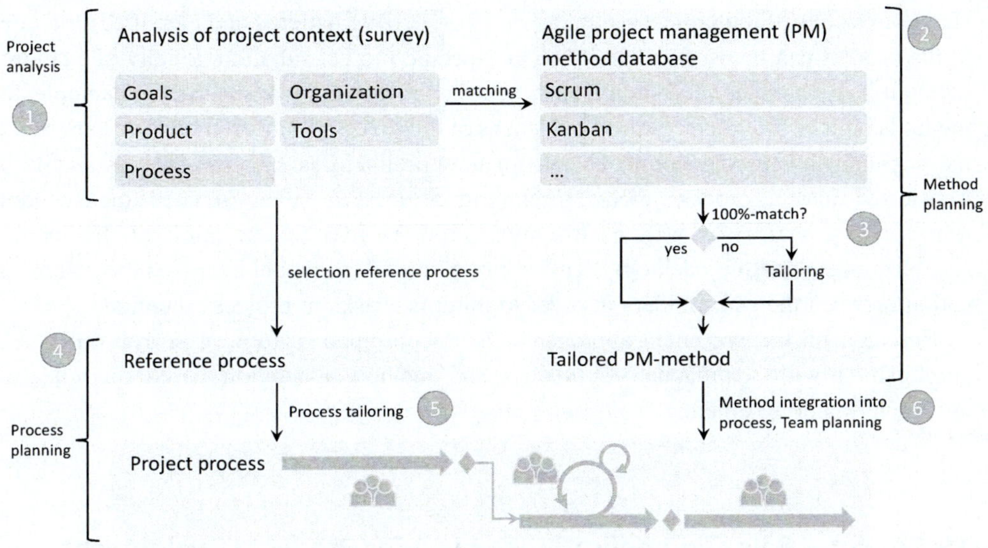

Fig. 15.5 Concept for hybrid project management in design processes (referenced steps are indicated in circles)

mechatronic engineering (approximately four to six for each thematic area). Then, the selected factors have been assigned to the thematic areas.

For each context factor there are guiding questions to identify their characteristics. An example illustrates this procedure: one of the factors assigned to the topic "product" is the context factor "degree of novelty". Here, for example, it is asked whether the product is to be developed requires a new design, an adapted design or only individual components to be designed newly. The answers of all questions allow a typing of the project and an objective decision on how to proceed.

To match suitable agile project management methods (step 2), a list of methods has been created (e.g. Scrum, Kanban, Extreme Programming). A description that defines the characteristics and requirements is prepared for each method. If an answer from the guiding questions matches the requirements of a method, this is noted. The methods with the most matches are suggested as suitable.

In most cases, there will not be a hundred per cent match (step 3). In this case, the differences have to be compensated by tailoring the method. Here, different operations can be used such as scaling the method, combining of various methods or adding, deleting and focusing on activities and roles. Before the tailored method can be integrated into the project process, a suitable reference process has to be selected (step 4). The reference process is adapted from a superordinate model (e.g. V-Model (Graessler and Hentze 2020), V-Model XT etc.) to specific enterprise requirements and is established in the enterprise. Several reference processes are sometimes offered within an enterprise, which can be selected according to their use (e.g. classic and hybrid approaches with agile sequences). After selection, the reference process has also to be tailored to the projects context (step 5).

Here, the results of the project typing as well as recurring patterns and identified structures in the process can be used to tailor it to the specific project situation. Finally, the tailored method(s) will be integrated situationally into the project process (step 6). For example, the method could be integrated in the early stages of the process if uncertainties are expected at the beginning of the development (e.g. for new product design). The process should be transferred into a process model, including at least a superordinate time schedule (e.g. Gantt diagram) as well as essential activities for the execution of the method (e.g. meetings, creating backlogs). The common use of the model increases the communication between the stakeholders in order to minimize risks in project execution.

Research on the presented approach will be continued in future research projects in combination with approaches to process and method adaptation based on reflection approaches, among others.

15.3.4 Potentials and Implications of I4.0 for Product Development

As a result of the global industry facing significant economic challenges (Erol et al. 2016) a growing customer demand for improved product functions, higher quality standards, increased product variety and opportunities for product customisation can be observed. (Reik et al. 2006; Wichmann et al. 2019) The latest and most innovative technologies, known as Industry 4.0, are used to meet these challenges on the shop floor in order to increase the exchange of information flows and the integration of virtual and physical structures across the entire value chain (Drath and Horch 2014; Erol et al. 2016). However, especially small and medium-sized enterprises (SMEs) are lagging behind in implementing Industry 4.0 technologies and are facing several hurdles such as an adequate estimation of the effects of Industry 4.0 deployment (Matt et al. 2020; Schneider et al. 2020). This issue is currently addressed by research in general, including by the approach introduced in the following, which aims at supporting SMEs to overcome these implementation hurdles.

Basically, a diffuse understanding of Industry 4.0 can be identified in both research and industry (Inkermann et al. 2019), whereby the definition of the term varies greatly depending on the perspective and individual field. For the purpose of standardisation, the following definition of Industry 4.0 was developed within the framework of the *Innovation Network Synus*[4]—a recently finished research project:

"In the project Synus Industry 4.0 is understood as the networking of individual heterogeneous automation (physical world) and information components (cyber world) within a production system and product development. These interact with the objective of increasing the value of business processes and products. The special focus here is on data

[4]The research project "Synus" (Methods and tools for the synergetic conception and evaluation of Industry 4.0 technologies) was funded (10/2017 to 12/2020) by the European Regional Development Fund (EFRE I ZW 6- 85012454) and managed by the project management agency NBank.

acquisition, methods of data processing and provision, communication and interaction between machines as well as between humans and machines." (Institut für Konstruktionstechnik 2020). The basis for the common Industry 4.0 understanding within the Synus project is the combined expertise of the project partners from the fields of product development, tooling and manufacturing technology, automotive management and industrial production, control technology and vehicle mechatronics, and applied systems engineering.

The current challenge for the industry, according to our research, is in particular the development of a sustainable economic strength that addresses global technological as well as societal change (Schneider et al. 2020). In this context, Industry 4.0 technologies provide the necessary assistance for the implementation of highly flexible enterprise structures as well as for horizontal and vertical networking of added-value chains to create added-value networks (Geissbauer et al. 2014). Despite the necessity of a widespread implementation of Industry 4.0 technologies, industry generally tends to react with restraint, as especially small and medium-sized enterprises (SMEs) are confronted with major hurdles in Industry 4.0 implementation (Matt et al. 2020; Schneider et al. 2020). These hurdles are based on insufficient ability to assess the effects of individual Industry 4.0 technologies in conjunction with high initial investments and an increasing overall system complexity through the use of Industry 4.0 as well as insufficiently fulfilled basic infrastructural requirements by industrial enterprises (Andelfinger and Hänisch 2017; Matt et al. 2020). In this context, SMEs are at an increased disadvantage compared to large companies due to their lower liquidity and availability of resources for assessing the impact of individual Industry 4.0 technologies (Schneider et al. 2020).

In the past decade, a large number of research projects have been initiated in the field of Industry 4.0 in order to assist industry in responding to and preparing for the social, technological and economic changes that are currently apparent and can be foreseen in the future (Schneider et al. 2021). Based on systematic literature research we were able to identify the research focal points shown in Fig. 15.6 as well as their inherent specification (Schneider et al. 2021). In the course of this survey, 30 representative Industry 4.0 projects were identified and analysed. Each red dot in Fig. 15.6 represents a project. The allocation to the individual areas of the figure visualizes the distribution of the research projects across the addressed research focus areas. Accordingly, the current research efforts can basically be assigned to the development of novel Industry 4.0 technologies or the development of methodical tools for maturity analysis, Industry 4.0 technology assessment or roadmap design (Schneider et al. 2021).

In the course of our research, we are currently developing a comprehensive methodical approach (blue bar, Fig. 15.6) that is intended to accelerate the initiation and simplification of the implementation of I4.0 technologies in SMEs. The overall approach, referred to as the "Industry 4.0 Method and Knowledge Platform", thereby incorporates the specifications of method development in the field of Industry 4.0 (Fig. 15.6) and addresses existing research gaps such as the insufficient integration of product development

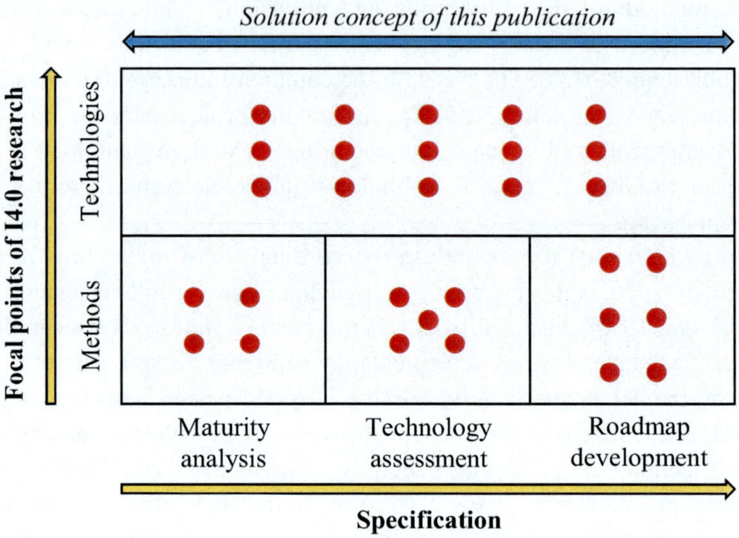

Fig. 15.6 Current focal points of research in the field of Industry 4.0 (Schneider et al. 2021)

characteristics and aspects into evaluation processes and perspectives (Schneider et al. 2021).

The Industry 4.0 Method and Knowledge Platform is intended to answer the following questions from a scientific as well as from an industrial point of view (Schneider et al. 2021):

- What is Industry 4.0 about and what are central terms?
- Which fundamental Industry 4.0 technologies do currently exist?
- How does the current enterprise state in terms of Industry 4.0 look like?
- Which Industry 4.0 implementation potentials exist?
- Which specific I4.0 technology can exploit these potentials to the best possible extend?
- How could the impact of this technology look like on a rough-quantitative level?
- Which strategic steps are to be made to implement Industry 4.0 and which experts can be contacted for this purpose?

As part of current research work, the Industry 4.0 Method and Knowledge Platform is implemented and will be industrially validated and subsequently finalised within 2021.

15.4 Conclusion and Future Work

This chapter presented an overview of four methodological approaches to support the development of complex multidisciplinary systems. These approaches are results or interim results of current and ongoing research and partly also the subject of doctoral research projects (Tarik Şahin (Release Planning), Julian Baschin (Project Management), David Schneider (Industry 4.0)). For a more detailed presentation of the methodological approaches developed, their classification in relation to the state of research and (initial) validations, reference is made to the publications of the authors and previously named collaborators cited in the sections.

Only with a combination of different development methods, such as those presented in this article, the requirements for future product development as well as for future technical systems can be addressed. Research on design methodology can have a major impact on promoting the concept of the Circular Economy and enabling its implementation. Research on the methodological approaches presented will continue and be part of various future projects.

References

Albers A, Bursac N, Wintergerst E (2015) Product generation development—importance and challenges from a design research perspective. In: New developments in mechanics and mechanical engineering : Proceedings of the international conference on mechanical engineering (ME 2015), 15–17 Mar 2015, Vienna, Austria: Proceedings of the international conference on theoretical mechanics and applied mechanics (TMAM 2015), 15–17 Mar 2015, Vienna, Austria, Mastorakis NE (ed), vol 13. s.l, pp 16–21

Albers A, Heimicke J, Spadinger M, Reiss N, Breitschuh J, Richter T, Bursac N, Marthaler F (2019) A systematic approach to situation-adequate mechatronic system development by ASD—agile systems design. Procedia CIRP 84:1015–1022. https://doi.org/10.1016/j.procir.2019.03.312

Andelfinger VP, Hänisch T (2017) Industrie 4.0. Springer Fachmedien Wiesbaden, Wiesbaden

Baschin J, Huth T, Vietor T (2020) Context-specific agile process design to support the planning of product development projects. Proc Des Soc Des Conf 1:491–500. https://doi.org/10.1017/dsd.2020.109

Brinker J, Gembarski PC, Hagen S, Thomas O (2020) Anwendungspotenziale von Additive Repair und Refurbishment für Service-orientierte Geschäftsmodelle. In: Lachmayer R, Rettschlag K, Kaierle S (eds) Konstruktion für die Additive Fertigung 2019. Springer, Berlin, pp 43–54

Browning TR, Fricke E, Negele H (2006) Key concepts in modeling product development processes. Syst Eng 9:104–128. https://doi.org/10.1002/sys.20047

Carlshamre P (2002) Release planning in market-driven software product development: provoking an understanding. Requir Eng 7:139–151. https://doi.org/10.1007/s007660200010

Dömges R, Pohl K (1998) Adapting traceability environments to project-specific needs. Commun ACM 41:54–62. https://doi.org/10.1145/290133.290149

Drath R, Horch A (2014) Industrie 4.0: hit or hype? [Industry Forum]. EEE Ind Electron Mag 8:56–58. https://doi.org/10.1109/MIE.2014.2312079

Eigner M, Roubanov D, Zafirov R (2014) Modellbasierte virtuelle Produktentwicklung. Springer, Berlin

Ellen MacArthur Foundation (2017) Concept—what is a circular economy? A framework for an economy that is restorative and regenerative by design. https://www.ellenmacarthurfoundation.org/circular-economy/concept. Accessed 21 Jan 2021

Erol S, Schuhmacher A, Sihn W (2016) Strategic guidance towards Industry 4.0—a three-stage process model. In: Dimitrov D, Oosthuizen T (eds) International conference on competitive manufacturing: COMA '16, 27–29 Jan 2016, Stellenbosch, South Africa: Proceedings. Global Competitiveness Centre in Engineering Department of Industrial Engineering Stellenbosch University, Stellenbosch, South Africa, pp 495–500

Estefan JA (2008) Survey of model-based systems engineering (MBSE) methodologies 25

European Commission (2018) Reports in critical raw materials and the circular economy. https://ec.europa.eu/commission/publications/report-critical-raw-materials-and-circular-economy_en. Accessed 21 Jan 2021

European Environment Agency (2017) Circular by design: Products in the circular economy. EEA report, no 2017, 6. Publications Office of the European Union, Luxembourg

Fraunhofer-Institut für Zuverlässigkeit und Mikrointegration IZM (2019) MoDeSt: Produktzirkularität durch modulares Design—Strategien für langlebige Smartphones. https://modest-projekt.de/. Accessed 4 Feb 2021

Gausemeier J, Dumitrescu R, Steffen D, Czaja A, Wiederkehr O, Tschirner C (2015) Systems engineering in industrial practice. Paderborn, Germany. Accessed 25 Feb 2019

Geissbauer R, Schrauf S, Koch V, Kuge S (2014) Industrie 4.0 - Chancen und Herausforderungen der vierten industriellen Revolution. https://www.kantaremnid.de/studien/pdf/pwc_studie_industrie_4.0.pdf

Gotel OCZ, Finkelstein CW (1994) An analysis of the requirements traceability problem. In: Proceedings of IEEE international conference on requirements engineering, pp 94–101

Graessler I, Hentze J (2020) The new V-model of VDI 2206 and its validation. at-Automatisierungstechnik 68:312–324. https://doi.org/10.1515/auto-2020-0015

Graessler I, Oleff C, Scholle P (2020) Method for systematic assessment of requirement change risk in industrial practice. Appl Sci 10:8697. https://doi.org/10.3390/app10238697

Groenveld P (1997) Roadmapping integrates business and technology. Res Technol Manag 40:48–55. https://doi.org/10.1080/08956308.1997.11671157

Hüsselmann C, Dönges S, Karpf S (2019) Zielgerichtete Adaption des Projektmanagements-Verschwendung vermeiden und Wertschöpfung erhöhen durch Projekttypisierung. WI-[Reports] – Arbeitspapiere Wirtschaftsingenieurwesen

Inkermann D, Huth T, Vietor T (2018) Towards cross-domain release engineering-potentials and challenges for automotive industry. In: Rausch A, Knieke C, Schranz M (eds) ADAPTIVE 2018: The tenth international conference on adaptive and self-adaptive systems and applications, 18–22 Feb 2018, Barcelona, Spain. IARIA, Wilmington, DE, pp 93–99

Inkermann D, Schneider D, Martin NL, Lembeck H, Zhang J, Thiede S (2019) A framework to classify Industry 4.0 technologies across production and product development. Procedia CIRP 84:973–978. https://doi.org/10.1016/j.procir.2019.04.218

Institut für Konstruktionstechnik (2020) https://www.innovationsverbund-synus.eu/orientierung-sensibilisierung/. Accessed 21 January 2021

ISO/IEC/IEEE 24765 (2017) ISO/IEC/IEEE 24765:2017(E)—systems and software engineering-vocabulary. ISO/IEC/IEEE International Standard

Kleiner S, Kramer C (2013) Model based design with systems engineering based on RFLP using V6. In: Abramovici M, Stark R (eds) Smart product engineering. Springer, Berlin, pp 93–102

Komus K (2019) Status Quo (Scaled) Agile 2019/2020. Koblenz, Germany. https://www.hs-koblenz.de/bpm-labor/status-quo-scaled-agile-2020

Krause D, Vietor T, Inkermann D, Hanna M, Richter T, Wortmann N (2021) Produktarchitektur. In: Bender B, Gericke K (eds) Pahl/Beitz Konstruktionslehre. Springer, Berlin, pp 335–393

Matt DT, Modrák V, Zsifkovits H (2020) Industry 4.0 for SMEs. Springer, Cham

Negele H, Fricke E, Igenbergs E (1997) ZOPH—a systemic approach to the modeling of product development systems. INCOSE Int Symp 7:266–273. https://doi.org/10.1002/j.2334-5837.1997.tb02181.x

Object Management Group (2017a) OMG Systems Modeling Language (OMG SysML™). https://www.omg.org/spec/SysML/1.5/. Accessed 8 Feb 2019

Object Management Group (2017b) OML® Unified Modeling Language® (OMG UML®). http://www.omg.org/spec/UML/2.5.1. Accessed 8 Feb 2019

Pigosso DC, Zanette ET, Filho AG, Ometto AR, Rozenfeld H (2010) Ecodesign methods focused on remanufacturing. J Cleaner Prod 18:21–31. https://doi.org/10.1016/j.jclepro.2009.09.005

Pohl K (2012) Model-based engineering of embedded systems: The SPES 2020 methodology. Springer, Heidelberg

Prakash S, Dehoust G, Gsell M, Schleicher T, Stamminger R (2016) Einfluss der Nutzungsdauer von Produkten auf ihre Umweltwirkung: Schaffung einer Informationsgrundlage und Entwicklung von Strategien gegen „Obsoleszenz". Dessau-Roßlau, Dessau-Roßlau. https://www.umweltbundesamt.de/publikationen/einfluss-der-nutzungsdauer-von-produkten-auf-ihre-1

Reik MP, McIntosh RI, Owen GW, Mileham AR, Culley SJ (2006) Design for changeover (DFC). In: Blecker T, Friedrich G (eds) Mass customization: challenges and solutions, vol 87. Kluwer Academic, Boston, pp 111–136

Riesener M, Rebentisch E, Doelle C, Kuhn M, Brockmann S (2019) Methodology for the design of agile product development networks. Procedia CIRP 84:1029–1034. https://doi.org/10.1016/j.procir.2019.04.172

Ropohl G (ed) (1975) Systemtechnik—Grundlagen und Anwendung. Hanser, München

Ruhe G (2010) Product release planning. Auerbach, Boca Raton, FL

Rupp C (2014) Requirements-Engineering und -Management: Aus der Praxis von klassisch bis agil, 6th edn. Hanser, München

Şahin T, Inkermann D, Vietor T (2019) Towards consistent value orientation in release planning. In: Volume 7: 31st international conference on design theory and methodology. American Society of Mechanical Engineers

Şahin T, Huth T, Axmann J, Vietor T (2020) A Methodology for value-oriented strategic release planning to provide continuous product upgrading. In: 2020 IEEE international conference on industrial engineering and engineering management (IEEM). IEEE, pp 1032–1036

Şahin T, Köster L, Huth T, Vietor T (2021) How to Upgrade Vehicles? Release Planning in the Automotive Industry. In: Bargende M, Reuss H-C, Wagner A (eds) 21. Internationales Stuttgarter Symposium. Springer Fachmedien Wiesbaden, Wiesbaden, pp 155–173

Schneider D, Huth T, Vietor T, Schumacher P, Weckenborg C, Spengler T (2020) Development of a potential model to support the assessment and introduction of Industry 4.0 technologies. Proc Des Soc Des Conf 1:707–716. https://doi.org/10.1017/dsd.2020.85

Schneider D, Huth T, Vietor T (2021) Development of an Industry 4.0 method and knowledge platform for strategic technology implementation. Procedia CIRP 100:613–618. https://doi.org/10.1016/j.procir.2021.05.132

Schuh G, Riesener M (2018) Produktkomplexität managen: Strategien—Methoden—Tools, 3rd edn. Hanser, München

Schuh G, Rebentisch E, Riesener M, Diels F, Dolle C, Eich S (2017) Agile-waterfall hybrid product development in the manufacturing industry—introducing guidelines for implementation of parallel use of the two models. In: 2017 IEEE international conference on industrial engineering and engineering management (IEEM). IEEE, pp 725–729

Schuh G, Dölle C, Kantelberg J, Menges A (2018) Identification of agile mechanisms of action as basis for agile product development. Procedia CIRP 70:19–24. https://doi.org/10.1016/j.procir.2018.02.007

Smit K, Zoet M, Berkhout M (2016) A framework for traceability of legal requirements in the Dutch governmental context. In: BLED 2016 proceedings

Stahel WR (2016) The circular economy. Nature 531:435–438. https://doi.org/10.1038/531435a

Svahnberg M, Gorschek T, Feldt R, Torkar R, Saleem SB, Shafique MU (2010) A systematic review on strategic release planning models. Inf Softw Technol 52:237–248. https://doi.org/10.1016/j.infsof.2009.11.006

Umeda Y, Daimon T, Kondoh S (2007) Life cycle option selection based on the difference of value and physical lifetimes for life cycle design. In: Bocquet J-C (ed) DS 42 Proceedings of ICED 2007, the 16th international conference on engineering design, 28–31 July 2007, Paris, France. The Design Society, Glasgow

Vietor T, Herrmann C, Spengler TS (eds) (2015) Synergetische Produktentwicklung: Unternehmensübergreifend erfolgreich zusammenarbeiten; Ergebnisse des Verbundprojekts SynProd. Berichte aus dem Maschinenbau. Shaker, Aachen

Weber M, Weisbrod J (2003) Requirements engineering in automotive development: experiences and challenges. IEEE Softw 20:16–24. https://doi.org/10.1109/MS.2003.1159025

Webster K (2017) The circular economy: A wealth of flows. Ellen MacArthur Foundation, Cowes, Isle of Wight, UK

Weilkiens T (2014) Systems Engineering mit SysML/UML: Anforderungen, Analyse, Architektur, 3rd edn. dpunkt.verl., Heidelberg

Werkzeugmaschinenlabor WZL der RWTH Aachen (2020) ReLIFE—Adaptives Remanufacturing zur Lebenszyklusoptimierung vernetzter Investitionsgüter: Projekthomepage. http://relife.wzl.rwth-aachen.de/de/default.html. Accessed 4 Feb 2021

Wichmann RL, Eisenbart B, Gericke K (2019) The direction of industry: a literature review on Industry 4.0. Proc Int Conf Eng Des 1:2129–2138. https://doi.org/10.1017/dsi.2019.219

Wits WW, García JRR, Becker JMJ (2016) How additive manufacturing enables more sustainable end-user maintenance, repair and overhaul (MRO) strategies. Procedia CIRP 40:693–698. https://doi.org/10.1016/j.procir.2016.01.156